MOLECULAR AND CELLULAR GERONTOLOGY

ANNALS OF THE NEW YORK ACADEMY OF SCIENCES
Volume 908

MOLECULAR AND CELLULAR GERONTOLOGY

Edited by Olivier Toussaint, Heinz D. Osiewacz, Gordon J. Lithgow, and Christine Brack

The New York Academy of Sciences
New York, New York
2000

Library of Congress Cataloging-in-Publication Data

Molecular and cellular gerontology / editors, Olivier Toussaint ... [et al.].
       p. ; cm. — (Annals of the New York Academy of Sciences, ISSN 0077-8923 ; v. 908)
    Includes bibliographical references and index.
       ISBN 1-57331-271-1 (cloth : alk. paper) — ISBN 1-57331-272-X (paper : alk. paper)
       1. Aging—Molecular aspects—Congresses. 2. Cells—Aging—Congresses. 3.
    Longevity—Congresses. I. Toussaint, Olivier. II. EMBO Workshop on Molecular and
    Cellular Gerontology (1999 : Ticino, Switzerland) III. Series.
       [DNLM: 1. Aging—physiology—Congresses. 2. Cell Aging—Congresses. 3. Molecular
    Biology—Congresses. WT 104 M7172 2000]
    Q11 .N5 vol. 908
    [QP86]
    500 s—dc21
    [612.6'7]   00-034861

GYAT / PCP
Printed in the United States of America
ISBN 1-57331-271-1 (cloth)
ISBN 1-57331-272-X (paper)
ISSN 0077-8923

ANNALS OF THE NEW YORK ACADEMY OF SCIENCES
Volume 908
June 2000

MOLECULAR AND CELLULAR GERONTOLOGY[a]

Editors
OLIVIER TOUSSAINT, HEINZ D. OSIEWACZ, GORDON J. LITHGOW, AND
CHRISTINE BRACK

Conference Organizers
CHRISTINE BRACK, OLIVIER TOUSSAINT, HEINZ D. OSIEWACZ, AND
GORDON J. LITHGOW

CONTENTS

Preface. *By* CHRISTINE BRACK, GORDON J. LITHGOW, HEINZ D. OSIEWACZ, AND
OLIVIER TOUSSAINT ix

Some New Directions for Research on the Biology of Aging. *By* GEORGE M.
MARTIN 1

Molecular Gerontology. Bridging the Simple and the Complex. *By* THOMAS
B.L. KIRKWOOD 14

Metabolic Control and Gene Dysregulation in Yeast Aging. *By* S. MICHAL
JAZWINSKI 21

Mitochondrial Oxidative Stress and Aging in the Filamentous Fungus
Podospora anserina. By HEINZ D. OSIEWACZ AND CORINA BORGHOUTS   31

Molecular Genetic Mechanisms of Life Span Manipulation in *Caenorhabditis elegans. By* SHIN MURAKAMI, PATRICIA M. TEDESCO, JAMES R.
CYPSER, AND THOMAS E. JOHNSON 40

Genetics of Human Aging. The Search for Genes Contributing to Human
Longevity and Diseases of the Old. *By* P. ELINE SLAGBOOM, BASTIAAN
T. HEIJMANS, MARIAN BEEKMAN, RUDI G.J. WESTENDORP, AND
INGRID MEULENBELT 50

Genetics of Survival. *By* FRANÇOIS SCHÄCHTER 64

[a]This volume is the result of a conference entitled **EMBO Workshop on Molecular and Cellular Gerontology**, held in Serpiano, Ticino, Switzerland, September 18–22, 1999.

Melanin Accumulation Accelerates Melanocyte Senescence by a Mechanism Involving p16^{INK4a}/CDK4/pRB and E2F1. *By* DEBDUTTA BANDYOPADHYAY AND ESTELA E. MEDRANO 71

Stress-Induced Premature Senescence. Essence of Life, Evolution, Stress, and Aging. *By* OLIVIER TOUSSAINT, PATRICK DUMONT, JEAN-FRANÇOIS DIERICK, THIERRY PASCAL, CHRISTOPHE FRIPPIAT, FLORENCE CHAINIAUX, FRANCIS SLUSE, FRANÇOIS ELIAERS, AND JOSÉ REMACLE .. 85

Role of Oxidative Stress in Telomere Length Regulation and Replicative Senescence. *By* THOMAS VON ZGLINICKI 99

Replicative Senescence and Oxidant-Induced Premature Senescence. Beyond the Control of Cell Cycle Checkpoints. *By* QIN M. CHEN 111

Poly(ADP-Ribosyl)ation, Genomic Instability, and Longevity. *By* ALEXANDER BÜRKLE 126

Aging and Longevity. A Paradigm of Complementation between Homeostatic Mechanisms and Genetic Control? *By* CHARIKLIA PETROPOULOU, NIKI CHONDROGIANNI, DAVINA SIMOES, GEORGIA AGIOSTRATIDOU, NATALIA DROSOPOULOS, VIOLETTA KOTSOTA, AND EFSTATHIOS S. GONOS 133

Protein Degradation by the Proteasome and Its Implications in Aging. *By* BERTRAND FRIGUET, ANNE-LAURE BULTEAU, NIKI CHONDROGIANNI, MARIANGELA CONCONI, AND ISABELLE PETROPOULOS 143

Fibroblast Responses to Exogenous and Autocrine Growth Factors Relevant to Tissue Repair. The Effect of Aging. *By* DIMITRIS KLETSAS, HARRIS PRATSINIS, IRENE ZERVOLEA, PANAGIOTIS HANDRIS, ELENI SEVASLIDOU, ENZO OTTAVIANI, AND DIMITRI STATHAKOS 155

The Werner Syndrome. A Model for the Study of Human Aging. *By* JAN O. NEHLIN, GUNHILD LANGE SKOVGAARD, AND VILHELM A. BOHR 167

Calorie Restriction and Age-Related Oxidative Stress. *By* B.J. MERRY 180

Role of Mitochondrial DNA Mutations in Disease and Aging. *By* D.A. COTTRELL, E.L. BLAKELY, G.M. BORTHWICK, M.A. JOHNSON, G.A. TAYLOR, E.J. BRIERLEY, P. INCE, AND D.M. TURNBULL 199

Inherited Variability of the Mitochondrial Genome and Successful Aging in Humans. *By* G. DE BENEDICTIS, G. CARRIERI, O. VARCASIA, M. BONAFÈ, AND C. FRANCESCHI 208

Mitochondrial Oxidative Stress. Physiologic Consequences and Potential for a Role in Aging. *By* SIMON MELOV 219

Tissue Mitochondrial DNA Changes. A Stochastic System. *By* GEORGE KOPSIDAS, SERGEY A. KOVALENKO, DAMIEN R. HEFFERNAN, NATALIA YAROVAYA, LUDMILLA KRAMAROVA, DIANE STOJANOVSKI, JUDY BORG, MONIRUL ISLAM, APHRODITE CARAGOUNIS, AND ANTHONY W. LINNANE 226

Inflamm-aging. An Evolutionary Perspective on Immunosenescence. *By* CLAUDIO FRANCESCHI, MASSIMILIANO BONAFÈ, SILVANA VALENSIN, FABIOLA OLIVIERI, MARIA DE LUCA, ENZO OTTAVIANI, AND GIOVANNA DE BENEDICTIS 244

The CA1 Region of the Human Hippocampus Is a Hot Spot in Alzheimer's Disease. *By* M.J. WEST, C.H. KAWAS, L.J. MARTIN, AND J.C. TRONCOSO 255

Transgenic Mouse Models of Alzheimer's Disease. *By* KLAUS D.
BORNEMANN AND MATTHIAS STAUFENBIEL 260

Molecular Misreading. A New Type of Transcript Mutation in Gerontology.
By FRED W. VAN LEEUWEN, DAVID F. FISCHER, ROB BENNE, AND
ELLY M. HOL 267

Biogerontology: The Next Step. *By* SURESH I.S. RATTAN 282

Nuclear-Mitochondrial Interactions Involved in Aging in *Podospora
anserina*. *By* CORINA BORGHOUTS AND HEINZ D. OSIEWACZ 291

Inherited Frailty. ApoE Alleles Determine Survival after a Diagnosis of
Heart Disease or Stroke at Ages 85+. *By* E.H. CORDER, H. BASUN,
L. FRATIGLIONI, Z. GUO, L. LANNFELT, M. VIITANEN, L.S. CORDER,
K.G. MANTON, AND B. WINBLAD 295

MtDNA Deletions in Aging and in Nonmitochondrial Pathology. *By*
A. CORMIO, A.M.S. LEZZA, J. VECCHIET, G. FELZANI, L. MARANGI,
F.W. GUGLIELMI, A. FRANCAVILLA, P. CANTATORE, AND
M.N. GADALETA 299

Transcriptome and Proteome Analysis in Human Senescent Fibroblasts and
Fibroblasts Undergoing Premature Senescence Induced by Repeated
Sublethal Stresses. *By* JEAN-FRANÇOIS DIERICK, THIERRY PASCAL,
FLORENCE CHAINIAUX, FRANÇOIS ELIAERS, JOSÉ REMACLE, PETER
MOSE LARSEN, PETER ROEPSTORFF, AND OLIVIER TOUSSAINT 302

Human Diploid Fibroblasts Display a Decreased Level of *c-fos* mRNA at 72
Hours after Exposure to Sublethal H_2O_2 Stress. *By* PATRICK DUMONT,
MAGGI BURTON, QIN M. CHEN, CHRISTOPHE FRIPPIAT,
THIERRY PASCAL, JEAN-FRANÇOIS DIERICK, FRANÇOIS ELIAERS,
FLORENCE CHAINIAUX, JOSÉ REMACLE, AND OLIVIER TOUSSAINT   306

Replicative Senescence of Interleukin-2–Dependent Human T Lympho-
cytes. Flow Cytometric Characteristics of Phenotype Changes. *By*
EWA JARUGA, JANUSZ SKIERSKI, EWA RADZIESZEWSKA, AND
EWA SIKORA 310

Age-Related Changes in Irradiation-Induced Apoptosis and Expression of
p21 and p53 in Crypt Stem Cells of Murine Intestine. *By* KAREEN
MARTIN, CHRISTOPHER S. POTTEN, AND THOMAS B.L. KIRKWOOD 315

Testing Evolutionary Theories of Aging. *By* GAWAIN MCCOLL, NICOLE L.
JENKINS, DAVID W. WALKER, AND GORDON J. LITHGOW 319

A Novel *in Vitro* Model of Conditionally Immortalized Human Vascular
Smooth Muscle Cells. A Tool for Aging Studies. *By* HARRIS PRATSINIS,
CATHERINE DEMOLIOU-MASON, ALUN HUGHES, AND
DIMITRIS KLETSAS 321

Using Stress Resistance to Isolate Novel Longevity Mutations in *Caenorhabditis elegans*. *By* JAMES N. SAMPAYO, NICOLE L. JENKINS, AND
GORDON J. LITHGOW 324

Telomere Length As a Marker of Oxidative Stress in Primary Human Fibroblast Cultures. *By* VIOLETA SERRA, TILMAN GRUNE, NICOLLE SITTE,
GABRIELE SARETZKI, AND THOMAS VON ZGLINICKI 327

Programmed Cell Death and Senescence in Skeletal Muscle Stem Cells. *By*
KATHRYN WOODS, ANNA MARRONE, AND JANET SMITH 331

Study of the H1 Linker Histone Variant, H1o, during the *in Vitro* Aging of
    Human Diploid Fibroblasts. *By* D.S. TSAPALI, K.E. SEKERI-PATARYAS,
    AND T.G. SOURLINGAS 336

Aging-Related Muscle Dysfunction. Failure of Adaptation to Oxidative
    Stress? *By* S. SPIERS, F. MCARDLE, AND M.J. JACKSON 341

A Theoretical Model for "*in machina*" Experiments on Immunosenescence.
    By SILVANA VALENSIN AND GIANNI DI CARO 344

INDEX OF CONTRIBUTORS 349

Financial assistance was received from:

- AETAS, FONDATION POUR LA RECHERCHE SUR LE VIEILLISSEMENT,
  GENEVA
- "BIOGERONTOLOGY," KLUWER ACADEMIC PUBLISHERS
- CLONTECH AG, BASEL
- CROSSAIR LTD. CO. FOR REGIONAL EUROPEAN AIR TRANSPORT,
  BASEL
- F. HOFFMANN-LA ROCHE AG, BASEL
- FREIWILLIGE AKADEMISCHE GESELLSCHAFT, BASEL
- GERIATRIC UNIVERSITY CLINIC, BASEL
- NOVARTIS PHARMA AG, BASEL
- PUK - PSYCHIATRIC UNIVERSITY CLINIC, BASEL
- RIA - RESEARCH INTO AGEING, LONDON
- SAMW - SWISS ACADEMY OF MEDICAL SCIENCES, BASEL
- VERUM, FOUNDATION FOR BEHAVIOUR AND ENVIRONMENT,
  MUNICH

Preface

Aging is not only an important biological issue, but also an important social and emotional issue that affects an increasing number of elderly people in industrialized countries.

Understanding the molecular mechanisms underlying the physiological aging process may ultimately lead to the understanding and prevention of age-related diseases.

The EMBO Workshop on Molecular and Cellular Gerontology, held in Serpiano, Switzerland, in September 1999 brought together delegates from 14 different countries, representing 21 nationalities. The sessions covered a wide range of complementary research areas: molecular and cellular aspects and theories of aging, molecular basis of the loss of homeostatic maintenance mechanisms, lower and higher model organisms including transgenic animal models, genetics of human longevity, clinical and molecular aspects of age-related diseases, and integrated systems such as brain and immune system aging.

Biological gerontology is clearly a multidisciplinary field, and a platform such as the one presented by this EMBO Workshop is a unique chance to stimulate further international collaboration in this field. The current volume contains 26 plenary papers and 15 extended abstracts presented at this first EMBO Workshop on Molecular and Cellular Gerontology. This volume presents an excellent picture of the state-of-the-art as well as the outlook for future developments in molecular gerontology.

The editors are most grateful to EMBO (European Molecular Biology Organisation) for supporting this workshop and to the New York Academy of Sciences.

We also acknowledge additional support from the following sponsors (in alphabetic order) who helped to make this meeting successful: Aetas Foundation, Geneva; "Biogerontology," Kluwer Academic Publishers; Clontech AG, Basel; Crossair AG, Basel; F. Hoffmann-La Roche AG, Basel; Freiwillige Akademische Gesellschaft (F.A.G.), Basel; Geriatrische Universitätsklinik, Basel; Novartis Pharma AG, Basel; Psychiatrische Universitätsklinik, Basel; Research into Aging, London; SAMW Schweizerische Akademie der Medizinischen Wissenschaften, Basel; Verum Foundation, Munich, Germany.

CH. BRACK, *University of Basel, Switzerland*
G. LITHGOW, *University of Manchester, U.K.*
H. OSIEWACZ, *University of Frankfurt, Germany*
O. TOUSSAINT, *University of Namur, Belgium*

Some New Directions for Research on the Biology of Aging

GEORGE M. MARTIN[a]

Departments of Pathology and Genetics, University of Washington, Seattle, Washington 98195, USA

ABSTRACT: A highly selective, eclectic, and personal view of new directions and new opportunities for research on the biology of aging is briefly outlined. Some concern is raised regarding the present emphasis on the use of centenarians for the definition of genetic loci responsible for unusually robust retention of structure and function. More progress is likely to be made were we to focus on the genetic basis for "elite" aging in middle-aged subjects examined for very specific phenotypes, as these are likely to be far less polygenic. Descriptive gerontology is entering a renaissance, given such new clinical tools as functional MRI and basic science tools such as functional genomics and proteomics. Advances in genomics should expedite answers to such questions as why some avian species have exceptionally long lifespans despite unusual loads of oxidative stress. One hopes to see renewed mechanistic studies, using such tools, at the systems levels. New methodologies are permitting the evaluation of stochastic alterations in gene structure and function in postreplicative cells. The exciting work on molecular misreading should prompt us to reexplore the Orgel hypothesis as it applies to such cell types. Epigenetic shifts in gene expression that occur in association with sexual maturation and the cessation of growth may have deleterious consequences late in the life course. It will therefore be important for gerontologists to investigate the molecular biology of pubescence. Finally, our community should investigate the impact of environmental "gerontogens," agents that accelerate specific processes of aging and specific senescent phenotypes.

INTRODUCTION

Much of the excitement of doing science comes from the fact that we cannot easily predict the most rewarding new directions in almost any field of investigation. That dictum probably applies more to the study of aging than to most biological problems. Biogerontology interacts with all domains of biology at all levels of analysis, from molecules to populations. Add to that menu the mix of psychosocial avenues of research (which I shall make no pretense of addressing in this brief essay), and we have an almost impossible task. I can only offer a *potpourri* of my personal views of what phenotypes, which experimental approaches, and which mechanisms are especially neglected, promising, or important aspects of our science and which therefore deserve more attention as candidates for new directions for aging research.

[a]gmmartin@u.washington.edu

1

PHENOTYPES

Biogerontologists who work with model systems such as *Caenorhabditis elegans* and *Drosophila melanogaster* have reasonably concentrated almost exclusively upon life table parameters, particularly maximum lifespan. (Clearly, one new direction is the systematic definition of anatomic and pathophysiologic aspects of aging in these poorly understood models of aging.) We have only sampled a small subset of such model systems amenable to genetic analysis. Given the rapid advances in genomics, we should soon be able to choose from a much wider array of organisms.[1] Comparative studies across phyla can be very powerful in identifying genetic modulators of aging (and, hence, basic mechanisms) that might obtain in essentially all species ("pan-specific public modulators") or in a group of related animals such as mammals ("group-specific public modulators").[2] Medical types like me, however, have been preoccupied with the biochemical mechanisms underlying a range of late life diseases and disabilities. "Longevity genes" for people who live to be 100 but who have been demented for 15 years are not doing such individuals much good. These different approaches, of course, are not mutually exclusive. We need both.

Great progress is being made in the genetics of dementias of the Alzheimer type, but many other clinically important senescent phenotypes are crying out for more research. I shall highlight a few of these "orphan" phenotypes.[3] In terms of new directions, however, I shall emphasize a third approach, investigations of the biochemical genetic basis for unusually robust retention of structure and function. Such approaches have been neglected, with the exception of the recent flurry of interest in centenarians. Although I am happy to see this new direction of research, I am not very optimistic about our ability to interpret what I guess will be a mountain of polymorphic associations and linkage assignments, perhaps often different for different populations. Many more of us should be focusing upon rates of change of highly specific physiologic domains in middle-aged subjects. I give some arguments to support this point of view below. A recent request for applications from the U.S. National Institute on Aging highlights the importance of such research (http://grants.nih.gov/grants/guide/rfa-files/RFA-AG-99-007.html).

SOME ORPHAN PHENOTYPES OF CLINICAL INTEREST

We can pat ourselves on the back for the good progress we are making in discovering why some individuals are especially sensitive to dementias of the Alzheimer type. Not surprisingly, most of our solid information comes from the rare, "private" autosomal dominant mutations leading to early onset of disease. The hope is that this will lead us to an understanding of the vastly more common late onset "sporadic" forms of the disease. The three genes so far identified tend to support the amyloid cascade hypothesis of pathogenesis.[4] This might prove to be the case for what could be regarded as the clinically more significant "public" polymorphisms, of which only the *APOE* polymorphism has been well established.[5] It is virtually certain that other polymorphic susceptibility alleles will eventually be documented in a number of independent populations. Good candidates include the alpha 2 macroglobulin polymorphism[6] and the FE65 polymorphism.[7] Meanwhile, there are many patients waiting for help with a great variety of other geriatric ailments that impair the quality of life. I have referred to some of these as "orphan" phenotypes because of their comparative neglect by funding agencies and investigators.[3] Examples include idiopathic normal pressure hydrocephalus, trigeminal neuralgia, loss of olfactory sense, peripheral neuropathy, aging of the germ line, and degenerative intervertebral disk disease. Whereas the latter commonly presents in middle age, I consider any disability that is prevalent by about age 45 as a geriatric disorder, as any phenotype that does not reach a phenotypic threshold of expression until then will have escaped the force of natural selection.[2] I will make only a few comments about some of these research opportunities.

I was surprised to learn recently that there is not a single NIA Program-Project grant dealing with the theme of germ-line aging. I pointed out the importance of understanding ovarian aging in a review published some 22 years ago.[8] I am sure many others before and after have made the same point. If anything, it is now of greater clinical significance than ever before, because so many couples are choosing to postpone reproduction. The result has been a booming business in prenatal diagnosis to detect aneuploid pregnancies in older mothers. There is less concern for older fathers, but they too are at increased risk for the production of mutant sperm.[9] Given the recent excitement about the Bloom, Werner, and Cockayne syndromes,[10–12] all of which are likely to impact upon DNA transactions in meiosis as well as in mitosis, one would have expected a renaissance of interest in the fidelity of meiosis as it relates to aging, but I have seen no evidence of such a movement.

The apparent de-repression of integrated viral agents in late life is one of the more fascinating of the relatively neglected gerontological pathologies. Reactivation of childhood chicken pox is the classical example, but I wonder how many other such entities are waiting to be discovered. The trigeminal neuralgia form of *Herpes zoster* is among the most debilitating disorders of older people. Postherpetic neuralgia is in fact the leading cause of intractable pain in the elderly and is the single major cause of suicide in patients over the age of 70 who suffer from chronic pain.[13] Recent evidence indicates that T-cell recognition of an immediate early gene protein (IE63) may be one important component in controlling herpes reactivation from latency.[14] One would hope for more engagement of this problem by immunologists.

Normal pressure hydrocephalus is characterized by the clinical triad of gait disturbance, urinary incontinence, and dementia. A recent review came to the conclusion that "The causes and pathophysiology of idiopathic normal-pressure hydrocephalus (NPH) still remain poorly understood."[15] There are reasons to believe that the underlying pathogenesis is vascular, as arterial hypertension is a significant risk factor[16,17] and periventricular arteriosclerosis has been documented.[18] An interesting clue is that NPH can be associated with systemic lupus erythematosus.[19]

APPROACHES

Discovery Science and Hypothesis-Testing Science

Ever since I can remember, gerontologists have had trouble getting their grants approved by peer-review study groups, because they have proposed descriptive studies that were not sufficiently hypothesis-driven. But the tools for descriptive science (my colleague Lee Hood prefers the term "discovery" science) have improved dramatically. The spectacular advances in genomics will permit the rapid accumulation of our understanding of unusual gene action in new model organisms. Among these, I would give a high priority to avian genomics. Unfortunately for gerontology, the only avian species currently receiving significant attention by molecular geneticists is the domestic chicken. Evolutionary biologists have argued that species that have evolved or retained more efficient flight would be of greater interest for the study of the biology of aging.[20] These species have evolved remarkably long lifespans despite high body temperatures, high blood sugar levels, and high lifetime utilization of calories as compared to comparably sized mammals. Given these considerations, my colleague Steven N. Austad and I have embarked upon a program of avian genomics, with a focus on budgerigars (parakeets) (*Melopsittacus undulatus*) as the model organism.

We will soon have complete genomic sequences for all of the more traditional model organisms (*H. sapiens*, *M. domesticus*, *D. melanogaster*, and *C. elegans*). We will therefore see much more rapid progress on such projects as the linkage analysis of monogenic disorders of relevance to the pathobiology of human aging and the quantitative trait analysis of longevities of worms and flies. The new developments in functional genomics are also very exciting. The most recent initiative is in "proteomics," the goal being to comprehensively define networks of interacting proteins. Meanwhile, great progress has been made in the development and application of massive gene expression array technologies. The decision of *Science* magazine to publish such a study — involving a screening of changes in gene expression with aging and caloric restriction — is an example of the new kind of descriptive science that will help our field develop.[21]

The Importance of Single Cell Analytic Techniques

Studies such as the one just mentioned need to be accompanied, eventually, by single cell analytic techniques in order to address the age-old gerontological confound of shifts in population heterogeneity. For example, the aforementioned study[21] utilized skeletal muscle. In my experience, aging muscles may have more interstitial connective tissue, more adipocytes, more histiocytes, more lymphocytes, more fat cells, and relatively less skeletal muscle. Old muscle may also exhibit multiple foci of satellite cell-mediated muscle regeneration and neuritic sprouting at neuromuscular junctions. The microarray study just referred to[21] in fact provided molecular evidence of what histopathologists have observed. Techniques such as in situ polymerase chain reaction (PCR),[22] laser-assisted microdissection,[23] single cell cytoplasmic aspiration,[24] immunocytochemistry (especially accompanied by confocal microscopy),[25] and quantitative stereology[26] may help sort out these difficulties. Further studies with such tissues as skeletal muscle will also have to disentangle the relative contributions of disuse atrophy and intrinsic biological aging. Both are important to understand at the molecular levels. One way to get around this problem when studying skeletal muscle would be to include muscles like those of the tongue and diaphragm, as they are in constant use. This is also an argument for using myocardium as a model for aging in nondividing populations of cells.

A Revival of Somatic Cell Genetics

I first became involved in the study of cell aging as a frustrated postdoctoral fellow in the laboratory of Guido Pontecorvo at Glasgow University. "Ponte" had pioneered research on the parasexual cycle of fungi, using such genetic recombinational mechanisms as haploidization, mitotic nondisjunction, and mitotic crossing-over to map genes to chromosomes and to define the linear order of chromosomal genes.[27] The lab was in the process of switching from *Aspergillus nidulans* to cultivated human somatic cells. While I was eventually successful in providing some modest cytogenetic evidence for the existence of a parasexual cycle in human somatic cells,[28] my efforts at carrying out serial subcloning and replicative plating were dismal failures. This was 1961. Hayflick and Moorhead[29] had just published their famous paper on the limited replicative lifespans of such cells. But we can now establish cultures of various types of human somatic cells from individual human donors via the transgenic expression of the catalytic subunit of human telomerase.[30] We therefore now have the opportunity to revisit somatic cell genetic approaches to interesting biological questions, including those relevant to gerontology. In addition to exploitation of a parasexual cycle for genetic analysis, one can induce a variety of point mutations, conditionally express a variety of transfected genetic constructs, and fuse cells with a variety of cell types. An example of a specific application would be the use of such cultures to establish genetic haplotypes. The *APOE* chromosomal domain would be a good place to start. Another example would be the development of isogenic strains bearing various mutations for the study of the mitotic cell cycle, apoptosis, terminal differentiation, dedifferentiation, etc. One of my dreams is to turn skin fibroblasts into neuroblasts and to then terminally differentiate them into various types of glial and neural cells. This would give us immediate access to gene action in human CNS cells from individual subjects without the necessity of brain biopsy.

Some Difficulties with the Centenarian Approach

The first conceptual difficulty one faces when one initiates a genetic association study or a genetic segregation study to find genetic alleles that provide major contributions to the probability of becoming a centenarian is the fact that stochastic events clearly make a major impact upon lifespan. First of all, this is predicted by the evolutionary biological theory of aging. According to that theory, senescence is nonadaptive. It is not the result of determinative, sequential gene action. It can be viewed as a byproduct of gene action selected for entirely different purposes, the enhancement of reproductive fitness; some senescent phenotypes may also be the result of idiosyncratic constitutional mutations that escape the force of natural selection.[31,32]

Secondly, human lifespan is likely to be modulated by a large number of disease-associated genetic loci.[33] We may be misled by the reports of enhanced lifespans resulting from single gene mutations or the transfection of a single genetic construct in model organisms. These experiments[34,35] have utilized highly inbred laboratory stocks, each of which can presumed to have its special set of vulnerabilities.

A third concern is the obvious difficulty in getting DNA from members of three generations of a pedigree in order to sort out phase relations of markers.

Fourth, one would be hard put to carry out longitudinal studies in centenarians in order to establish an intrinsically slow rate of change of a putatively relevant biological marker. Moreover, any such physiologic studies would be obscured by the fact that centenarians are likely to suffer from a number of comorbid processes.

Finally, attempts to make generalizations to multiple populations should be difficult and expensive. A priori, given the likelihood of important roles for many gene-gene interactions and environmental-gene interactions, one would expect different subsets of polymorphisms to emerge in different populations.

The Genetic Basis of "Elite Aging"

I will arbitrarily define, as examples of elite aging, rates of decline of specific physiologic parameters that are slower than those observed in 99% of the population of interest. Here is one research strategy for a genetic epidemiology study that might be productive[36]: (1) Decide upon a specific physiologic function, the more specific the better, as a narrow phenotype is more likely to lead to the identification of a few loci with major effects. The assay should be relatively cheap, relatively noninvasive, highly reproducible, and, of great importance, extremely sensitive. (2) Use the assay(s) to screen a very large number of middle-aged individuals in at least two contrasting populations, preferably those that are relatively ethnically homogeneous (e.g., Japanese and Finnish). Why do I choose middle-aged subjects? First of all, as briefly noted above, by the age of 45, quantitative estimates of the rate of decline of the Malthusian parameter in age-structured human populations[2] indicate that the force of natural selection has essentially disappeared. Therefore, significant rates of functional and structural decline in various body systems should be underway by middle age. Second, people are still generally healthy in middle age, so that we have fewer concerns about comorbidity. Third, there are vast numbers of such individuals in most human populations. Fourth, one will often be able to obtain DNA from three generations (the parents and children of the middle-aged index cases). Fifth, sibs of about the same age (say within two years of the index case) should be easy to find, permitting genetic studies on sib-pairs. (3) Choose individuals who score in the upper one percentile of the population. These are the elite individuals who have remarkably robust retention of function in a given assay or assays. Carry out assays in sibs of these index cases. Define a subset of sib pairs with extreme concordance or extreme discordance for the phenotype of interest. Such pairs would provide enhanced power for genetic analysis.[37] (4) Carry out a genome wide screen for markers closely associated with the phenotype of interest. (5) Identify major genetic loci via positional cloning. Such research will be expensive and time consuming but should greatly increase our understanding of the molecular biology of aging in human beings. The challenge will be the definition of the phenotypes to be investigated and the choice of suitably sensitive assays. Perhaps this challenge will help to revive more traditional physiologic research in departments of physiology and medicine. A growing proportion of faculty members in those departments have become biochemists and molecular biologists.

Neuroimaging

Gerontologists have only begun to tap the vast potential of neuroimaging methodologies for the study of brain aging. These include magnetic resonance imaging

(MRI), functional MRI,[38] MRI spectroscopy,[39] and positron emission tomography (PET).[40]

MECHANISMS

Neuroendocrine Cascades

The concept of neuroendocrine cascades in aging was initially championed by Caleb Finch.[41] There are several reasons why I believe gerontologists will be driven to devote comparatively more effort to research on the neuroendocrine system. First, there is the exciting work in *C. elegans* on the regulation of the dauer pathway. Mutations in this pathway can lead to marked increases in lifespan. The trigger for the switch in gene expression is a pheromone related to conditions of larval crowding. This signal is mediated by a neurosecretory function.[42] An insulin-like receptor is an early player in the cascade and a downstream forkhead transcription factor is the key to the regulation of a suite of other yet-to-be-defined genes.[43] The parallels of the effects on lifespan of metabolic controls of fuel utilization in nematodes and in calorically restricted rodents[44] are striking. Second, there is increasing recognition of type 2 diabetes mellitus as a segmental progeroid syndrome. Third, there are the long-lived dwarf mice with lesions in the pituitary and associated hypothyroidism.[45] Fourth, there is the well-documented role of the thyroid hormone on metabolic rate and mitochondrial function. Fifth, there is the clinically important question of sleep disorders of the elderly, the association of sleep with lowering of core body temperatures, and the mysterious "restorative" role of sleep. The neuroendocrine basis for the setting of core body temperature deserves a great deal more research. It is still conceivable that virtually all of the reports of enhancements of animal lifespans are based on a lowering of core body temperatures or metabolic rates. This has long been known for the case of poikilotherms like fish, drosophila, and nematodes. More recent evidence has come from the study of lifespan mutations in *C. elegans,*[46] dwarf mice,[45] and calorically restricted rodents.[47]Sixth, the beautiful research from Cynthia Kenyon's lab on the coupling of reproductive biology with lifespan[48] also argues for more attention to such questions in mammals. Finally, we will probably be faced with an avalanche of clinical trials of such agents as selective estrogen and testosterone receptor agonists, growth hormone, and growth hormone release hormone. We shall need to develop the basic science underpinnings for clinicians.

Late Effects of Early Epigenetic Switches in Gene Expression

The laboratory mouse attains its full adult size at around 6 months of age, just about the time Karen Swisshelm found evidence for the beginning of a selective inactivation of a specific chromosomal site of ribosomal RNA sythesis.[49] One can assume that this mechanism evolved to partially downregulate protein synthesis after the major epoch of somatic growth. But once such a mechanism becomes engaged, it appears to have a "life of its own." The result could be a slowly progressive and insidious course of decline in protein synthesis and turnover into senescent phases of the life course.

IGF2 undergoes parental genomic imprinting. A switch from monoallelic to bi-allelic methylation of its promoter is said to occur "during aging" of human beings.[50] Although the existing data are scarce, they are consistent with a spreading of methylation to the promoter of the other parental allele beginning in the second decade of life. I therefore conclude that such a switch evolved as a result of strong selective pressure. Once initiated, the inactivation continues unabated, with almost complete methylation occurring at advanced ages. Here is yet another example of an adaptive switch without any built-in protection of a "brake" or "reverse gear." Given the importance of insulin-like growth factors in the pathobiology of aging, once can imagine that the late-life consequences of this strategy are substantial.

A third example is the downregulation of the expression of the NR2B subunit of the NMDA receptor at around the time of sexual maturity.[51] This epigenetic switch appears to be one of the causes of the declines in the efficiencies of memory and learning in aging mammals; the constitutive expression of additional copies of that gene leads to superior performances in several assays of memory and learning in transgenic mice.[51] But how would the downmodulation of such an apparently advantageous trait lead to enhanced reproductive fitness? Some fascinating trade-off effects remain to be discovered. We could in fact learn a great deal about the biology of aging by studying the biology of pubescence.

Alterations in Proliferative Homeostasis

Here are additional examples of phenotypes of great interest to biogerontologists that I believe also have their origins via an epigenetic switch in gene expression during early development. Unlike nematodes and fruit flies, many aged tissues from mammalian species (perhaps all mammalian species), including humans,[52] exhibit a mix of atrophy, hyperplasia, benign neoplasia, and, with considerable frequency, malignant neoplasia. Much progress has been made in the dissection of somatic mutational alterations leading to the malignant phenotype.[53] Here we have a concrete example of a stochastic mechanism leading to an important senescent phenotype. Is the hyperplasia that is found in aging tissues a precursor of neoplastic proliferation? This is a plausible scenario, as it could serve as a tumor promoter mechanism. We have almost no knowledge of how such altered homeostasis occurs, although we can guess it has to do with a loss of the fidelity of cell-cell communications and, perhaps, cell-matrix communications. One hypothesis worth testing would be that these aberrations result from asynchronous clonal attenuations[54] of cell types that participate in maintaining proliferative homeostasis.

The vast majority of research on replicative senescence has involved skin fibroblast-like cells. By contrast, there has been a paucity of research on the senescence of cells of the vascular wall (medial smooth muscle and endothelial cells).[55,56] This is unfortunate, as it may prove to be highly relevant to the pathogenesis of atherosclerosis, the major cause of morbidity and mortality in aging populations of western societies.

The question arises as to how clonal attenuation evolved. Most authors have emphasized the notion that it evolved as a mechanism to protect cells from undergoing neoplastic transformation.[57] I have argued elsewhere that perhaps a more plausible explanation is that the switch from stem cell modes of replication to tangential modes of replication was selected to permit finer control of organogenesis.[58] There is very little information as to when and how, during embryogenesis, stem cell modes of replication are abandoned and tangential types of replication are initiated.

Gerontogens

Many biologists are gainfully employed pursuing research on mutagens, carcinogens, and teratogens, but the concept of gerontogens has not yet caught on. I have defined gerontogens as exogenous agents that can accelerate aspects of biological aging.[59] One might classify these into two groups in the same way one might classify deleterious genetic alleles that impact upon aging.[33,60] Thus, there may be "segmental gerontogens" and "unimodal gerontogens." In the real world, exposure to ultraviolet light is inescapable and, therefore, various nature-nurture interactions produce different degrees of ultraviolet-induced aging of the skin. Ultraviolet-induced skin aging can therefore be considered to be an example of a unimodal gerontogen. Ultraviolet light accelerates the rates of development of such phenotypes as epidermal atrophy, basophylic degeneration of collagen, fragmentation of elastin, senile keratosis, Bowen's disease, basal cell carcinoma, squamous cell carcinoma, and malignant melanoma. These phenotypes can be presumed to result from both protein and DNA damage. Oxidative stress is widely believed to be among the more general mechanisms of aging. It will be important, therefore, to have more detailed research on the molecular biology of UV-mediated skin pathologies. An example of such relevant research is the discovery that tandem CC to TT tandem mutations are commonly found in the p53 gene of neoplasms that develop in sun-exposed skin.[61] These tandem mutations are candidate "signature mutations" for oxidative damage.[62]

A good candidate for a segmental gerontogen is tobacco smoke. It will be important to ascertain which of its numerous genotoxic agents produces DNA damage that most closely resembles that which can be observed spontaneously in various aging cell types. A study of parallel effects on proteins will also be of interest. The proneoplastic effects of tobacco smoke are exceedingly well established. Perhaps of even greater relevance to the pathobiology of aging, however, are its effects upon the vascular system, particularly the microvasculature.[63]

The Aging Brain and the Tyranny of Plaques and Tangles

Virtually every laboratory in the world with major interests in the pathogenesis, prevention, or treatment of dementia of the Alzheimer type works on aspects of the metabolism of the beta amyloid precursor protein. Small peptide derivatives of that protein are thought to be responsible for the genesis of one of the two principle histopathological lesions, the neuritic plaque. Judging from our publications (for example), our lab would be counted among the *baptists* of the scientific world. The exceptions are mostly those who are devoted to unraveling neurofibrillary tangles, the second major histopathological lesion. These colleagues are enamored of the microtubular associated protein tau, and hence are referred to as *tauists*. Hardly anyone is examining the molecular biology of the aging brain parenchyma *in between* the famous plaques and tangles. The work on molecular misreading referred to below is an exception. We need more such exceptions.

The Microvasculature

The microvasculature (capillaries, venules, and arterioles) are arguably the most important structures between the plaques and tangles discussed above. After years of relative neglect (except for the observations of congophilic angiopathy), the microvasculature of the brain has finally been getting some attention.[64] Almost nothing is known, however, about changes, in the aging organism, in the kinetics of accumulation of altered structure and function of mitochondria relative to what has been observed in unfractionated brain tissues.[65] We also need more research on the susceptibility of endothelial cells to apoptotic and necrotic injuries, the decline in the proliferative potentials of endothelial cells, and the changes in structure and function of basement membranes. Although there is a significant body of research on the blood brain barrier (BBB) in aging mammals, I am unaware of any systematic studies of individual variations in the degree to which the integrity of the BBB is maintained during aging among populations of human subjects.

Gene Instability in Postreplicative Cells

Powerful new methods for the study of mutation *in vivo* will provide us with a great deal more information about the frequencies and spectra of mutations in various cell types, including postreplicative cells such as neurons.[66] In addition to mutation, we will need to understand the basis for such processes as DNA amplifications, including increases in ploidy. The latter occurs in the aging myocardium, particularly in association with hypertension.

Molecular Misreading in Postreplicative Cells: Should We Revisit the Orgel Hypothesis?

Fred van Leewen and his colleagues[67] have provided us with exciting new evidence for the occurrence of transcriptional errors in aging neurons. There may be a general vulnerability of the genome to develop frameshifts when the complex machinery of DNA-dependent RNA polymerase engages sequences such as GAGAG. We should not ignore, however, molecular misreading at the translational level. It will be particularly important to search for age-related increases in transcriptional and translational fidelity for the case of proteins that are themselves involved in protein synthesis, as this provides the possibility of a positive feedback loop, as first envisaged by Leslie Orgel.[68] Orgel's hypothesis has been "pronounced dead" by gerontologists a number of times, but continues to be discussed.[69,70] But this conclusion was mainly based on experiments with the *in vitro* model of replicative senescence.[71] Perhaps the only test of that idea carried out in postreplicative cells *in vivo* was an experiment performed by Peter Rabinovitch and myself some 17 years ago, in which we challenged young and old mice with the encephalomyocarditis virus, an agent that replicates in neurons and myocardial cells.[72] There were two major predictions: (1) that the yield of virus should decrease in old tissues because of the theorized Orgelian exponential rise in the infidelity of protein synthesis and (2) that the frequencies of mutant viral progeny would rise for the same reasons. Prediction one was falsified, but we were unable to test prediction two because of technical reasons, including the intrinsically high mutation rates of such viral agents. Given the new experiments of van Leeuwen and colleagues, I suggest that the Orgel hypothesis was buried prematurely and that the body should be exhumed for another autopsy.

REFERENCES

1. MARTIN, G.M. & M.S. TURKER. 1988. Model systems for the genetic analysis of mechanisms of aging. J.Gerontol. **43**: B33–B39.
2. MARTIN, G.M., S.N. AUSTAD & T.E. JOHNSON. 1996. Genetic analysis of ageing: role of oxidative damage and environmental stresses. Nat.Genet. **13**: 25–34.
3. MARTIN, G.M. 1993. FRAR course on laboratory approaches to aging. "Orphan" phenotypes in gerontological research. Aging (Milano) **5**: 325–332.
4. HARDY, J.A. & G.A. HIGGINS. 1992. Alzheimer's disease: the amyloid cascade hypothesis. Science **256**: 184–185.
5. SWARTZ, R.H., S.E. BLACK & P. GEORGE-HYSLOP. 1999. Apolipoprotein E and Alzheimer's disease: a genetic, molecular and neuroimaging review. Can. J. Neurol. Sci. **26**: 77–88.
6. BLACKER, D., M.A. WILCOX, N.M. LAIRD et al. 1998. Alpha-2 macroglobulin is genetically associated with Alzheimer disease [see comments]. Nat.Genet. **19**: 357–360.
7. HU, Q., W.A. KUKULL, S.L. BRESSLER et al. 1998. The human FE65 gene: genomic structure and an intronic biallelic polymorphism associated with sporadic dementia of the Alzheimer type. Hum.Genet. **103**: 295–303.
8. MARTIN, G.M. 1977. Cellular aging-postreplicative cells. A review (Part II). Am. J. Pathol. **89**: 513–530.
9. VOGEL, F. & R. RATHENBERG. 1975. Spontaneous mutation in man. Adv. Hum. Genet. **5**: 223–318.
10. BOHR, V.A., G. DIANOV, A. BALAJEE et al. 1998. DNA repair and transcription in human premature aging disorders. J. Invest. Dermatol. Symp. Proc. **3**: 11–13.
11. FRY, M. & L.A. LOEB. 1998. The three faces of the WS helicase [news; comment]. Nat. Genet. **19**: 308–309.
12. YAMAGATA, K., J. KATO, A. SHIMAMOTO et al. 1998. Bloom's and Werner's syndrome genes suppress hyperrecombination in yeast sgs1 mutant: implication for genomic instability in human diseases. Proc. Natl. Acad. Sci. USA **95**: 8733–8738.
13. HESS, T.M., L.J. LUTZ, L.A. NAUSS & T.J. LAMER. 1990. Treatment of acute herpetic neuralgia. A case report and review of the literature [see comments]. Minn. Med. **73**: 37–40.
14. SADZOT-DELVAUX, C., P.R. KINCHINGTON, S. DEBRUS et al. 1997. Recognition of the latency-associated immediate early protein IE63 of varicella-zoster virus by human memory T lymphocytes. J. Immunol. **159**: 2802–2806.
15. MORI, K. & T. MIMA. 1998. To what extent has the pathophysiology of normal-pressure hydrocephalus been clarified? Crit. Rev. Neurosurg. **8**: 232–243.
16. KRAUSS, J.K., J.P. REGEL, W. VACH et al. 1996. Vascular risk factors and arteriosclerotic disease in idiopathic normal- pressure hydrocephalus of the elderly. Stroke **27**: 24–29.
17. PENAR, P.L., W.D. LAKIN & J. YU. 1995. Normal pressure hydrocephalus: an analysis of aetiology and response to shunting based on mathematical modeling. Neurol. Res. **17**: 83–88.
18. AKAI, K., S. UCHIGASAKI, U. TANAKA & A. KOMATSU. 1987. Normal pressure hydrocephalus. Neuropathological study. Acta Pathol. Jpn. **37**: 97–110.
19. UHL, M.D., B.E. WERNER, T.J. ROMANO & B.L. ZIDAR. 1990. Normal pressure hydrocephalus in a patient with systemic lupus erythematosus. J. Rheumatol. **17**: 1689–1691.
20. HOLMES, D.J. & S.N. AUSTAD. 1995. Birds as animal models for the comparative biology of aging: a prospectus. J. Gerontol. A Biol. Sci. Med. Sci. **50**: B59–B66.
21. LEE, C.K., R.G. KLOPP, R. WEINDRUCH & T.A. PROLLA. 1999. Gene expression profile of aging and its retardation by caloric restriction. Science **285**: 1390–1393.
22. LONG, A.A. 1998. In-situ polymerase chain reaction: foundation of the technology and today's options. Eur. J. Histochem. **42**: 101–109.

23. BANKS, R.E., M.J. DUNN, M.A. FORBES et al. 1999. The potential use of laser capture microdissection to selectively obtain distinct populations of cells for proteomic analysis: preliminary findings. Electrophoresis **20:** 689–700.
24. KACHARMINA, J.E., P.B. CRINO & J. EBERWINE. 1999. Preparation of cDNA from single cells and subcellular regions. Methods Enzymol. **303:** 3–18.
25. BROWN, N.L. 1999. Imaging gene expression using antibody probes. Methods Mol. Biol. **122:** 75–91.
26. WEST, M.J., L. SLOMIANKA & H.J. GUNDERSEN. 1991. Unbiased stereological estimation of the total number of neurons in the subdivisions of the rat hippocampus using the optical fractionator. Anat. Rec. **231:** 482–497.
27. PONTECORVO, G. 1958. Trends in Genetic Analysis. Columbia University Press. New York.
28. MARTIN, G.M. & C.A. SPRAGUE. 1969. Parasexual cycle in cultivated human somatic cells. Science **166:** 761–763.
29. HAYFLICK, L. & P.S. MOORHEAD. 1961. The serial cultivation of human diploid cell strains. Exp. Cell Res. **25:** 585–621.
30. BODNAR, A.G., M. OUELLETTE, M. FROLKIS et al. 1998. Extension of life-span by introduction of telomerase into normal human cells [see comments]. Science **279:** 349–352.
31. KIRKWOOD, T.B. & M.R. ROSE. 1991. Evolution of senescence: late survival sacrificed for reproduction. Philos. Trans. R. Soc. Lond. B Biol. Sci. **332:** 15–24.
32. ROSE, M.R. 1991. Evolutionary Biology of Aging. Ref Type: Serial (Book, Monograph). Oxford University Press.
33. MARTIN, G.M. 1978. Genetic syndromes in man with potential relevance to the pathobiology of aging. Birth Defects Orig. Art. Ser. **14:** 5–39.
34. LIN, Y.J., L. SEROUDE & S. BENZER. 1998. Extended life-span and stress resistance in the Drosophila mutant methuselah [see comments]. Science **282:** 943–946.
35. PARKES, T.L., A.J. ELIA, D. DICKINSON et al. 1998. Extension of Drosophila lifespan by overexpression of human SOD1 in motorneurons [comment] [see comments]. Nat. Genet. **19:** 171–174.
36. MARTIN, G.M. 1998. Toward a genetic analysis of unusually successful neural aging. In Handbook of the Aging Brain. E. Wang & D.S. Snyder, Eds.: 125–134. Ref Type: Serial (Book, Monograph). Academic Press. New York.
37. ZHANG, H. & N. RISCH. 1996. Mapping quantitative-trait loci in humans by use of extreme concordant sib pairs: selected sampling by parental phenotypes [published erratum appears in Am. J. Hum. Genet. 1997. **60:** 748–749]. Am. J. Hum. Genet. **59:** 951–957.
38. CORKIN, S. 1998. Functional MRI for studying episodic memory in aging and Alzheimer's disease. Geriatrics **53** (Suppl. 1)**:** S13–S15.
39. ROTHMAN, D.L., N.R. SIBSON, F. HYDER et al. 1999. In vivo nuclear magnetic resonance spectroscopy studies of the relationship between the glutamate-glutamine neurotransmitter cycle and functional neuroenergetics [In Process Citation]. Philos. Trans. R. Soc. Lond. B Biol. Sci. **354:** 1165–1177.
40. CASELLI, R.J., N.R. GRAFF-RADFORD, E.M. REIMAN et al. 1999. Preclinical memory decline in cognitively normal apolipoprotein E-epsilon4 homozygotes. Neurology **53:** 201–207.
41. FINCH, C.E. 1976. The regulation of physiological changes during mammalian aging. Q. Rev. Biol. **51:** 49–83.
42. AILION, M., T. INOUE, C.I. WEAVER et al. 1999. Neurosecretory control of aging in Caenorhabditis elegans. Proc. Natl. Acad. Sci. USA **96:** 7394–7397.
43. LIN, K., J.B. DORMAN, A. RODAN & C. KENYON. 1997. daf-16: an HNF-3/forkhead family member that can function to double the life-span of Caenorhabditis elegans [see comments]. Science **278:** 1319–1322.
44. SOHAL, R.S. & R. WEINDRUCH. 1996. Oxidative stress, caloric restriction, and aging. Science **273:** 59–63.
45. BARTKE, A., H.M. BROWN-BORG, A.M. BODE et al. 1998. Does growth hormone prevent or accelerate aging? Exp. Gerontol. **33:** 675–687.
46. VAN VOORHIES, W. 1999. Genetic and environmental conditions that increase longevity in Caenorhabditis elegans decrease metabolic rate. Proc. Natl. Acad. Sci. USA **96:** 11399–11403.

47. Turturro, A. & R.W. Hart. 1991. Longevity-assurance mechanisms and caloric restriction. Ann. N.Y. Acad. Sci. **621:** 363–372.
48. Hsin, H. & C. Kenyon. 1999. Signals from the reproductive system regulate the lifespan of *C. elegans* [see comments]. Nature **399:** 362–366.
49. Swisshelm, K., C.M. Disteche, J. Thorvaldsen *et al.* 1990. Age-related increase in methylation of ribosomal genes and inactivation of chromosome-specific rRNA gene clusters in mouse. Mutat. Res. **237:** 131–146.
50. Issa, J.P., P.M. Vertino, C.D. Boehm *et al.* 1996. Switch from monoallelic to biallelic human IGF2 promoter methylation during aging and carcinogenesis. Proc. Natl. Acad. Sci. USA **93:** 11757–11762.
51. Tang, Y.-P., E. Shimizu, G.R. Dube *et al.* 1999. Genetic enhancement of learning and memory in mice. Nature **401:** 63–69.
52. Martin, G.M. 1979. Proliferative homeostasis and its age-related aberrations. Mech.Ageing Dev. **9:** 385–391.
53. Fearon, E.R. 1995. Molecular genetics of colorectal cancer. Ann. N.Y. Acad. Sci. **768:** 101–110.
54. Martin, G.M., C.A. Sprague, T.H. Norwood & W.R. Pendergrass. 1974. Clonal selection, attenuation and differentiation in an *in vitro* model of hyperplasia. Am. J. Pathol. **74:** 137–154.
55. Martin, G.M. & C.A. Sprague. 1972. Clonal senescence and atherosclerosis. Lancet **2:** 1370–1371.
56. Chang, E. & C.B. Harley. 1995. Telomere length and replicative aging in human vascular tissues. Proc. Natl. Acad. Sci. USA **92:** 11190–11194.
57. Campisi, J. 1997. The biology of replicative senescence. Eur. J. Cancer **33:** 703–709.
58. Martin, G. M. 1993. Clonal attenuation: causes and consequences [editorial]. J. Gerontol. **48:** B171–B172.
59. Martin, G.M. 1987. Interactions of aging and environmental agents: the gerontological perspective. Prog. Clin. Biol. Res. **228:** 25–80.
60. Martin, G.M. 1982. Syndromes of accelerated aging. Natl. Cancer Inst. Monogr. **60:** 241–247.
61. Queille, S., S. Seite, S. Tison *et al.* 1998. p53 mutations in cutaneous lesions induced in the hairless mouse by a solar ultraviolet light simulator. Mol. Carcinog. **22:** 167–174.
62. Reid, T.M. & L.A. Loeb. 1993. Tandem double CC-->TT mutations are produced by reactive oxygen species. Proc. Natl. Acad. Sci. USA **90:** 3904–3907.
63. Monfrecola, G., G. Riccio, C. Savarese *et al.* 1998. The acute effect of smoking on cutaneous microcirculation blood flow in habitual smokers and nonsmokers. Dermatology **197:** 115–118.
64. de la Torre, J. C. 1997. Hemodynamic consequences of deformed microvessels in the brain in Alzheimer's disease. Ann. N.Y. Acad. Sci. **826:** 75–91.
65. Cortopassi, G.A. & A. Wong. 1999. Mitochondria in organismal aging and degeneration. Biochim. Biophys. Acta **1410:** 183–193.
66. Dolle, M.E., H.J. Martus, M. Novak *et al.* 1999. Characterization of color mutants in lacZ plasmid-based transgenic mice, as detected by positive selection [In Process Citation]. Mutagenesis **14:** 287–293.
67. van Leeuwen, F.W., J.P. Burbach & E.M. Hol. 1998. Mutations in RNA: a first example of molecular misreading in Alzheimer's disease. Trends Neurosci. **21:** 331–335.
68. Orgel, L.E. 1970. The maintenance of the accuracy of protein synthesis and its relevance to ageing: a correction. Proc. Natl. Acad. Sci. USA **67:** 1476.
69. Holliday, R. 1996. The current status of the protein error theory of aging. Exp. Gerontol. **31:** 449–452.
70. Gallant, J., C. Kurland, J. Parker *et al.* 1997. The error catastrophe theory of aging. Point counterpoint [letter]. Exp. Gerontol. **32:** 333–346.
71. Harley, C.B., J.W. Pollard, J.W. Chamberlain *et al.* 1980. Protein synthetic errors do not increase during aging of cultured human fibroblasts. Proc. Natl. Acad. Sci.USA **77:** 1885–1889.
72. Rabinovitch, P.S. & G.M. Martin. 1982. Encephalomyocarditis virus as a probe of errors in macromolecular synthesis in aging mice. Mech. Ageing Dev. **20:** 155–163.

Molecular Gerontology

Bridging the Simple and the Complex

THOMAS B.L. KIRKWOOD

University of Newcastle, Department of Gerontology, Wolfson Research Centre, Newcastle General Hospital, Westgate Road, Newcastle upon Tyne NE4 6BE, UK

ABSTRACT: It is clear, both empirically and theoretically, that the mechanisms of aging are multiple and complex. Nevertheless, single gene mutations and simple interventions such as calorie restriction have broad effects on the senescent phenotype. The major challenge is to unite highly reductionist analysis of molecular components with integrative model systems that can "put it all together." Two themes are developed. In the first, biochemical models are described that show how the network concept of cellular aging can be used to integrate multiple biochemical mechanisms that contribute to cellular instability. In the second theme, the role of intrinsic developmental chance is examined as a major factor contributing, in addition to genes and environment, to the divergence of the senescent phenotype. The implications of these themes for research strategies in molecular gerontology are discussed.

INTRODUCTION

Aging is the progressive decline in an organism's internal condition, leading to an increasing risk of death. Although it is now commonly accepted that aging evolved because of the declining force of natural selection with age and because of the metabolic costs of longevity assurance, there is as yet no general consensus about the exact biochemical mechanisms that are involved. Evolution theory suggests that aging is caused by a life-long accumulation of random damage in somatic cells and tissues due to an evolved limitation in the levels of key maintenance functions. This idea, termed the disposable soma theory, predicts a central role for cell maintenance and stress response mechanisms in regulating the duration of life.[1-5] There is potentially a large number of such mechanisms, and individual theories have focused, in particular, on the roles of free radicals and oxidative damage,[6-8] aberrant proteins,[9,10] defective mitochondria,[11-13] and somatic mutations.[14-17]

Two important implications of the disposable soma theory are (1) the need to consider integrated approaches to the study of cell maintenance, and (2) the explicit prediction that the mechanisms of cell aging are intrinsically stochastic.

The concept of a network of maintenance functions brings coherence to the study of individual reactions and processes that contribute to the overall process of senescence, but it also exposes some of the difficulties. In particular, it is necessary to understand how each component process affects the function and viability of the network as a whole and to examine their potentially synergistic interactions. The recognition that the mechanisms of aging are stochastic has broad implications for the investigation of aging at the molecular and cellular levels and for understanding the limitations on genomic specification of the senescent phenotype.[18]

THE SCOPE OF NETWORKS

To examine the implications of the network concept for intracellular mechanisms of aging, a series of theoretical models of candidate molecular mechanisms of aging has been developed.[19–21] The last of these, the so-called MARS model (for mitochondria, aberrant proteins, radicals, scavengers), incorporates the following features: accumulation of defective mitochondria; accumulation of aberrant proteins in protein synthesis; the damaging actions of oxygen-free radicals and the protective role of antioxidant enzymes such as superoxide dismutase; and the turnover of proteins by proteolytic scavengers.

The MARS model successfully predicted many of the experimental findings in cells and organisms. These include: a sharp rise with age in the fraction of inactive proteins, explaining the frequently observed loss of specific enzyme activity; the development with age of only a slight rise in the fraction of erroneous proteins, explaining why a decline in enzyme specificity is only rarely observed; a significant increase with age in protein half-life; a decrease with age in the mitochondrial population; an increase with age in the fraction of defective mitochondria; an increase with age in the average rate of free radical production per mitochondrion; and a decrease with age in the average level of ATP generation per mitochondrion.

Nevertheless, even a model with the predictive power of the MARS model is a compromise between complexity and simplicity. It provides a framework for further conceptual development as new concepts and data emerge. In particular, the MARS model did not consider the potentially important effect of cell division on candidate mechanisms of aging. An example is the recent extension of the mitochondrial theory of aging proposed by de Grey[22] and modeled by Kowald and Kirkwood.[23] The plausibility of the mitochondrial theory rests crucially on the kinetics of accumulation of defective mitochondria within cells. The most common suggestion is that defective mitochondria have some kind of replication advantage, perhaps because they have a smaller genome after a deletion event. This concept, however, has several problems.

At least in mammalian cells, all genes encoded by the mitochondrial genome are necessary for some step of the energy generation process. Therefore, damage to the mtDNA is likely to result in impaired ATP production. Several energy-dependent steps are needed for mitochondrial replication, namely, (1) the respiratory enzymes have to be doubled by protein synthesis within the mitochondrion; (2) all other proteins have to be doubled by import from the cytoplasm; and (3) the mtDNA has to be replicated. Considering this, it is hard to see how a defective mitochondrion can achieve an accelerated division rate. The idea is also inconsistent with an important experimental finding: damaged mitochondria accumulate more in postmitotic tissues such as brain and muscle than in mitotically active tissues such as liver or skin.[24–26] Whenever a cell divides, its mitochondrial population has to double. Mitochondria in rapidly dividing tissues therefore have higher growth rates than do mitochondria in postmitotic tissues. But if defective mitochondria indeed proliferate faster than intact ones, an overall increase in the mitochondrial growth rate should mean that the defective mitochondria accumulate faster; in other words, defective mitochondria should accumulate more quickly in dividing cells. Nevertheless, several studies have shown that muscle fibers with abnormalities of the electron transport system are apparently taken over by mitochondria of a single mutant mtDNA genotype,[27,28] suggesting that defective mitochondria somehow outcompete the wild-type.

The paradox is thus that the experimental findings point to clonal expansion of defective mitochondria but at the same time argue against a faster division rate of these mitochondria.

The resolution by de Grey of this paradox was to suggest that mutant and wild-type mitochondria differ not in the rates of replication but in their rates of degradation. A mitochondrion typically remains only a few weeks within the cell, after which it is degraded. The net growth rate of a population of mitochondria is the difference between the birth (replication) and death (degradation) rates. Mutants can therefore accumulate in a population either by increasing their division rate or by lowering their rate of degradation. If damaged mitochondria accumulate because they have a slower degradation rate, this would explain clonal expansion while avoiding the need for defective mitochondria to divide faster than wild-type.

The power of this kind of theoretical modeling is that it shows how apparently contradictory experimental data can be explained through the interplay of more than one process, namely, that damaged mitochondria are degraded slower than intact ones *and* have a growth disadvantage. The growth disadvantage reflects the impaired energy generation and reduced proton gradient of defective mitochondria. The delayed degradation gives damaged mitochondria the necessary selection advantage. This leads to clonal expansion of damaged organelles even though they have a lower replication rate than do healthy ones.

The network approach has important implications for other aspects of molecular gerontology. An important prediction of the model is that rapid cell division is an efficient mechanism not only for the rejuvenation of the mitochondrial population but also for the dilution of molecular waste products such as protease-resistant aberrant proteins. This means that tissue culture experiments studying *in vitro* aging may underestimate the role of these kinds of damage-accumulation mechanisms. It is now widely accepted that the phenomenon of cell replicative senescence (the Hayflick limit) results from the progressive shortening of telomeres during cell division.[29] However, while telomere reduction must eventually result in loss of cell proliferative capacity, this process is unlikely to be the sole contributor to cell deterioration with age. In populations of cells dividing more slowly than under the forced conditions of *in vitro* culture, other mechanisms probably also contribute to cell aging. The network concept reminds us not only that these mechanisms may interact, as for example when oxidative stress accelerates telomere loss,[30] but also that individual mechanisms may make very different relative contributions *in vivo* and *in vitro*.

THE PLAY OF CHANCE

Individual lifespans in inbred laboratory organisms (e.g., nematodes, fruitflies, and mice) show marked variance.[31] These variations are at least as great, as a percentage of the species average lifespan, as those of human identical twins who experience far more diverse environments during life. Reproductive senescence in inbred mice also varies widely, epitomized by inbred mice, which differ extensively in the time of onset of the lengthening and eventual loss of estrous cycles.[32] Because reproductive aging in female mammals ultimately results from loss of ovarian oocytes, the individual differences in reproductive aging are most likely linked to observed wide variations in oocyte numbers formed during development.[33,34] The nervous system also contains variable numbers of postmitotic and irreplaceable neurons,[35,36] which may be relevant to variations in the timing of neurodegenerative losses of function in later life.[37]

Variations in cell numbers are partially corrected or constrained during development. The nematode is an extreme example, with its tightly controlled number of cells. However, the positions of cell bodies and the contacts between cells are less precisely determined in certain body regions. Variations in the fates of individual cells arise during development and in adult life and may contribute to the diversity of individual senescent phenotypes. The basic mechanisms contributing to chance variations in cell fate are asymmetry in cell division, randomness in cell differentiation and cell death, and variations in cell migration. Cell differentiation is determined by ligand-gated receptor activation both at the gene level by binding of transcription factors and at the membrane level, by signal cascades initiated by binding of external and internal ligands to membrane receptor proteins. In both cases, a large number of essentially random molecular interactions are required before a ligand fortuitously collides with and binds to the appropriate receptor site.

Multicellular organisms comprise complex hierarchies and lineages of cells. They are effectively clonal mosaics of cells. A cell may differentiate down alternative pathways according to which pathway is activated first. Mammalian tissues vary in clonal heterogeneity, that is, the incorporation of separate, cross-differentiating cell clones versus a single clone. In the developing cerebral cortex, for example, retroviral marker studies showed that neurons from a single clone can disperse across wide distances and participate in different cytoarchitectonic areas.[38,39] Other tissues, particularly those with continual renewal from stem cells such as the hematopoietic system and the intestinal epithelium, are maintained through the repeated expansion of varied clonal sub-populations. After clones expand, the cells are eventually discarded or die, so that a continuous process of clonal succession, or attenuation, occurs. With advancing age, stem cells may decline in some tissues, resulting in greater chance fluctuations in the clonal composition of the tissues. Skewing of X-chromosome inactivation patterns in blood lymphocytes increase with age in women, suggesting stochastic clonal loss, possibly with stem cell decline.

Cultured diploid somatic cells show marked variation in clonal division potential.[40] Within a mass population individual cells vary extensively in the numbers of cell divisions they can support. The *average* number of divisions declines progressively with advancing cell culture age, but even early passage cultures exhibit wide variance in clonal division potential. Even the two daughter cells from a single division may vary greatly in their division potential. This example provides incontrovertible evidence that the determination of cell replicative senescence is stochastic.

A longstanding hypothesis suggests that somatic genome instability, in particular, somatic mutation, contributes causally to aging processes. There is now much evidence in mammals for the accumulation throughout life of somatic cell DNA mutations, epigenetic defects such as loss of DNA methylation patterns, and abnormalities of chromosomal number and structure.[17,41,42] Together, these lesions result in a general drift away from the unique genomic identity of the zygote.

At a molecular level, most mRNA and proteins of adult mammals are replaced through ongoing biosynthesis. Intracellular proteins have half-lives ranging from minutes to days. The constant turnover removes most damaged or erroneous molecules, but some extracellular proteins are not replaced in adults and can therefore also accumulate damage. Molecules that are particularly long-lived include collagen and elastin of tendon and connective tissue, crystallins of the lens, and dentin in teeth.

In summary, there is huge scope for chance variations arising during development and in adult life to influence the aging phenotype.[18] Recognition that chance plays an important role in modulating outcomes of aging, in addition to the roles played by genes and external environment, may also help gain insight into the so-called 'mortality plateaus' of advanced ages.[43–45] In humans after about age 90 and in several invertebrates, the Gompertzian exponential increase in age-specific mortality rates eventually slows down and may even decrease at advanced age. Genetic heterogeneity is suspected as a cause of mortality plateaus. However, genetic heterogeneity does not explain the mortality plateaus seen in *Drosophila,* because mortality plateaus were similar in genetically homogeneous and heterogeneous stocks.[43] *Caenorhabditis* also had mortality plateaus, although isogenic worms showed less of a plateau than does a genetically heterogeneous population.[46] A nonexclusive alternative to genetic heterogeneity to explain the mortality plateaus is that variance in frailty arises through *intrinsic* chance variations during development and adult life.

CONCLUSIONS

Recognition that a network of mechanisms contributes to aging and that the components of this network are themselves intrinsically stochastic needs to be firmly embedded in the experimental approaches to investing molecular gerontology. This is not to deny the importance of conventional reductionist analyses. Indeed, the challenge of identifying the contributions to aging from heterogeneous fine scale molecular events should provide an impetus to further refine the reductionist techniques at our disposal. Nevertheless, we need to recognize that the era of individual, competing hypotheses about causes of aging is over.

The disposable soma theory provides a framework to consider how the structure and function of an organism reflect design features for the tolerance of chance variations that arise during development. Natural selection only requires the soma to function during the normal life expectancy in the wild environment. By analogy with engineering, some tolerance is built into the living machinery, which is seen in the chance variations affecting cell numbers and in the random faults affecting molecules and higher order structures.

If we are to understand the full complexity of how networks and chance influence the aging process, there is considerable need for multidiscplinary approaches, including mathematical modeling. The value of mathematical models as predictive and analytic tools for the understanding of complex processes is amply shown in carcinogenesis, chemotaxis, development, immunology, neurophysiology, wound healing, as well as aging. The challenge for molecular gerontology is to show how a range of often minor alterations in the populations of molecules and cells that make up the organism can explain the major alterations in phenotype that lead to increased frailty, morbidity, and mortality with advancing age.

ACKNOWLEDGMENT

These ideas have been developed in collaboration with Caleb Finch, Claudio Franceschi, and Axel Kowald.

REFERENCES

1. KIRKWOOD, T.B.L. 1977. Evolution of ageing. Nature **270:** 301–304.
2. KIRKWOOD, T.B.L. 1981. Evolution of repair: survival versus reproduction. *In* Physiological Ecology: An Evolutionary Approach to Resource Use. C.R. Townsend & P. Calow, Eds. : 165–189. Blackwell Scientific Publications. Oxford.
3. KIRKWOOD, T.B.L. & R. HOLLIDAY. 1979. The evolution of ageing and longevity. Proc. R. Soc. Lond., Ser. B **205:** 531–546.
4. KIRKWOOD, T.B.L. & C. FRANCESCHI. 1992. Is aging as complex as it would appear? Ann. N.Y. Acad. Sci. **663:** 412–417.
5. RATTAN, S.I.S. & O. TOUSSAINT. 1996. Molecular Gerontology: Research Status and Strategies. Plenum Press. New York.
6. HARMAN, D. 1956. A theory based on free radical and radiation chemistry. J. Gerontol. **11:** 298–300.
7. HARMAN, D. 1981. The aging process. Proc Natl. Acad. Sci. USA **78:** 7124–7128.
8. SOHAL, R.S. & W.C. ORR. 1995. Is oxidative stress a causal factor in aging? *In* Molecular Aspects of Aging. K. Esser & G.M. Martin, Eds. : 109–127. J. Wiley & Sons. Chichester.
9. ORGEL, L.E. 1963. The maintenance of the accuracy of protein synthesis and its relevance to ageing. Proc. Natl. Acad. Sci. USA **49:** 517–521.
10. ROSENBERGER, R.F. 1991. Senescence and the accumulation of altered proteins. Mutat. Res. **256:** 255–262.
11. RICHTER, C. 1988. Do mitochondrial fragments promote cancer and aging? FEBS Lett. **241:** 1–5.
12. LINNANE, A.W., S. MARZUKI, T. OZAWA & M. TANAKA. 1989. Mitochondrial DNA mutations as an important contributor to ageing and degenerative diseases. Lancet **I:** 642–645.
13. WALLACE, D.C. 1992. Mitochondrial genetics: a paradigm for aging and degenerative diseases? Science **256:** 628–632.
14. SZILARD, L. 1959. On the nature of the aging process. Proc. Natl. Acad. Sci. USA **45:** 35–45.
15. CURTIS, H.J. 1966. Biological Mechanisms of Aging. CC Thomas. Springfield, IL.
16. BURNET, F.M. 1974. Intrinsic Mutagenesis: A Genetic Approach to Ageing. J. Wiley & Sons. New York.
17. VIJG, J. 1990. DNA sequence changes in aging: how frequent, how important? Aging **2:** 227–229.
18. FINCH, C.E. & T.B.L. KIRKWOOD. 2000. Chance, Development and Aging. Oxford University Press. New York.
19. KOWALD, A. & T.B.L. KIRKWOOD. 1993. Mitochondrial mutations, cellular instability and ageing: modelling the population dynamics of mitochondria. Mutat. Res. **295:** 93–103.
20. KOWALD, A. & T.B.L. KIRKWOOD. 1994. Towards a network theory of ageing: a model combining the free radical theory and the protein error theory. J. Theor. Biol. **168:** 75–94.
21. KOWALD, A. & T.B.L. KIRKWOOD. 1996. A network theory of ageing: the interactions of defective mitochondria, aberrant proteins, free radicals and scavengers in the ageing process. Mutat. Res. **316:** 209–236.
22. DE GREY, A.D.N.J. 1997. A proposed refinement of the mitochondrial free radical theory of aging. BioEssays **19:** 161–166.
23. KOWALD, A. & T.B.L. KIRKWOOD. 2000. Accumulation of defective mitochondria through delayed degradation of damaged organelles and its possible role in the ageing of post-mitotic and dividing cells. J. Theor. Biol. **202:** 145–160.

24. CORTOPASSI, G.A., D. SHIBATA, N.W. SOONG et al. 1992. A pattern of accumulation of a somatic deletion of mitochondrial DNA in aging human tissues. Proc. Natl. Acad. Sci. USA **89:** 7370–7374.

25. SUGIYAMA, S., M. TAKASAWA, M. HAYAKAWA et al. 1993. Changes in skeletal muscle, heart and liver mitochondrial electron transport activities in rats and dogs of various ages. Biochem. Mol. Biol. Intl. **30:** 937–944.

26. LEE, H.C., C.Y. PANG, H.S. HSU et al. 1994. Differential accumulations of 4977 bp deletion in mitochondrial DNA of various tissues in human ageing. Biochim. Biophys. Acta **1226:** 37–43.

27. MÜLLER-HÖCKER, J., P. SEIBEL, K. SCHNEIDERBANGER et al. 1993. Different in situ hybridization patterns of mitochondrial DNA in cytochrome c oxidase-deficient extraocular muscle fibres in the elderly. Virchows Arch. A **422:** 7–15.

28. BRIERLEY, E.J., M.A. JOHNSON, R.N. LIGHTOWLERS et al. 1998. Role of mitochondrial DNA mutations in human aging: implications for the central nervous system and muscle. Ann. Neurol. **43:** 217–223.

29. BODNAR, A.G., M. OUELLETTE, M. FROLKIS et al. 1998. Extension of life-span by introduction of telomerase into normal human cells. Science **279:** 349–352.

30. VON ZGLINICKI, T., G. SARETZKI, W. DOCKE et al. 1995. Mild hyperoxia shortens telomeres and inhibits proliferation of fibroblasts. A model for senescence. Exp. Cell Res. **220:** 186–193.

31. FINCH, C.E. & R.E. TANZI. 1997. The genetics of aging. Science **278:** 407–411.

32. FINCH, C.E., L.S. FELICIO, K. FLURKEY et al. 1980. Studies on ovarian-hypothalamic-pituitary interactions during reproductive aging in C57BL/6J mice. In Brain-Endocrine Interaction. IV. Neuropeptides in Development and Aging. D. Scott & J.L. Sladek, Jr. Peptides, vol 1, suppl. 1. :163–175. ANKHO International. Fayetteville, NY.

33. GOSDEN, R.G., S.C. LAING, L.S. FELICIO et al. 1983. Imminent oocyte exhaustion and reduced follicular recruitment mark the transition to acyclicity in aging C57Bl/6J mice. Biol. Reprod. **12:** 2483–2488.

34. RICHARDSON, S.J., V. SENIKAS, J.F. NELSON. 1987. Follicular depletion during the menopausal transition: evidence for accelerated loss and ultimate exhaustion. J. Clin. Edocrinol. Metab. **65:** 1231–1237.

35. WEST, M.J. 1993. Regionally specific loss of neurons in the subdivisions of the rat hippocampus using the optical fractionator. Neurobiol. Aging **14:** 287–293.

36. RASMUSSEN, T., T. SCHLIEMANN, J.C. SORENSEN et al. 1996. Memory impaired rats: no loss of principal hippocampal and subicular neurons. Neurobiol. Aging **17:** 143–147.

37. FINCH, C.E. 1997. Longevity: is everything under control? Non-genetic and non-environmental sources of variation. In Longevity: to the Limits and Beyond. J.-M. Robine, J. Vaupel, B. Jeune & M. Allard , Eds. : 165–178. Springer-Verlag. Heidelberg.

38. GROVE, E.A., T.B.L. KIRKWOOD & J. PRICE. 1992. Neuronal precursor cells in the rat hippocampal formation contribute to more than one cytoarchitectonic area. Neuron **8:** 217–229.

39. WALSH, C. & C.L. CEPKO. 1992. Widespread dispersion of neuronal clones across functional regions of the cerebral cortex. Science **255:** 434–440.

40. SMITH, J.R. & R.G. WHITNEY. 1980. Intraclonal variation in proliferative potential of human diploid fibroblasts: stochastic mechanism for cellular aging. Science **207:** 82–84.

41. HOLLIDAY, R. 1987. The inheritance of epigenetic defects. Science **238:** 163–170.

42. DOLLE, M.E.T., H. GESE, G.L. HOPKINS et al. 1997. Rapid accumulation of genome rearrangements in liver but not in brain of old mice. Nature Genet. **17:** 431–434.

43. CURTSINGER, J.W., H.H. FUKUI, D.R. TOWNSEND. & J.W. VAUPEL. 1992. Demography of genotypes: failure of the limited lifespan paradigm in Drosophila melanogaster. Science **258:** 461–463.

44. VAUPEL, J.W., J.R. CAREY, K. CHRISTENSEN et al. 1998. Biodemographic trajectories of longevity. Science **280:** 855–860.

45. KIRKWOOD, T.B.L. 1999. Evolution, molecular biology and mortality plateaus. In Molecular Biology of Aging, Alfred Benzon Symposium 44. V.A. Bohr, B.F.C. Clark & T. Stevnsner, Eds. : 383–390. Munksgaard. Copenhagen.

46. BROOKS, A., G.J. LITHGOW & T.E. JOHNSON. 1994. Mortality rates in a genetically heterogeneous population of Caenorhabditis elegans. Science **263:** 688–671.

Metabolic Control and Gene Dysregulation in Yeast Aging

S. MICHAL JAZWINSKI[a]

Department of Biochemistry and Molecular Biology, Louisiana State University Health Sciences Center, New Orleans, Louisiana 70112, USA

ABSTRACT: Life span in the yeast *Saccharomyces cerevisiae* is usually measured by the number of divisions individual cells complete. Four broad physiologic processes that determine yeast life span have been identified: metabolic control, resistance to stress, chromatin-dependent gene regulation, and genetic stability. A pathway of interorganelle communication involving mitochondria, the nucleus, and peroxisomes has provided a molecular mechanism of aging based on metabolic control. This pathway functions continuously, rather than as an on-off switch, in determining life span. The longevity gene *RAS2* modulates this pathway. *RAS2* also modulates a variety of other cellular processes, including stress responses and chromatin-dependent gene regulation. An optimal level of Ras2p activity is required for maximum longevity. This may be due to the integration of life maintenance processes by *RAS2*, which functions as a homeostatic device in yeast longevity. Loss of transcriptional silencing of heterochromatic regions of the genome is a mark of yeast aging. It is now clear that the functional status of chromatin plays an important role in aging. Changes in this functional status result in gene dysregulation, which can be altered by manipulation of the histone deacetylase genes. Silencing of ribosomal DNA appears to be of particular importance. Extrachromosomal ribosomal DNA circles are neither sufficient nor necessary for yeast aging.

INTRODUCTION

Aging of the yeast *Saccharomyces cerevisiae* is measured by the number of divisions that individual cells of this unicellular eukaryote complete before they die,[1,2] rather than by any phenotypic changes that occur during this replicative life span. More recently, aging of this organism has also been studied as the passage of time before cells in a stationary culture lose viability.[3] These two modes of aging may have more in common than meets the eye, as will be discussed.

Genetic analysis of aging has yielded a plentiful bounty of 17 longevity genes that play a role in determining the replicative life span. These genes participate in four broad physiologic processes: metabolic control, stress resistance, chromatin-dependent gene regulation, and genetic stability. Interestingly, some of these genes, such as *RAS2*,[4–6] appear to function in more than one of these processes, suggesting that they may play an integrative role. The genetics of survival in stationary phase is also

[a]Address for correspondence: Dr. S. Michal Jazwinski, Department of Biochemistry and Molecular Biology, LSU Health Sciences Center, 1901 Perdido St., Box P7-2, New Orleans, LA 70112, USA.
sjazwi@lsumc.edu

a well-developed area.[7] The key feature of the adaptation to stationary phase is the induction of stress responses. It was found recently that oxidative stress is one of these, because of the reduced survival in the absence of Cu, Zn-superoxide dismutase.[3] This article deals with replicative life span. At the end, replicative life span and survival in stationary phase will be juxtaposed.

THE RETROGRADE RESPONSE

Several years ago, an intracellular signaling pathway from the mitochondrion to the nucleus was discovered in yeast and named retrograde regulation.[8] In the retrograde response, a signal from the mitochondrion leads to activation of a heterodimeric transcription factor,[9,10] which translocates into the nucleus and binds to the promoters of a variety of genes. These genes encode various metabolic enzymes that are located in the mitochondrion, the cytoplasm, and the peroxisome.[11–13] Thus, the retrograde response constitutes a mode of interorganelle communication. The retrograde response has typically been induced in petite yeast. These are yeast strains that lack fully functional mitochondria, because of the loss of mitochondrial DNA sequences. It was also found that mutations in a nuclear gene, *COX4*, that disrupt the electron transport chain also induce the retrograde response.[4] Petite yeast are not small; rather, the colonies that they form on the medium used to differentiate them are diminutive. The nature of the physiologic signal that may normally induce the retrograde response is currently unknown.

The retrograde response has been in search of a role in yeast biology since its discovery. It has now been shown to constitute a molecular mechanism of yeast longevity.[4] The induction of this pathway in petite yeast results in an extension of life span. This is a very robust response; it has been demonstrated in four different yeast strains. Significantly, differences exist in the details of induction of the retrograde response in these strains; however, whenever this response is induced, extension of life span occurs.[4] In fact, the retrograde response can be induced by both genetic and environmental means, resulting in life span extension. This constitutes very strong evidence for a broad-based mechanism of aging in yeast.

Deletion of the *RTG2* gene, a downstream effector of the retrograde response, abrogates the increase in life span seen in petite yeast.[4] The Rtg2 protein is involved in the formation of the active heterodimer Rtg1p-Rtg3p,[14] which regulates transcription of retrograde responsive genes. These genes encode metabolic enzymes, as mentioned earlier. Their induction represents activation of peroxisomal function and the shift from utilization of glucose to acetate as a carbon source. This is coupled to stimulation of gluconeogenesis and activation of the glyoxylate cycle, which provides Krebs cycle intermediates. These rather profound metabolic changes resemble some of those found in mutants of *Caenorhabditis elegans* that display an extended life span as well as in *Drosophila* selected for extended longevity (reviewed in ref. 15). Furthermore, they bear a resemblance to the changes that are found on caloric restriction in rodents,[4,15] which also increases life span. This similarity extends to the enhanced resistance of petite yeast to lethal heat shock.[4] It is not certain, however, which of the many changes that occur on induction of the retrograde response are actually responsible for extension of life span.

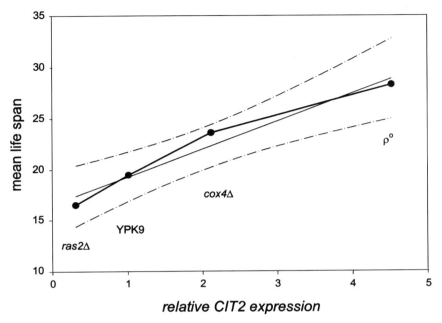

FIGURE 1. The retrograde response functions as a rheostat in determining life span. The *S. cerevisiae* strain YPK9 was the wild-type control. Three additional yeast strains were generated from YPK9 by deleting *RAS2* or *COX4* or by eliminating mitochondrial DNA (ρ°). Each of these strains was grown exponentially in rich YPD medium, and RNA was prepared. RNA was analyzed on Northern blots for expression of *CIT2*, a diagnostic for induction of the retrograde response.[4] The replicative life spans of individual cells of each of these strains were determined, as described.[4] Mean life spans are plotted as a function of *CIT2* expression. The *thin line* depicts linear regression ($r^2 = 0.96$) and the *dashed lines* are the 95% confidence intervals.

There are two fundamental ways in which the retrograde response may function to determine longevity. First, it may behave as a simple on-off switch. Induction of this response above a certain threshold level may, in this case, trigger an invariable increase in life span. Alternatively, the retrograde response may function as a rheostat. Above a certain induction threshold, an increase in the intensity of the retrograde response results in a commensurate increase in life span. It is the second of these two alternatives that may actually pertain (FIG. 1). There are several genetic manipulations that result in induction of the retrograde response. These manipulations result in varying levels of induction. Importantly, the increase in life span is directly proportional to the level of induction. This suggests that the normal function of the retrograde response is to react in a measured way, by changes in longevity, to a signal or signals that can vary in intensity under different circumstances. This has interesting consequences, and it places certain restrictions on the nature of the physiologic signal that normally triggers the retrograde response. In sum, the retrograde response signals the status of the mitochondrion to the nucleus, and it is an example of metabolic control that plays a role in determining longevity.

RAS2 FUNCTIONS AS A HOMEOSTAT IN YEAST AGING

A wide array of processes are important in maintaining the integrity of the cell and its potential to divide, in other words its replicative life span. By definition, the cell cycle itself plays a direct role. The role of metabolism and its control in determining longevity is on a firm footing with the discovery that the retrograde response determines life span. The importance of resistance to heat stress and ultraviolet radiation (UV) has also been well documented.[5,16] Genetic stability is also a factor in maintaining yeast life span.[17–19]

Two other cellular processes are likely candidates for a role in determining life span. They are chromatin-dependent gene regulation and cellular spatial organization. There are changes that occur during aging in both of these areas. Loss of chromatin-dependent transcriptional silencing was found both at subtelomeric loci[20] and at the silent mating-type loci.[21] A loss of spatial organization (cell polarity) occurs with age in yeast that can be recognized as an increase in the frequency of random budding pattern.[6] There is support for the role of transcriptional silencing and cell polarity in aging. Two genes, *CDC7*[22] and *SIR4*,[23] that are important for transcriptional silencing have been implicated in yeast aging. A genetically induced delay in the increase in random budding is correlated with an increase in longevity.[6,24] The evidence, however, must be considered circumstantial, given the pleiotropic nature of the effects of these genes involved in transcriptional silencing and cell polarity.

RAS2 was clearly documented as a longevity-determining gene in yeast.[24,25] This gene is involved in sensing the nutritional status of the cell.[26] It is directly involved in cell division cycle control.[27,28] *RAS2* was recently shown to modulate the retrograde response,[4] and it is clearly involved in the response to stress.[29] The response to heat stress,[5,30] which requires *RAS2*, plays a role in determining life span. *RAS2* also modulates transcriptional silencing, enhancing it, and cell polarity, delaying its loss during aging, in a manner consistent with its effect on life span.[6] Thus, *RAS2* appears to impinge on the cellular processes that are known or are suspected of playing a role in yeast aging.

Participation of *RAS2* in the response to heat stress deserves some comment. *RAS2* downregulates transcription of stress response genes that have a stress response regulatory element (STRE) in their promoters.[29] Because of this activity, this gene tends to decrease survival when yeasts are exposed to a lethal heat shock. On the other hand, *RAS2* has a protective effect when yeasts are exposed to chronic bouts of sublethal heat shock, due to its effective downregulation of stress genes and upregulation of growth-promoting genes.[5] *RAS2*, along with *RAS1*, is required for the increase in life span seen when yeasts are conditioned by a transient, sublethal heat stress.[30] These various effects on the response to heat stress appear to be related to the different situations encountered by dividing versus nondividing yeasts. The former can reestablish the population from only a small fraction of surviving cells after a lethal heat shock, and the ability to maintain their replicative life span during chronic stress may represent a balance they strike between the responses to these differing heat stresses. Nondividing yeasts, on the other hand, must be able to withstand lethal stress, because they do not divide to constantly renew the population.

The several cellular processes that are modulated by *RAS2* present competing demands on cellular energy and substrate resources. Thus, *RAS2* is perfectly positioned to coordinate or integrate these processes in a shifting external and internal environment. It is therefore not by accident that *RAS2* plays a role in assessing the nutritional status of the cell, although the mechanisms involved in this process are not well understood.[26] This may be its central role in determining life span. *RAS2* can be considered a homeostatic device in yeast longevity, on these grounds. Consistent with this role, there is an optimal level of *RAS2* activity associated with maximum longevity.[24,25,31] Thus, dependence of life span on *RAS2* expression is biphasic. The genetic background, environment, and other epigenetic factors are likely to determine where on this biphasic map a yeast finds itself. The effect of environment has been demonstrated for the conditioning effect of transient, sublethal heat stress that extends life span.[30]

There are two distinct ways in which *RAS2* might function as a homeostatic device in yeast longevity. It could function as an adaptation gene. In this capacity, it would participate in pathways that compensate for changes that occur during aging. This would be a passive role. *RAS2*, on the other hand, might function as a regulatory gene by participating in pathways that regulate cellular processes that are important for life maintenance. This, in turn, could be categorized as an active role. These two modes of action are not mutually exclusive. Obviously, the latter would provide more opportunity for life span-enhancing interventions. Indeed, the fact that overexpression of *RAS2* extends life span[24] supports this active role, because it indicates that this manipulation can overcome all of the deficits that are limiting for longevity under a particular set of circumstances. A corollary is the fact that *RAS2* modulates multiple processes that have been implicated in longevity.

GENE DYSREGULATION

Loss of transcriptional silencing in heterochromatic regions of the yeast genome occurs with age, and circumstantial evidence suggests that this may be causally associated with aging, as indicated earlier. Additional evidence providing strong support for this association was recently obtained.[32] This evidence has come from identification of the histone deacetylase genes, *RPD3* and *HDA1*, as longevity genes.

The deletion of *RPD3* strengthens silencing at all three, known heterochromatic loci in yeast: subtelomeric genes, silent mating-type locus, and ribosomal DNA (rDNA) locus.[32,33] This change in the functional status of chromatin is correlated with a substantial increase in life span.[32] The Sir protein silencing complex is not required for either the silencing change or the increase in life span. In contrast, deletion of *HDA1* has no effect on life span. However, when it is coupled to a deletion of *SIR3*, there is an increase in longevity.[32] The strengthening of silencing, in this case, was only seen in rDNA. *SIR3* itself had no effect on life span. Deletion of both *RPD3* and *HDA1* curtailed life span, because of an increase in initial mortality of the population indicating some overlap in function of the two deacetylases.

SIR2 is required for silencing at the rDNA locus, and *SIR3* and *SIR4* are dispensable.[34,35] Deletion of *RPD3* suppresses the loss of silencing in rDNA caused by the

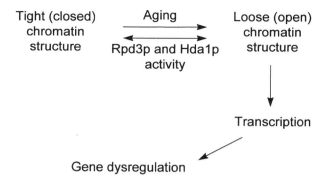

FIGURE 2. Gene dysregulation caused by loosening of chromatin structure with age can be reversed by altering histone acetylation patterns. This model depicts the effect of age on chromatin structure and transcriptional status. This effect of aging can be reversed by manipulation of the histone deactylase genes *RPD3* and *HDA1*, returning the chromatin to a "young" state with a concomitant increase in life span.

loss of *SIR2*.[32] However, the silencing observed does not reach the level seen when *RPD3* alone is deleted, and there is no recovery of the decline in life span caused by *sir2* mutation. All of the data point to the importance for longevity of silencing of rDNA. Nevertheless, the possibility exists that other regions of chromatin may also be important, because *RPD3* is known to exert an effect at other loci. *RPD3* and *HDA1* do not overlap entirely in function. Regardless of whether other chromatin regions are involved, these studies show that the functional state of chromatin is important for life span. The opening of chromatin domains that occurs during aging, resulting in inappropriate gene transcription, can be reversed by manipulating the acetylation of histones, with an associated increase in longevity (FIG. 2). This provides evidence that gene dysregulation is a cause of aging.

Given the focal position of the rDNA locus in studies of the deacetylase genes, it is tempting to speculate on the manner in which life span is affected by the transcriptional status of this locus. It is known that cellular rRNA content increases with age, but this does not keep up with the increase in cell size, resulting in an overall decrease in rRNA concentration in the cell.[36,37] These changes in rRNA levels are accompanied by a decline in protein synthesis with age.[37] One can speculate that the increase in rRNA synthesis during aging occurs in the absence of an increase in ribosomal protein synthesis. This would lead to an imbalance and defective assembly of active ribosomes. The silencing of rDNA by deletion of histone deactylases would reduce rRNA production and maintain the balance and continued assembly of functional ribosomes. The induction of the retrograde response results in the generation of extrachromosomal rDNA circles (ERCs), with an associated capacity for increased synthesis of rRNA.[38] Yet, the retrograde response extends life span. I, therefore, predict that the retrograde response also results in the induction of other components of the protein synthetic machinery, providing the balance needed to maintain active ribosome production for a prolonged period. This may be one of the ways in which the induction of the retrograde response contributes to the extension of life span.

As just noted, the retrograde response induces copious production of ERCs and at the same time extends life span substantially. The production of ERCs has been found to be a cause of aging in yeast.[17,18] Together, these results indicate that ERCs are not a cause of aging under all circumstances; in fact, the retrograde response predominates by extending life span in the face of ERC production. As suggested earlier, the generation of ERCs in amounts balanced by ribosomal protein production may actually be salutary for longevity. Reduction of ERC production[18] may achieve the same ends, as the yeast ages. The *sir2* mutation opens up rDNA chromatin and reduces its silencing. It also results in an increase in recombination in rDNA that favors ERC production.[32] As expected, the *sir2* mutant displays a curtailed life span, but no increase in ERCs was detected.[32] Here again is another uncoupling of yeast aging from ERC production. Thus, ERCs are neither sufficient nor necessary for yeast aging.

DIVIDING AND NONDIVIDING CELLS: A METABOLIC CONTINUUM

A distinction is often made between the aging of dividing and nondividing cells and between organisms that contain predominantly postmitotic cells and those that also have mitotic cells. The telomere attrition that attends each round of DNA replication during cell division has been taken as a distinguishing feature of most mitotic cells. This loss of telomeric repeats has been postulated to affect the transcriptional status of subtelomeric genes during cellular senescence.[39] Any distinction between the two cell types, based on the effects of telomere shortening on expression of subtelomeric genes, has been shown to be artificial.[20] Not only was the first evidence for changes in transcriptional state of subtelomeric genes with aging provided in this study, but also these changes were shown to occur without any telomere attrition.[20] Thus, mechanisms of aging involving gene dysregulation will be similar in cells in which telomere shortening occurs and in those in which it does not.

Is there any other basis for separately classifying dividing and nondividing cells from the point of view of aging? Both cell types are metabolically active. Some nondividing cells, such as neurons, are even more active than dividing cells. Thus, mechanisms of aging involving metabolic control will be similar in both types of cells, although the intensity of metabolic activity may differ from tissue to tissue, perhaps affecting the level of stress with which the cell must cope.

A qualitatively different situation has been proposed to exist between stationary and exponentially growing populations of yeast.[3] However, it was pointed out earlier that in both cases the "metabolic clock" is running, albeit at a different pace.[40,41] Indeed, the metabolic activity of stationary phase yeast cells is a fraction of that of logarithmic phase yeast,[42] and even yeast spores that can survive in a dormant state almost indefinitely respire at a low rate.[43] Stationary and dividing yeasts are therefore part of a metabolic continuum. Because metabolic activity is more intense, it should be easier to discern mechanisms of aging related to metabolic control in dividing yeasts. It is likely that many other aspects of aging will also be similar in the two cell types, but it is also likely that many will differ. It is already clear that the significance of certain stress responses may differ between dividing and nondividing yeasts, based on analysis of the role of *RAS2* during heat stress that was discussed earlier.

If the metabolic continuum discussed here exists, it should be possible to observe an effect of metabolic activity while yeast cells are not dividing on their subsequent life span when they are allowed to divide. In fact, just such a dependence has been observed.[44] Anecdotal evidence indicated that yeast cells from stationary cultures have shorter life spans than do those from exponential cultures (J.B. Chen, N.P. D'mello, and S.M. Jazwinski, unpublished observations).

It is of interest to speculate about the nature of the effects of metabolic activity on yeast aging. Metabolic activity brings in tow stress. This stress includes oxidative stress, under appropriate growth conditions. This stress can lead to damage, and, in the case of oxidative stress, this would include damage to mitochondria in particular. In dividing cells, any damage that accumulates in organelles and cell structures that are retained in mother cells or "filtered" for damaged copies, such as mitochondrial DNA may be,[45] would constitute the senescence factor[22] and cause aging in yeast. Similar processes might operate in animal cells. However, apart from stem cells and certain other dividing cell types, there is no asymmetry between the two products of mitosis. In dividing cells that do not exhibit this asymmetry, undamaged clones may be constantly selected. In nondividing cells, both yeast and animal, damage would consistently accumulate, putting a premium on stress response and damage repair processes.

ACKNOWLEDGMENTS

I thank Dr. Paul A. Kirchman in my laboratory for providing the data in FIGURE 1. This work was supported by grants from the National Institute on Aging of the National Institutes of Health (USPHS).

REFERENCES

1. MORTIMER, R.K. & J.R. JOHNSTON. 1959. Life span of individual yeast cells. Nature **183:** 1751–1752.
2. MULLER, I., M. ZIMMERMANN, D. BECKER & M. FLOMER. 1980. Calendar life span versus budding life span of *Saccharomyces cerevisiae*. Mech. Ageing. Dev. **12:** 47–52.
3. LONGO, V.D., E.B. GRALLA & J.S. VALENTINE. 1996. Superoxide dismutase activity is essential for stationary phase survival in *Saccharomyces cerevisiae*. Mitochondrial production of toxic oxygen species *in vivo*. J. Biol. Chem. **271:** 12275–12280.
4. KIRCHMAN, P.A., S. KIM, C.-Y. LAI & S.M. JAZWINSKI. 1999. Interorganelle signaling is a determinant of longevity in *Saccharomyces cerevisiae*. Genetics **152:** 179–190.
5. SHAMA, S., P.A. KIRCHMAN, J.C. JIANG & S.M. JAZWINSKI. 1998. Role of *RAS2* in recovery from chronic stress: effect on yeast life span. Exp. Cell Res. **245:** 368–378.
6. JAZWINSKI, S.M., S. KIM, C.-Y. LAI & A. BENGURIA. 1998. Epigenetic stratification: the role of individual change in the biological aging process. Exp. Gerontol. **33:** 571–580.
7. WERNER-WASHBURNE, M., E. BRAUN, G.C. JOHNSTON & R.A. SINGER. Stationary phase in the yeast *Saccharomyces cerevisiae*. Microbiol. Rev. **57:** 383–401.
8. PARIKH, V.S., M.M. MORGAN, R. SCOTT *et al.* 1987. The mitochondrial genotype can influence gene expression in yeast. Science **235:** 576–580.
9. LIAO, X. & R.A. BUTOW. 1993. *RTG1* and *RTG2*: two yeast genes required for a novel path of communication from mitochondria to the nucleus. Cell **72:** 61–71.
10. JIA, Y., B. ROTHERMEL, J. THORNTON & R.A. BUTOW. 1997. A basic helix-loop-helix-leucine zipper transcription complex in yeast functions in a signaling pathway from mitochondria to the nucleus. Mol. Cell. Biol. **17:** 1110–1117.

11. SMALL, W.C., R.D. BRODEUR, A. SANDOR *et al.* 1995. Enzymatic and metabolic studies on retrograde regulation mutants of yeast. Biochemistry **34:** 5569–5576.
12. CHELSTOWSKA, A. & R.A. BUTOW. 1995. RTG genes in yeast that function in communication between mitochondria and the nucleus are also required for expression of genes encoding peroxisomal proteins. J. Biol. Chem. **270:**18141–18146.
13. VÉLOT, C., P. HAVIERNIK & G.J.-M. LAUQUIN. 1996. The *Saccharomyces cerevisiae RTG2* gene is a regulator of aconitase expression under catabolite repression conditions. Genetics **144:** 893–903.
14. ROTHERMEL, B.A., J.L. THORNTON & R.A. BUTOW. 1997. Rtg3p, a basic helix-loop-helix/leucine zipper protein that functions in mitochondrial-induced changes in gene expression, contains independent activation domains. J. Biol. Chem. **272:** 19801–19807.
15. JAZWINSKI, S.M. 2000. Coordination of metabolic activity and response to stress in yeast longevity. *In* Molecular Biology of Ageing. S. Hekimi, Ed.: 21–44. Springer. Heidelberg.
16. KALE, S.P. & S.M. JAZWINSKI. 1996. Differential response to UV stress and DNA damage during the yeast replicative life span. Dev. Genet. **18:** 154–160.
17. SINCLAIR, D.A. & L. GUARENTE. 1997. Extrachromosomal rDNA circles: a cause of aging in yeast. Cell **91:** 1033–1042.
18. DEFOSSEZ, P.-A., R. PRUSTY, M. KAEBERLEIN *et al.* 1999. Elimination of replication block protein Fob1 extends the life span of yeast mother cells. Mol. Cell **3:** 447–455.
19. PARK, P.U., P.-A. DEFOSSEZ & L. GUARENTE. 1999. Effects of mutations in DNA repair genes on formation of ribosomal DNA circles and life span in *Saccharomyces cerevisiae*. Mol. Cell. Biol. **19:** 3848–3856.
20. KIM, S., B. VILLEPONTEAU & S.M. JAZWINSKI. 1996. Effect of replicative age on transcriptional silencing near telomeres in *Saccharomyces cerevisiae*. Biochem. Biophys. Res. Commun. **219:** 370–376.
21. SMEAL, T., J. CLAUS, B. KENNEDY *et al.* 1996. Loss of transcriptional silencing causes sterility in old mother cells of *S. cerevisiae*. Cell **84:** 633–642.
22. EGILMEZ, N.K. & S.M. JAZWINSKI. 1989. Evidence for the involvement of a cytoplasmic factor in the aging of the yeast *Saccharomyces cerevisiae*. J. Bacteriol. **171:** 37–42.
23. KENNEDY, B.K., N.R. AUSTRIACO, J. ZHANG & L. GUARENTE. 1995. Mutation in the silencing gene *SIR4* can delay aging in *S. cerevisiae*. Cell **80:** 485–496.
24. SUN, J., S.P. KALE, A.M. CHILDRESS *et al.* 1994. Divergent roles of *RAS1* and *RAS2* in yeast longevity. J. Biol. Chem. **269:**18638–18645.
25. CHEN, J.B., J. SUN & S.M. JAZWINSKI. 1990. Prolongation of the yeast life span by the v-Ha-*RAS* oncogene. Mol. Microbiol. **4:** 2081–2086.
26. TATCHELL, K. 1993. *RAS* genes in the budding yeast *Saccharomyces cerevisiae*. *In* Signal Transduction: Prokaryotic and Simple Eukaryotic Systems. T. Kurjan, Ed. : 147–188. Academic Press. San Diego.
27. BARONI, M.D., P. MONTI & L. ALBERGHINA. 1994. Repression of growth-regulated G1 cyclin expression by cyclic AMP in budding yeast. Nature **371:** 339–342.
28. TOKIWA, G., M. TYERS, T. VOLPE & B. FUTCHER. 1994. Inhibition of G1 cyclin activity by the Ras/cAMP pathway in yeast. Nature **371:** 342–345.
29. MARCHLER, G., C. SCHULLER, G. ADAM & H. RUIS. 1993. A *Saccharomyces cerevisiae* UAS element controlled by protein kinase A activates transcription in response to a variety of stress conditions. EMBO J. **12:** 1997–2003.
30. SHAMA, S., C.-Y. LAI, J.M. ANTONIAZZI *et al.* 1998. Heat stress-induced life span extension in yeast. Exp. Cell Res. **245:** 379–388.
31. JAZWINSKI, S.M. 1999. Longevity, genes, and aging: a view provided by a genetic model system. Exp. Gerontol. **34:** 1–6.
32. KIM, S., A. BENGURIA, C.-Y. LAI & S.M. JAZWINSKI. 1999. Modulation of life-span by histone deacetylase genes in *Saccharomyces cerevisiae*. Mol. Biol. Cell. **10:** 3125–3136.
33. SMITH, J.S., E. CAPUTO & J.D. BOEKE. 1999. A genetic screen for ribosomal DNA silencing defects identifies multiple DNA replication and chromatin-modulating factors. Mol. Cell. Biol. **19:** 3184–3197.
34. SMITH, J.S. & J.D. BOEKE. 1997. An unusual form of transcriptional silencing in yeast ribosomal DNA. Genes Dev. **11:** 241–254.

35. SMITH, J.S., C.B. BRACHMANN, L. PILLUS & J.D. BOEKE. 1998. Distribution of a limited Sir2 protein pool regulates the strength of yeast rDNA silencing and is modulated by Sir4p. Genetics **149:** 1205–1219.
36. JAZWINSKI, S.M. 1996. Longevity-assurance genes and mitochondrial DNA alterations: yeast and filamentous fungi. *In* Handbook of the Biology of Aging, 4ᵗʰ Ed. E.L. Schneider & J.W. Rowe, Eds. : 39–54. Academic Press. San Diego.
37. MOTIZUKI, M. & K. TSURUGI. 1992. The effect of aging on protein synthesis in the yeast *Saccharomyces cerevisiae*. Mech. Ageing Dev. **64:** 235–245.
38. CONRAD-WEBB, H. & R.A. BUTOW. 1995. A polymerase switch in the synthesis of rRNA in *Saccharomyces cerevisiae*. Mol. Cell. Biol. **15:** 2420–2428.
39. HARLEY, C.B. 1991. Telomere loss: mitotic clock or genetic time bomb? Mutat. Res. **256:** 271–282.
40. JAZWINSKI, S.M. 1999. Molecular mechanisms of yeast longevity. Trends Microbiol. **7:** 247–252.
41. JAZWINSKI, S.M. 1990. An experimental system for the molecular analysis of the aging process: the budding yeast *Saccharomyces cerevisiae*. J. Gerontol. **45:** B68–B74.
42. LAGUNAS, R. & E. RUIZ. 1988. Balance of production and consumption of ATP in ammonium-starved *Saccharomyces cerevisiae*. J. Gen. Microbiol. **134:** 2507–2511.
43. MILLER, J.J. 1989. Sporulation in *Saccharomyces cerevisiae*. *In* The Yeasts. A.H. Rose & J.S. Harrison, Eds. : 489–550. Academic Press. San Diego.
44. ASHRAFI, K., D. SINCLAIR, J.I. GORDON & L. GUARENTE. 1999. Passage through stationary phase advances replicative aging in *Saccharomyces cerevisiae*. Proc. Natl. Acad. Sci. USA **96:** 9100–9105.
45. OKAMOTO, K., P.S. PERLMAN & R.A. BUTOW. 1998. The sorting of mitochondrial DNA and mitochondrial proteins in zygotes: preferential transmission of mitochondrial DNA to the medial bud. J. Cell Biol. **142:** 613–623.

Mitochondrial Oxidative Stress and Aging in the Filamentous Fungus *Podospora anserina*

HEINZ D. OSIEWACZ[a] AND CORINA BORGHOUTS

Molekulare Entwicklungsbiologie und Biotechnologie,
Johann Wolfgang Goethe-Universität, Botanisches Institut,
Marie Curie Strasse 9, D-60439 Frankfurt am Main, Germany

ABSTRACT: In the filamentous fungus *Podospora anserina*, mitochondrial oxidative stress is a major contributor to aging. Reactive oxygen species (ROS) generated as a result of electron leakage during respiration lead to damage of components of the electron transport chain. In aging wild-type cultures, damaged proteins cannot be replaced because the mitochondrial genes encoding some of the corresponding subunits gradually become deleted from the mitochondrial DNA (mtDNA). Consequently, these defects result in an increased generation of reactive oxygen species and respiration deficits leading to cell death. Analyses of wild-type strains and of different long-lived mutants of *P. anserina* provide strong evidence that molecular mechanisms controlling aging processes in this fungus are complex and act at different levels. A basic mechanism (e.g., damage by ROS) appears to be overlaid by prominent instabilities of the mtDNA.

SENESCENCE IN *PODOSPORA ANSERINA*

P. anserina is a filamentous ascomycete at the borderline between unicellular and multicellular organisms. The individuum is a mycelium consisting of branched and septated filaments, the so-called hyphae. Septae are cross walls perforated by a pore so that continuity between adjacent segments occurs. The pores are wide enough to allow all organelles to pass through. Thus, essentially the mycelium of *P. anserina* represents a huge multinucleate cell, a so-called coenocyte.

In contrast to almost all filamentous fungi, mycelia of *P. anserina* are characterized by a limited life span. After germination of an ascospore, the product of sexual reproduction, a juvenile mycelium develops. This mycelium grows at the hyphal tips. The growth rate is linear until it reaches a presenescent phase. At this phase, the growth rate decreases, the peripheral hyphae become undulate and slender, and pigmentation of the mycelium increases. Finally the culture stops growing and dies at the hyphal tips (FIG. 1).

Senescence in *P. anserina* was first reported in the early 1950s and subsequently extensively investigated.[1,2] It was found that life span is dependent on environmental factors (e.g., growth temperature, nutrition) and on nuclear as well as extranuclear genetic traits. In particular, the mitochondrial DNA (mtDNA) was found to play a major role.[3–5]

[a]Corresponding author. Phone:+49 69 79829264; fax: +49 69 79829363.
Osiewacz@em.uni-frankfurt.de

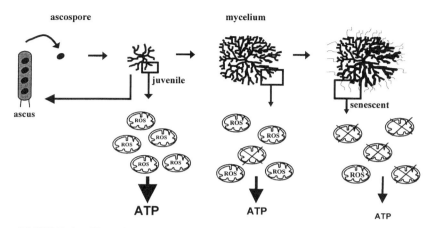

FIGURE 1. Life cycle of *Podospora anserina*. Germination of an ascospore formed as the product of sexual reproduction in a specific sporangium, termed the ascus, leads to the formation of a juvenile mycelium. The mycelium consists of branched filaments, the hyphae, and grows at the hyphal tips. After a strain-specific growth period, the growth rate of a mycelium decreases, the hyphae become undulate and slender, and they finally stop growing and die at the tips (senescence syndrome). During aging of a wild-type culture of *P. anserina*, the population of functional drifts towards compromised mitochondria and finally to nonfunctional mitochondria. Functional mitochondria produce high amounts of ATP and only low levels of ROS. In compromised mitochondria, high amounts of ROS and little ATP are generated. In senescent cultures, most mitochondria appear to be nonfunctional. Replacement of damaged mitochondria in later life phases is not possible, because in these stages the mitochondrial genes encoding different subunits of the respiratory chain are deleted from the mtDNA.

THE MITOCHONDRION, AN ESSENTIAL SEMIAUTONOMOUS CELLULAR COMPARTMENT

Mitochondria are organelles of eukaryotic cells that, in aerobic organisms, are essential for energy transduction (FIG. 2). The matrix space, in which different metabolic processes take place (e.g., tricarboxylic acid cycle, TCA), is surrounded by two biomembranes. The inner membrane contains the four protein complexes of the mitochondrial electron transport chain and the ATP-synthase complex (complex V) involved in the synthesis of adenosine triphosphate (ATP). Some of the subunits of the different complexes are encoded by the mtDNA (except for complex II), whereas others are encoded by the nuclear DNA. Nuclear encoded subunits are synthesized in the cytoplasm and become transported into the mitochondrion. Here they are assembled to functional protein complexes and are transported to the appropriate location in the inner membrane. Functional respiratory chains are dependent on cofactors (e.g., iron-containing heme, copper), that also need to be transported into the organelle (FIG. 2). The details of the underlying processes are not elucidated at the moment, but at least for copper (a cofactor of complex IV) it appears that copper-binding transporter proteins delivering the metal to cytochrome-c oxidase (COX) are involved.[6–10]

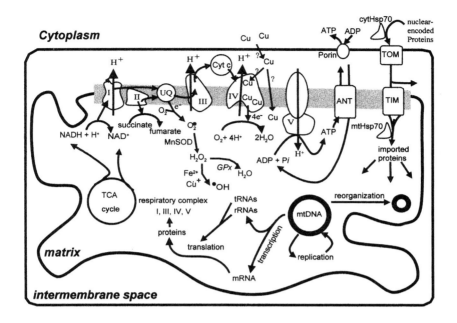

FIGURE 2. Scheme of a mitochondrion. In the matrix space, the mtDNA codes for components of the respiratory chain located in the mitochondrial inner membrane. Part of the energy-transducing apparatus is the tricarboxylic acid cycle (TCA), which is completely encoded by the nuclear DNA. The same holds true for protein complexes involved in the transport of proteins (TIM, TOM), molecular chaperones (e.g., hsp70), and metabolites (porin, ANT) into and out of mitochondria as well as those proteins involved in the maintenance and expression of the mitochondrial DNA (mtDNA). In different biologic systems, age-dependent mtDNA instabilities (reorganization) have been demonstrated. At the respiratory chain, electrons derived from reduced carbon sources are transported. In functional mitochondria the majority of electrons are finally transferred to oxygen at complex IV, leading to the generation of water. To some extent, electrons are leaking at different locations (e.g., at the ubiquinon pool, UQ), giving rise to formation of the superoxide anion ($O_2^{\bullet}$) that can lead to generation of the highly aggressive hydroxyl radical ($^{\bullet}OH$). The functionality of the mitochondrial system also depends on a controlled supply with metals such as iron and copper.

Additional essential components of the molecular machinery involved in the biogenesis of mitochondria are the complexes of the mitochondrial protein transport apparatus, the inner and the outer mitochondrial translocase (TIM and TOM), as well as molecular chaperones in the cytoplasm and in the mitochondrion. These components are all encoded by the nucleus. The same holds true for the proteins of the TCA cycle and all components of the mtDNA-maintaining system involved in the expression, replication, and reorganization of the mtDNA. From this brief description it is clear that the biogenesis and maintenance of mitochondria depends on well coordinated, complex nuclear–mitochondrial interactions.

MITOCHONDRIAL OXIDATIVE STRESS AS THE RESULT OF ELECTRON LEAKAGE AT THE RESPIRATORY CHAIN

The mitochondrial respiratory chain is prone to electron leakage.[11,12] That is, the electrons funneled into this system are not exclusively transferred in pairs to oxygen at complex IV. In particular, at the ubiquinone pool and at complex I, free radical intermediates are generated by transfer of a single electron to oxygen, resulting in superoxide. This reactive oxygen species (ROS) can lead to the generation of the highly aggressive hydroxyl radical, which is very hazardous because it can efficiently react with cellular components including nucleic acids, lipids, and proteins (FIG. 2).

During the life span of an organism, damaging the respiratory chain by mitochondrial ROS is thought to lead to compromised respiratory chains, generating increased amounts of ROS and less ATP. This kind of vicious cycle goes on further. If affected components of the respiratory chains cannot be replaced by undamaged proteins, the outcome is lethal and cells die because of energy deficits (FIG. 1). In cultures of *P. anserina* this scenario appears to take place because the mtDNA becomes rearranged during senescence and the genes coding for parts of the molecular machinery involved in mitochondrial energy transduction are deleted (see below).

AGE-RELATED mtDNA INSTABILITIES

The mtDNA of *P. anserina* is a circular molecule of 94 kb. It encodes 13 proteins of the respiratory chain, 2 subunits of mitochondrial ribosomes, 25 tRNAs, and 32 open reading frames, some of which are encoded by introns interrupting the coding sequence of genes.[13] The presence of introns is the reason that the mtDNA of *P. anserina* differs so markedly from the mtDNAs (about 16.5 kb in size) of humans and other mammals.

During aging, the mtDNA of *P. anserina* becomes reorganized. In wild-type strains, one specific rearrangement occurs regularly in the region of the first intron of the gene coding for subunit I (*CoI*) of the cytochrome oxidase complex. During aging, this intron, termed the pl-intron, becomes liberated and amplified as a covalently closed circular DNA molecule, the so-called plDNA (or asenDNA).[14–19] At the same time, those DNA fragments containing the integrated intron sequence disappear (FIG. 3, *Bgl*II restriction fragments 5 and 17).

The molecular mechanism leading to generation of the plDNA is thought to proceed via transposition of the pl-intron to the site at which one intron copy is already located.[19–21] This process ("homing-like" transposition) results in a duplication of the pl-intron sequence in the mtDNA. Finally, subsequent recombination processes between the duplicated sequences can explain the formation of circular plDNA molecules (FIG. 4).

In addition to the characteristic age-related reorganization processes occurring in the *CoI* gene, gross rearrangements in other parts of the mtDNA are observed that are also related to duplicated pl-intron sequences. In this case, intron duplication is the result of a pl-intron transposition to an ectopic site in the mtDNA. This site (FIG. 4; IBS*) is characterized by a short nucleotide sequence motif resembling the intron-binding site (IBS) of the pl-intron in the *CoI* gene.[22] The IBS* site is required

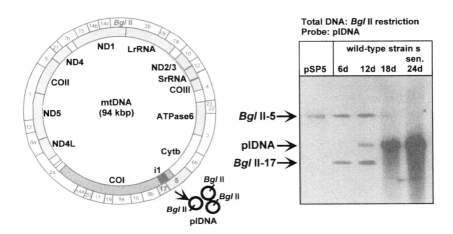

FIGURE 3. Age-related reorganizations of the mtDNA in the pl-intron (i1) region. The physical and genetic map of the mtDNA from a juvenile culture is shown on the *left*. The position of the pl-intron integrated into the coding sequence of the cytochrome oxidase I gene (*COI*) is indicated. Southern blot analysis of isolated mtDNA from wild-type cultures of *P. anserina* of different age, digested with *Bgl*II, was hybridized to the cloned sequence of the pl-intron (plDNA). During aging, the mtDNA fragments (*Bgl*II-5, *Bgl*II-17) containing the integrated intron copy disappear and the linearized plDNA accumulates as a fragment of 2.5 kb.

as an acceptor site for transposition of the intron. A reverse transcriptase involved in this process is encoded by the pl-intron.[23] The outcome of these rather complex molecular processes is the duplication of intron sequences in the mtDNA.

Another type of gross age-related mtDNA rearrangement in *P. anserina* is independent of the presence of pl-intron duplications. These rearrangements occur between disperse short direct repeats present in the mtDNA and result in the formation of circular molecules of various sizes. Circles containing an origin of replication accumulate during aging.[24,25]

The result of the different rearrangements is an age-related, almost quantitative reorganization of the mtDNA. Significantly, in senescent cultures of *P. anserina*, large parts of the information coding for the different essential components of the respiratory chain become deleted.

STABILIZATION OF THE mtDNA LEADS TO AN INCREASE IN LIFE SPAN

The picture of the molecular mechanisms involved in aging of *P. anserina* as just drawn is supported by data derived from analysis of various mutants. In particular, stabilization of the mtDNA was repeatedly reported to lead to an increased life span. In the long-lived mutant AL2-1, the liberation and/or amplification of plDNA is delayed due to the presence of a linear mitochondrial plasmid.[26–28]

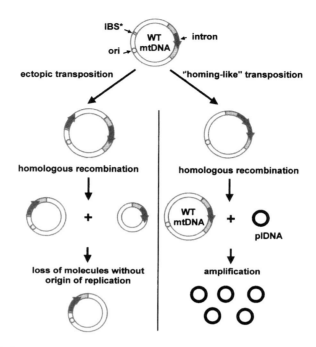

FIGURE 4. Model of processes leading to mtDNA reorganizations depending on duplicated pl-intron sequences (*black*). "Homing-like" transposition (*right*): transposition of the pl-intron to the position in the *CoI* gene (*gray*) in which an intron copy is already integrated leads to the generation of two tandemly oriented intron copies. Recombination between these copies leads to formation of circular plDNA, which is found amplified in senescent wild-type cultures. Ectopic transposition (*left*): transposition to an ectopic site in the mtDNA containing an IBS-like site (IBS*) leads also to duplication of the pl-intron. Intron sequences are separated by other mtDNA sequences. Recombination processes between the duplicated intron sequences lead to subcircles that subsequently may be lost if they do not contain an origin of replication (Ori).

In other mutants (ex and mex mutants), mtDNA stabilization is due to deletion of the *CoI* region including the whole pl-intron or parts of it.[29–32] Consequently, transposition of this mobile element cannot take place, and duplicated intron sequences involved in reorganization processes are not generated. As a result, these mutants display a highly significant increase in life span.

Finally, in long-lived mutant grisea, the loss-of-function mutation in a nuclear gene coding for a copper-modulated transcription factor leads to reduced uptake of copper from the growth medium. Interestingly, also in this mutant, the amplification of plDNA and the wild-type specific mtDNA reorganization are affected. Growing the mutant strain in medium supplemented with copper results in a reversion of the mutant phenotype to wild-type characteristics.[33–35] Copper thus appears to be of crucial significance in at least one of the different steps leading to generation of mtDNA deletions and amplification of plDNA.

Despite stabilization of the mtDNA, the analyzed mutants differ significantly in life span. Mutants, like grisea, are characterized by a rather moderate increase in life span, whereas mutants, like ex1, appear to be immortal. Thus, although instability of mtDNA clearly accelerates aging, this is not the only process affecting life span but rather part of a complex molecular network involved in life span control. Moreover, mutants like grisea clearly demonstrate that wild-type characteristic mtDNA reorganizations are not a prerequisite for aging in *P. anserina*. However, under wild-type conditions, extreme mtDNA instability may be the final or at least a major reason for bringing life to a close. For some unknown reasons, probably an adaptation to a specific natural niche, this accentuated mitochondrial etiology of aging has evolved in *P. anserina*. However, under protected laboratory conditions it is possible to select and to keep strains in which these mechanisms are affected. These strains may be unable to survive under natural conditions. Analysis of such mutant strains will help to unravel important additional details of the aging process at levels that in wild-type strains are overlaid by the typical processes just described. Probably, because there is also a clear mitochondrial basis of different degenerative processes including various diseases and aging in humans, novel clues arising from such investigations with *P. anserina* will be of significance in unraveling related processes in humans.

ACKNOWLEDGMENTS

We wish to thank the Deutsche Forschungsgemeinschaft (Bonn, Germany) for supporting the experimental work of the group.

REFERENCES

1. RIZET, G. 1953. Sur l'impossibilité d'obtenier la multiplication végétative ininterrompue et illimitée de l'ascomycète *Podospora anserina*. C.R. Hebd. Seances Acad. Sci. **237:** 838–840.
2. MARCOU, D. 1961. Notion de longévité et nature cytoplasmique du déterminant de sénescence chez quelques champignons. Ann. Sci. Natur. Bot. **11:** 653–764.
3. ESSER, K. & W. KELLER. 1976. Genes inhibiting senescence in the ascomycete *Podospora anserina*. Mol. Gen. Genet. **144:** 107–110.
4. TUDZYNSKI, P. & K. ESSER. 1977. Inhibitors of mitochondrial function prevent senescence in the ascomycete *Podospora anserina*. Mol. Gen. Genet. **153:** 111–113.
5. TUDZYNSKI, P. & K. ESSER. 1979. Chromosomal and extrachromosomal control of senescence in the ascomycete *Podospora anserina*. Mol. Gen. Genet. **173:** 71–84.
6. GLERUM, D.M., A. SHTANKO & A. TZAGOLOFF. 1996. Characterization of *COX17*, a yeast gene involved in copper metabolism and assembly of cytochrome oxidase. J. Biol. Chem. **271:** 14504–14509.
7. GLERUM, D.M., A. SHTANKO & A. TZAGOLOFF. 1996. *SCO1* and *SCO2* act as high copy suppressors of a mitochondrial copper recruitment defect in *Saccharomyces cerevisiae*. J. Biol. Chem. **271:** 20531–20535.
8. AMARAVADI, R., D.M. GLERUM & A. TZAGOLOFF. 1997. Isolation of a cDNA encoding the human homolog of *COX17*, a yeast gene essential for mitochondrial copper recruitment. Hum. Genet. **9:** 329–333.
9. BEERS, J., D.M. GLERUM & A. TZAGALOFF. 1997. Purification, characterization, and localization of yeast Cox17p, a mitochondrial copper shuttle. J. Biol. Chem. **272:** 33191–33196.

10. SRINIVASAN, C., M.C. POSEWITZ, G.N. GEORGE & D.R. WINGE. 1998. Characterization of the copper chaperone Cox17 of *Saccharomyces cerevisiae*. Biochemistry **37:** 7572–7577.
11. CADENAS, E., A. BOVERIS & C.L. RAGAN. 1977. Production of superoxide radicals and hydrogen peroxide by NADH-ubiquinone reductase and ubiquinone-cytochrome c reductase from beef heart mitochondria. Arch. Biochem. Biophys. **180:** 248–257.
12. TURRENS, T.F., A. ALEXANDRE & A.L. LEHINGER. 1985. Ubisemiquinone is the electron donor for superoxide formation by complex III of heart mitochondria. Arch. Biochem. Biophys. **237:** 408–414.
13. CUMMINGS, D.J., K.L. MCNALLY, J.M. DOMINICO & E.T. MATSUURA. 1990. The complete DNA sequence of the mitochondrial genome of *Podospora anserina*. Curr. Genet. **17:** 375–402.
14. STAHL, U., P.A. LEMKE, P. TUDZYNSKI *et al.* 1978. Evidence for plasmid like DNA in a filamentous fungus, the ascomycete *Podospora anserina*. Mol. Gen. Genet. **162:** 341–343.
15. CUMMINGS, D.J., L. BELCOUR & C. GRANDCHAMP. 1979. Mitochondrial DNA from *Podospora anserina*. II. Properties of mutant DNA and multimeric circular DNA from senescent cultures. Mol. Gen. Genet. **171:** 239–250.
16. BELCOUR, L., O. BEGEL, M.-O. MÓSSE & C. VIERNY. 1981. Mitochondrial DNA amplification in senescent cultures of *Podospora anserina*: variability between retained amplified sequences. Curr. Genet. **3:** 13–22.
17. KÜCK, U., U. STAHL & K. ESSER. 1981. Plasmid-like DNA is part of mitochondrial DNA in *Podospora anserina*. Curr. Genet. **3:** 151–156.
18. OSIEWACZ, H.D. & K. ESSER. 1984. The mitochondrial plasmid of *Podospora anserina*: a mobile intron of a mitochondrial gene. Curr. Genet. **8:** 299–305.
19. BELCOUR, L., A. SAINSARD-CHANET & C.H. SELLEM. 1994. Mobile group II introns, DNA circles, reverse transcriptase and senescence. Genetica **93:** 225–228.
20. OSIEWACZ, H.D. 1995. Molekulare Mechanismen biologischen Alterns. BIUZ **25:** 336–344.
21. OSIEWACZ, H.D. 1997. Genetic regulation of aging. J. Mol. Med. **75:** 715–727.
22. SELLEM, C.M., G. LECELLIER & L. BELCOUR. 1993. Transposition of a group-II intron. Nature **366:** 176–178.
23. FASSBENDER, S., K.H. BRÜHL, M. CIRIACY & U. KÜCK. 1994. Reverse transcriptase activity of an intron encoded polypeptide. EMBO J. **13:** 2075–2083.
24. CUMMINGS, D.J., I.A. MCNEIL, J.M. DOMENICO & E.T. MATSUURA. 1985. Excision amplification of mitochondrial DNA during senescence in *Podospora anserina*. DNA sequence of three unique plasmids. J. Mol. Biol. **185:** 659–680.
25. JAMET-VIERNY, C., J. BOULAY & J.F. BRIAND. 1997. Intramolecular cross-overs generate deleted mitochondrial DNA molecules in *Podospora anserina*. Curr. Genet. **31:** 162–170.
26. OSIEWACZ, H.D., J. HERMANNS, D. MARCOU *et al.* 1989. Mitochondrial DNA rearrangements are correlated with a delayed amplification of the mobile intron (plDNA) in a long-lived mutant of *Podospora anserina*. Mutat. Res. **219:** 9–15.
27. HERMANNS, J. & H.D. OSIEWACZ. 1992. The mitochondrial plasmid pAL2-1 of a long-lived *Podospora anserina* mutant is an invertron encoding a DNA and RNA polymerase. Curr. Genet. **22:** 491–500.
28. HERMANNS, J., A. ASSEBURG & H.D. OSIEWACZ. 1994. Evidence for a life span-prolonging effect of a linear plasmid in a longevity mutant of *Podospora anserina*. Mol. Gen. Genet. **243:** 297–307.
29. VIERNY, C., A.M. KELLER, O. BEGEL & L. BELCOUR. 1982. A sequence of mitochondrial DNA is associated with the onset of senescence in a fungus. Nature **297:** 157–159.
30. KÜCK, U., H.D. OSIEWACZ, U. SCHMID *et al.* 1985. The onset of senescence is affected by DNA rearrangements of a discontinuous mitochondrial gene in *Podospora anserina*. Curr. Genet. **9:** 373–382.
31. BELCOUR, L. & C. VIERNY. 1986. Variable DNA splicing sites of a mitochondrial intron: relationship to the senescence process in *Podospora anserina*. EMBO J. **5:** 609–614.

32. SCHULTE, E., U. KÜCK & K. ESSER. 1988. Extrachromosomal mutations from *Podospora anserina*: permanent vegetative growth in spite of multiple recombination events in the mitochondrial genome. Mol. Gen. Genet. **211:** 342–349.

33. MARBACH, K., J. FERNÁNDEZ-LARREA & U. STAHL. 1994. Reversion of a long-living undifferentiated mutant of *Podospora anserina* by copper. Curr. Genet. **26:** 184–186.

34. OSIEWACZ, H.D. & U. NUBER. 1996. Grisea, a putative copper activated transcription factor from *Podospora anserina* involved in differentiation and senescence. Mol. Gen. Genet. **252:** 115–124.

35. BORGHOUTS, C., E. KIMPEL & H.D. OSIEWACZ. 1997. Mitochondrial DNA rearrangements of *Podospora anserina* are under the control of the nuclear gene grisea. Proc. Natl. Acad. Sci. USA **94:** 10768–10773.

Molecular Genetic Mechanisms of Life Span Manipulation in *Caenorhabditis elegans*

SHIN MURAKAMI,[a] PATRICIA M. TEDESCO, JAMES R. CYPSER, AND THOMAS E. JOHNSON

Institute for Behavioral Genetics, University of Colorado, Boulder, Colorado 80309, USA

ABSTRACT: Aging and a limited life span are fundamental biological realities. Recent studies have demonstrated that longevity can be manipulated and have revealed molecular mechanisms underlying longevity control in the soil nematode *Caenorhabditis elegans*. Signals from both neurons and the gonad appear to negatively regulate longevity. One tissue-specific signal involves an insulin-like phosphatidylinositol 3-OH kinase pathway, dependent upon the DAF-16 forkhead transcription factor. These signals regulate mechanisms determining longevity that include the OLD-1 (formerly referred to as TKR-1) receptor tyrosine kinase. Interestingly, increased resistance to environmental stress shows a strong correlation with life extension.

Life span can be manipulated by both genetic and environmental factors. Genetic extension of life span has been accomplished in a variety of species.[1–5] Genes whose alteration causes life extension are referred to as gerontogenes.[3,4] Longevity in *Caenorhabditis elegans* has been extended by (1) polygenic alterations involving crosses between wild-type strains, (2) single-gene mutations, and (3) overexpression of life-extension genes. Most illuminating have been studies on the single-gene mutants, which have identified several gerontogenes and revealed molecular pathways, including an insulin-like signal transduction pathway that specifies life span.[5,6] Inactivating genes in this pathway leads to increased life expectancy. In contrast, genes that cause life extension when overexpressed represent another class: positive modulators of longevity. Thus, there are at least two ways to extend life span: inactivating negative regulators of life span and activating positive modulators of life span.

Interestingly, the life-extension mutants in *C. elegans* also have revealed a strong correlation between life extension and increased resistance to environmental stress.[2,7,8] Here, we will discuss the genetic manipulation of longevity, the molecular mechanisms underlying regulation of life span, and potential causes of longevity extension in *C. elegans*.

LIFE EXTENSION VERSUS LIFE SHORTENING

Life extension has been a good indicator of delayed aging and has been used to identify gerontogenes that regulate life span. By selecting for life extension rather

[a]Current address for correspondence: Shin Murakami, Division of Biological Sciences, University of Missouri, 310 Tucker Hall, Columbia, MO 65211.
MurakamiS@missouri.edu

than life shortening, one can be sure that these genes play an active role in regulating the length of life and the aging process. There are numerous ways to shorten life span that are nonspecific to aging; in many cases, it is very hard to demonstrate that shorter life is significantly indicative of faster aging in *C. elegans* and in other species. One problem with using the life-shortening phenotype is the lack of a convincing biological marker for aging in *C. elegans*. Therefore, in this review, we will mainly focus on studies using the life-extension phenotype and will be cautious about interpreting results from life-shortening alterations.

LIFE EXTENSION CAUSED BY MULTIPLE GENE ALTERATIONS

The first studies used crosses between the N2 and Bergerac "wild-type" strains followed by inbreeding of F2 progeny to generate recombinant-inbred strains that show life expectancies up to 63% longer than wild-type.[9] Using the RI strains, polygenes (called quantitative trait loci or QTLs) affecting life span and other life-history traits have been identified.[10–13]

SINGLE-GENE MUTATIONS: NEGATIVE REGULATORS OF LONGEVITY

To identify gerontogenes in *C. elegans,* forward genetics has commonly been used. Mutations have typically been generated using a mutagenic chemical, ethyl methanesulfate (EMS); so far, *age-1* and *age-2* are the only mutants originally identified in screens for increased longevity.[14–17] This method mainly generates reduction (hypomorphic) or loss-of-function (nullomorphic) mutations.[18] For example, EMS tends to generate point mutations, including stop codons, by G/C to A/T transitions and can also cause deletions of up to 10–20% of the mutagenized genome.[18–19]

age-1[b]

The first gerontogene mutant identified was *age-1*.[14,15] The *age-1(hx546)* reference allele has a life expectancy 65% longer than wild type and a maximum life span that is 105% longer.[15,20] *age-1* mutations have little effect on fertility, length of reproduction, or rate of development[15,16,20] but are dauer constitutive at the semi-lethal temperature of 27°C.[21] (The dauer is an alternative form of larvae produced under conditions of crowding or starvation. *daf* mutations affect the dauer-formation pathway[22] [see below].) *age-1(hx546)* is resistant to H_2O_2,[23] paraquat,[24] UV,[8] and heat[25,26] and has reduced frequency of deletions in mitochondrial DNA.[27] Three other alleles of *age-1* were independently isolated on the basis of longevity alone[16] and are also stress resistant but show subtle variations among themselves. Two more

---

[b]In *C. elegans,* genes are given names consisting of three italicized letters, a hyphen, and an italicized Arabic number (e.g., *age-1* and *old-1*). The protein product of a gene is referred to by the relevant gene name, written in nonitalic capitals, for example, the protein encoded by *old-1* is called OLD-1.

alleles of *age-1* have been isolated by selecting for increased thermotolerance (G. Lithgow, personal communication). All *age-1* alleles are also Daf-c at 27°C.

Mutations in another gene, previously called *daf-23*, also cause life-extension, dauer-constitutive (Daf-c), and stress-resistance phenotypes.[28] *daf-23* mutants are Daf-c at 25°C and *age-1* mutations fail to complement *daf-23* mutations.[21,29] Morris *et al.*[29] have cloned the *daf-23* locus, demonstrated that it shows structural homology with mammalian phosphatidylinositol-3-OH kinase (PI3 kinase), and suggested that *daf-23* and *age-1* are the same gene based on the failure to complement. To date, however, no mutation in the PI3 kinase structural locus has been found in any of the strains carrying the non-25°C-Daf-c alleles of *age-1* (Ref. 29; Murakami, Kliminskaya, and Johnson, unpublished). Thus, the formal possibility exists that *age-1* is not the same gene as *daf-23* and is not PI3K. Nevertheless, it is commonly assumed that PI3 kinase is the protein coded for by the *age-1* gene.

C. elegans *Gerontogenes That Specify Increased Longevity as Hypomorphs or Nullomorphs*

Except for *age-1* and *age-2* mutants, all life-extension mutants were isolated initially by screens for other phenotypes. These mutants can be classed according to their associated phenotypes. In all of these cases the Age (increased life span) phenotype results from hypomorphic or nullomorphic mutations.

Daf (*Dauer Larva Formation Abnormal*) Genes

age-1 is a member of this class. Mutants in *daf-2* are Daf-c at 25°C and result in twofold extension of life expectancy in the adult phase.[30] Additionally, *daf-2* interacts with *daf-12* to cause an almost fourfold increase in life expectancy.[28] *daf-2* bears structural homology with the human insulin receptor.[31]

The genes in the Daf class regulate dauer formation[22] and are the most well-characterized gerontogenes, but not all Daf genes cause life extension. AGE-1 PI3 kinase and DAF-2 insulin-like receptor form a genetic pathway,[8,28,30,32] dependent on the DAF-16 forkhead transcription factor.[33,34] Recently, PDK-1 (phosphatidylinositol dependent kinase-1) has been cloned and shown to be involved in this pathway.[35] Because the mammalian insulin receptor transduces a signal through PI3 kinase and PDKs,[36] it appears likely that a similar signal is transduced from the DAF-2 receptor, AGE-1 PI3 kinase and PDK-1 to the DAF-16 transcription factor (FIG. 1A). UNC-31 and UNC-64 appear to play a role in releasing a ligand of DAF-2,[36] perhaps specifying a neuronal signal for the DAF-2 receptor.[37] Mosaic analysis suggests that the function of *daf-2* in neuronal and/or other tissues originating from the AB cell is necessary for regulation of longevity.[37] Thus, the *age-1/daf-2* pathway is likely to transduce signal(s) from a limited spectrum of tissues that negatively regulate longevity.

The *age-1/daf-2* pathway also includes AKT-1 and AKT-2 (AKT: protein kinase B),[38] and DAF-18 (PTEN: PI3 phosphatase)[39,40] when regulating a metabolic shift, reproduction, and larval diapause (dauer stage). *daf-18* may antagonize AGE-1 PI3 kinase activity; however, it is unclear whether *akt-1* is also involved in regulation of longevity, since inactivation of *akt-1* suppresses the dauer-constitutive phenotype of *age-1* mutations but not the longevity phenotype.[38]

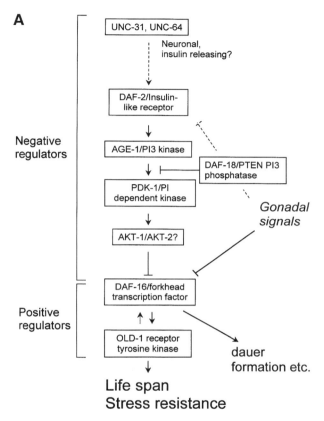

FIGURE 1. (A) Models for regulation of longevity. The *age-1/daf-2* insulin-like phosphatidylinositol 3-OH kinase pathway described within. Upon receipt of a signal from the DAF-2 insulin-like receptor,[31] the AGE-1 PI3 kinase is predicted to produce PIP3, which activates PDK-1 and AKT-1/AKT-2.[29,35,38] The AKTs may directly phosphorylate the DAF-16 forkhead transcription factor and may antagonize its function. The OLD-1 receptor tyrosine kinase appears to be downstream of these signal transduction proteins (Ref. 51; Murakami and Johnson, submitted). UNC-31 and UNC-64 are neuronal proteins and may be involved in releasing a ligand of the DAF-2 receptor.[36]

Spe (<u>Spe</u>rm Formation Defective) Genes

Two of the six mutant alleles of *spe-26*, a gene specifying proper segregation of cellular components affecting sperm activation, result in life extensions of about 65% for the hermaphrodite and the mated male,[41] although the details are contended.[42] An allele of *spe-10* also confers a modest (20%) life extension.[43]

Clk (<u>Cl</u>ock, Abnormal Biological Timing) Genes

Wong *et al.*[44] reported that two of four alleles of *clk-1*, both of which have altered cell cycle and developmental timing, also have increased life expectancy. The *clk-1*

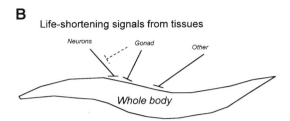

FIGURE 1. (B) Our proposed model. Negative signals from specific tissues repress molecular mechanisms for longevity in the whole body that may involve the DAF-16 forkhead transcription factor[33] and the OLD-1 receptor tyrosine kinase.[51] This life-extension mechanism also confers increased stress resistance. A life-shortening signal of neurons and other tissues (neurons etc.) involves the *age-1/daf-2* pathway, whereas a signal from the gonad interacts with the DAF-2 insulin-like receptor.

gene encodes a 187 amino acid protein showing some homology with yeast *CAT5p*, a gene required for transcriptional activation of several genes involved in gluconeogenesis in yeast.[45] They suggest that *clk-1* mutant alleles have reduced metabolic rates that are responsible for the extended longevity. Lakowski and Hekimi[46] have extended these studies to include *clk-2*, *clk-3*, and *gro-1,* all of which have modest (typically 20–30%) extensions of life span, but only for some alleles. Using double mutants, they have suggested that *clk-1* identifies a pathway distinct from that identified by *daf-2*.[46] Other studies using double mutants show that both *clk-1* and *daf-2* require a function of *daf-16*[7] or *ctl-1*, a cytosolic catalase gene.[47] However, all such interpretations run the risk that the life-shortening effects of *daf-16* and *ctl-1* mutants are nonspecific and result from an abnormal mode of death. Interestingly, the *clk-1;daf-2* double mutant extends life expectancy almost fivefold.

Eat Genes

Eat mutations cause reduced food intake, probably leading to dietary restriction. Some, but not all, of them show a life extension (*eat-2*, *eat-6*, *eat-13*, *eat-13*, *eat-18*, and *unc-26*).[48] Life extension in this class results from changes in eating behavior.

There are several additional gerontogene mutants that complement all known gerontogenes (e.g., Ref.16; Herndon *et al.,* in preparation), suggesting that there are as yet unidentified gerontogenes in *C. elegans*.

Gonadal Signals

It is likely that negative signals are transduced from multiple tissues to modulate life span. Although the postulated *age-1/daf-2* insulin-like pathway may be specific to neurons, gonadal signals have also been implicated in modulating life span.[49] Laser ablation of gonad precursor cells Z1–Z4 (Z1 and Z4 cells differentiate into the somatic gonad, and Z2 and Z3 precursor cells differentiate into the germ line) have revealed just such an effect. The signals of Z2 and Z3 shorten life span and are dependent on the DAF-16 forkhead transcription factor. It seems that those of Z1 and Z4 have little effect on life span but interact with *daf-2* mutants. Thus, it appears that tissue-specific signals can negatively regulate longevity. We extend our previously published model[8] to propose a molecular mechanism in which life extension is negatively regulated by such tissue-specific signals (FIG. 1B).

UPREGULATION OF POSITIVE REGULATORS OF LONGEVITY

We have used overexpression to upregulate genes positively modulating life span. Using a transgenic system, we identified a gerontogene that extends life span when overexpressed.[51] The gene encodes a novel tyrosine kinase receptor gene, *old-1* (overexpression longevity determinant).

Although positive modulators of longevity are poorly understood, *old-1* increases stress resistance and longevity (from 40 to 100%) when overexpressed in transgenic animals. These effects of *old-1* are comparable to those of the life-extension mutation *age-1* and are larger than those observed in *clk-1* and *spe-26* mutants. Importantly, the transgenic animals overexpressing *old-1* do not show altered developmental rates and show normal induction of dauer larvae. Nevertheless, the stress resistance and longevity of these animals is suppressed by mutations in *daf-16* (as is observed for the genes involved in the *age-1/daf-2* pathway, including *age-1, daf-2, pdk-1, unc-31,* and *unc-64*), therefore formally placing *old-1* in a common pathway with other gerontogenes. These results strongly support the correlation of stress resistance with life extension in *C. elegans* and illustrate the importance of (1) establishing the molecular mechanisms by which *old-1* overexpression induces stress resistance and life extension, and (2) identifying, in general, genes involved in regulating and enacting the stress response in *C. elegans*.

PROPOSED MOLECULAR MECHANISM OF LIFE EXTENSION

Organisms are continuously exposed to intrinsic environmental stress from inside the cell (e.g., oxidative stress) as well as extrinsic stress from outside environments (e.g., radiation and thermal stress). Such continuous exposure often promotes macromolecular damage, leading to deleterious effects that we call aging and senescence. Our proposed stress-resistance theory of aging hypothesizes that the ability to resist these environmental insults either by higher initial resistance to the stressor(s) and/or by more effective repair of critical damage is the rate-determining event leading to increased life expectancy.

Stress Resistance

Resistance to a variety of both intrinsic and extrinsic environmental stressors is strongly correlated with life extension in many species. In *C. elegans,* all gerontogene mutants tested show increased resistance to oxidative stress (*age-1* and *daf-2*),[23,24,52] thermal stress (*age-1, clk-1, eat-2, daf-2, spe-10,* and *spe-26*),[25,26,43] and UV stress (*age-1, clk-1, daf-2, daf-28, spe-10,* and *spe-26*).[8,43] Interestingly, the postulated *age-1/daf-2* insulin-like pathway regulates UV and oxidative stress resistance.[8,52] Overexpression of *old-1* also causes life extension and increased resistance to UV and heat stress (Ref. 51; Murakami and Johnson, submitted).

The increased resistance to a variety of environmental stressors and the putative roles of *age-1/daf-23*, *daf-2*, and *daf-16* in a signal-transduction pathway involving the phosphoinositides are consistent with a model in which life span is determined by the induction of a number of transcripts causing altered sensitivity to the environment and increased ability to resist or repair environmental damage.[2,8] The facts that the *spe-26* and *clk-1* Age (life-extension) alleles (but not the non-Age alleles) are also resistant to UV and that the UV resistance of these mutants is suppressed by *daf-16* without altering fertility[8] are consistent with increased resistance playing a causal role in the increased life expectancy of these mutants.

Other Possible Causes

Although extensive evidence supports increased stress resistance as a central mechanism generating the long-life phenotype of Age mutations, competing (if not necessarily mutually exclusive) hypotheses have been proposed. These include the concept of a central "life span" clock that can be genetically reset by mutation,[53] as well as the view that Age mutations convey longevity primarily by altering cellular metabolism.[33,54] However, few direct metabolic or biochemical measurements have been made to critically test these assumptions.

The sequence similarities of *daf-2* to the human insulin receptor and *age-1* to human PI3 kinase have led to the interpretation that these genes regulate an insulin response pathway regulating glucose metabolism.[31] Indeed, *age-1* and *daf-2* mutants cause a metabolic shift toward fat accumulation.[33] Similar metabolic shifts have also been observed in *daf-4* and *daf-7,* Daf mutants with wild-type life spans, suggesting that fat accumulation is associated with but not necessary for prolonged life span.

The *clk-1* gene appears to play a role in respiratory function in mitochondria. *clk-1* animals are longer-lived and have altered cell cycles and altered rhythmic behaviors.[45] The *clk-1* gene is probably orthologous to COQ7, which regulates yeast metabolism[55] through the biosynthesis of coenzyme Q. *clk-1* mutations slightly impair mitochondrial function[56] and uncouple energy production and consumption.[57]

Reduced rate of metabolism has also been investigated as a candidate for causing longevity by several labs. *clk-1* mutants consistently show large (70%) reductions in metabolism. Although controversial, long-lived mutants such as *age-1* and *daf-2* show little[57,58] or modest reduction in respiration.[59] In other assays, these mutants show no reduction in ATP levels.[57,58]

SPECULATIONS

Regulation of aging and longevity remains a compelling biological problem. Attractive, but largely untestable, evolutionary models argue against direct selection for an adaptive effect of aging. Recent studies of single-gene mutants have revealed tissue-specific signals that negatively regulate longevity. Such signals from neuronal tissues or gonads are sufficient for modulation of longevity. These signals are mediated, at least in part, by the DAF-2 insulin-like receptor, the AGE-1 PI3 kinase, and PDK-1. This pathway also appears to regulate a positive molecular mechanism for longevity that involves *old-1*. Other regulators will likely be identified using the methods described here as well as new methods such as biochip analyses. The correlation between longevity and stress resistance could be used to isolate additional regulators of longevity through use of stress resistance as a surrogate phenotype.

Details of the molecular mechanisms underlying stress resistance in the geronto-gene mutants are unclear. Stress resistance can be achieved by reducing stress-induced damage or by increasing ability to repair the damage. One way to reduce the damage would be to reduce the rate of metabolism, leading to reduction of oxidative stress from respiratory reactions. This mechanism may be responsible for increased stress resistance in long-lived DR (dietary-restricted) rodents.[60] In addition, some regulators of metabolism play an active role in stress resistance or *vice versa*. For example, the *age-1/daf-2* pathway regulates both metabolism[33] and stress resis-tance.[2,8] The *age-1/daf-2* pathway also determines a dauer-formation path that can be induced by starvation.[22] Moreover, dietary restriction (including starvation) can manipulate stress-inducible genes such as chaperones.[61] Thus, we see potential links between stress resistance and metabolism that could causally determine longevity. The uncovering molecular mechanisms encoded in the pathway will contribute fur-ther to the understanding of aging.

ACKNOWLEDGMENTS

This work was supported by the American Federation for Aging Research, the Ja-pan Society for the Promotion of Science, the National Institute of Health (RO1-AG08322, P01-AG08761, RO1-AG16219, and KO1-AA00195), the Glenn Founda-tion for Medical Research, and the Ellison Medical Foundation.

REFERENCES

1. MARTIN, G.M. *et al.* 1996. Genetic analysis of aging: role of oxidative damage and environmental stresses. Nat. Genet. **13:** 25–34.
2. JOHNSON, T.E. *et al.* 1996. Hypothesis: interventions that increase the response to stress offer the potential for effective life prolongation and increased health. J. Ger-ontol. **51:** B392–B395.
3. JOHNSON, T.E. & G.J. LITHGOW. 1992. The search for the genetic basis of aging: the identification of gerontogenes in the nematode *Caenorhabditis elegans.* J. Am. Geri-atr. Soc. **40:** 936–945.
4. RATTAN, S.I. 1998. The nature of gerontogenes and vitagenes. Antiaging effects of repeated heat shock on human fibroblasts. Ann. N.Y. Acad. Sci. **854:** 54–60.
5. WOOD, W.B. 1998. Aging of *Caenorhabditis elegans*: mosaics and mechanisms. Cell **95:** 147–150.
6. JOHNSON, F.B. *et al.* 1999. Molecular biology of aging. Cell **96:** 291–302.
7. LITHGOW, G.J. & T.B.L. KIRKWOOD. 1996. Mechanisms and evolution of aging. Sci-ence **273:** 80.
8. MURAKAMI, S. & T.E. JOHNSON. 1996. A genetic pathway conferring life extension and resistance to UV stress in *Caenorhabditis elegans.* Genetics **143:** 1207–1218.
9. JOHNSON, T.E. & W.B. WOOD. 1982. Genetic analysis of life-span in *Caenorhabditis elegans.* Proc. Natl. Acad. Sci. USA **79:** 6603–6607.
10. JOHNSON, T.E. 1987. Aging can be genetically dissected into component processes using long-lived strains of *C. elegans.* Proc. Natl. Acad. Sci. USA **84:** 3777–3781.
11. EBERT, R.H. *et al.* 1996. Defining genes that govern longevity in *Caenorhabditis ele-gans.* Dev. Genet. **18:** 131–143.
12. SHOOK, D.R. *et al.* 1996. Mapping quantitative trait loci affecting life history traits in the nematode *Caenorhabditis elegans.* Genetics **142:** 801–817.

13. SHOOK, D.R. & T.E. JOHNSON. 1999. Quantitative trait loci affecting survival and fertility-related traits in *Caenorhabditis elegans* show genotype–environment interactions, pleiotropy, and epistasis. Genetics **153:** 1233–1243.
14. KLASS, M.R. 1983. A method for the isolation of longevity mutants in the nematode *Caenorhabditis elegans* and initial results. Mech. Ageing Dev. **22:** 279–286.
15. FRIEDMAN, D.B. & T.E. JOHNSON. 1988. A mutation in the *age-1* gene in *Caenorhabditis elegans* lengthens life and reduces hermaphrodite fertility. Genetics **118:** 75–86.
16. DUHON, S.A. *et al.* 1996. Direct isolation of longevity mutants in the nematode *Caenorhabditis elegans*. Dev. Genet. **18:** 144–153.
17. YANG, Y.L. & D.L. WILSON. 1999. Characterization of a life-extending mutation in age-2, a new aging gene in *Caenorhabditis elegans*. J. Gerontol. **54:** B137–B142.
18. ANDERSON, P. 1995. *Caenorhabditis elegans*: modern biological analysis of an organism. Meth. Cell Biol. **48:** 31–58.
19. JANSEN, G. *et al.* 1997. Reverse genetics by chemical mutagenesis in *Caenorhabditis elegans*. Nat. Genet. **17:** 119–121.
20. JOHNSON, T.E. 1990. Increased life-span of *age-1* mutants in *Caenorhabditis elegans* and lower Gompertz rate of aging. Science **249:** 908–912.
21. MALONE, E. *et al.* 1996. Genetic analysis of the roles of *daf-28* and *age-1* in regulating *Caenorhabditis elegans* dauer formation. Genetics **143:** 1193–1205.
22. RIDDLE, D.L. & P.S. ALBERT. 1997. Genetic and environmental regulation of dauer larva development. *C. elegans* II. 739–768.
23. LARSEN, P.L. 1993. Aging and resistance to oxidative damage in *Caenorhabditis elegans*. Proc. Natl. Acad. Sci. USA **90:** 8905–8909.
24. VANFLETEREN, J.R. 1993. Oxidative stress and ageing in *Caenorhabditis elegans*. Biochem. J. **292:** 605–608.
25. LITHGOW, G.J. *et al.* 1994. Thermotolerance of a long-lived mutant of *Caenorhabditis elegans*. J. Gerontol. Biol. Sci. **49:** B270–B276.
26. LITHGOW, G.J. *et al.* 1995. Thermotolerance and extended life-span conferred by single-gene mutations and induced by thermal stress. Proc. Natl Acad. Sci. USA **92:** 7540–7544.
27. MELOV, S. *et al.* 1995. Increased frequency of deletions in the mitochondrial genome with age of *Caenorhabditis elegans*. Nucleic Acids Res. **23:** 1419–1425.
28. LARSEN, P.L. *et al.* 1995. Genes that regulate both development and longevity in *Caenorhabditis elegans*. Genetics **139:** 1567–1583.
29. MORRIS, J.Z. *et al.* 1996. A phosphatidylinositol-3-OH kinase family member regulating longevity and diapause in *Caenorhabditis elegans*. Nature **382:** 536–539.
30. KENYON, C. *et al.* 1993. A *C. elegans* mutant that lives twice as long as wild type. Nature **366:** 461–464.
31. KIMURA, K. *et al.* 1997. *daf-2*, an insulin receptor-like gene that regulates longevity and diapause in *Caenorhabditis elegans*. Science **277:** 942–946.
32. DORMAN, J.P. *et al.* 1995. The *age-1* and *daf-2* genes function in a common pathway to control the lifespan of *Caenorhabditis elegans*. Genetics **141:** 1399–1406.
33. OGG, S. *et al.* 1997. The fork head transcription factor DAF-16 transduces insulin-like metabolic and longevity signals in *Caenorhabditis elegans*. Nature **389:** 994–999.
34. LIN, K. *et al.* 1997. *daf-16*: An HNF-3-forkhead family member that can function to double the lifespan of *Caenorhabditis elegans*. Science **278:** 1319–1322.
35. PARADISE, S. *et al.* 1999. A PDK homolog is necessary and sufficient to transduce AGE-1 PI3 kinase signals that regulate diapause in *Caenorhabditis elegans*. Genes Dev. **13:** 1438–1452.
36. ALLION, M. *et al.* 1999. Neurosecretory control of aging in *Caenorhabditis elegans*. Proc. Natl. Acad. Sci. USA **96:** 7394–7397.
37. APFELD, J. & C. KENYON. 1998. Cell nonautonomy of *Caenorhabditis elegans daf-2* function in the regulation of diapause and life span. Cell **95:** 199–210.
38. PARADISE, S. & G. RUVKUN. 1998. *Caenorhabditis elegans* Akt/PKB transduces insulin receptor-like signals from AGE-1 PI3 kinae to the DAF-16 transcription factor. Genes Dev. **12:** 2488–2498.
39. OGG, S. & G. RUVKUN. 1998. The *C. elegans* PTEN homolog, DAF-18, acts in the insulin receptor-like metabolic signaling pathway. Mol. Cell **2:** 887–893.

40. ROUAULT, J.P. *et al.* 1999. Regulation of dauer larva development in *Caenorhabditis elegans* by *daf-18*, a homologue of the tumour suppressor PTEN. Curr. Biol. **25:** 329–332.
41. VANVOORHIES, W.A. 1992. Production of sperm reduces nematode life span. Nature **360:** 456–458.
42. GEMS, D. & D. RIDDLE. 1996. Longevity in *Caenorhabditis elegans* reduced by mating but not gamete production. Nature **379:** 723–725.
43. CYPSER, J.R. & T.E. JOHNSON. 1999. The *spe-10* has longer life and increased stress resistance. Neurobiol. Aging **20:** 503–512.
44. WONG, A. *et al.* 1995. Mutants in the *clk-1* gene of *Caenorhabditis elegans* affect developmental and behavioral timing. Genetics **139:** 1247–1259.
45. EWBANK, J.J. *et al.* 1997. Structural and functional conservation of the *Caenorhabditis elegans* timing gene *clk-1*. Science **275:** 980–983.
46. LAKOWSKI, B. & S. HEKIMI. 1996. Determination of life-span in *Caenorhabditis elegans* by four clock genes. Science **272:** 1010–1013.
47. TAUB, J. *et al.* 1999. A cytosolic catalase is needed to extend adult lifespan in *C. elegans daf-C* and *clk-1* mutants. Nature **13:** 162–166.
48. LAKOWSKI, B. & S. HEKIMI. 1998. The genetics of caloric restriction in *Caenorhabditis elegans*. Proc. Natl. Acad. Sci. USA **95:** 3091–3096.
49. HSIN, H. & C. KENYON. 1999. Signals from the reproductive system regulate the lifespan of *C. elegans*. Nature **27:** 362–366.
50. MELLO, C.C. *et al.* 1991. Efficient gene transfer in *C. elegans*: extrachromosomal maintenance and integration of transforming sequences. EMBO J. **10:** 3959–3970.
51. MURAKAMI, S. & T.E. JOHNSON. 1998. Life extension and stress resistance in *Caenorhabditis elegans* modulated by the *tkr-1* gene. Curr. Biol. **8:** 1091–1094.
52. HONDA, S. & Y. HONDA. 1999. The *daf-2* gene network for longevity regulates oxidative stress resistance and Mn-superoxide dismutase gene expression in *Caenorhabditis elegans*. FASEB J. **13:** 1385–1395.
53. KENYON, C. 1994. Ponce d'elegans: genetic quest for the fountain of youth. Cell **84:** 501–504.
54. GUARENTE, L. 1997. Aging. What makes us tick? Science **275:** 943–944.
55. JONASSEN, T. *et al.* 1998. Yeast *clk-1* homologue (Coq7/Cat5) is a mitochondrial protein in coenzyme Q synthesis. J. Biol. Chem. **273:** 3351–3357.
56. FELKAI, S. *et al.* 1999. CLK-1 controls respiration, behavior and aging in the nematode *Caenorhabditis elegans*. EMBO J. **18:** 1783–1792.
57. BRECKMAN, B.P. *et al.* 1999. Apparent uncoupling of energy production and consumption in long-lived Clk mutants of *Caenorhabditis elegans*. Curr. Biol. **9:** 493–496.
58. VANFLETEREN, J.R. & A. DE VREESE. 1996. A rate of aerobic metabolism and superoxide production rate potential in the nematode *Caenorhabditis elegans*. J. Exp. Zool. **274:** 93–100.
59. VAN VOORHIES, W.A. & S. WARD. 1999. Genetic and environmental conditions that increase longevity in *Caenorhabditis elegans* decrease metabolic rate. Proc. Natl. Acad. Sci. USA **96:** 11399–11403.
60. LEE, C.K. *et al.* 1999. Gene expression profile of aging and its retardation by caloric restriction. Science **285:** 1390–1393.
61. KURTZ, S. *et al.* 1986. An ancient developmental induction: heat-shock proteins induced in sporulation and oogenesis. Science **231:** 1154–1157.

Genetics of Human Aging

The Search for Genes Contributing to Human Longevity and Diseases of the Old

P. ELINE SLAGBOOM,[a,b] BASTIAAN T. HEIJMANS,[b,c] MARIAN BEEKMAN,[d] RUDI G.J. WESTENDORP,[c] AND INGRID MEULENBELT[b]

[b]*Gaubius Laboratory, TNO Prevention and Health, Leiden, The Netherlands*

[c]*Section of Gerontology and Geriatrics, Department of General Medicine and* [d]*Department of Human Genetics, Leiden University Medical Centre, Leiden, The Netherlands*

ABSTRACT: An aging population of humans reflects early-onset morbidity and mortality as well as late-onset disease in the phase when the mortality rate doubles and, finally, longevity of extremely long-lived subjects. Genetic influences have been reported to be relevant for each of these three phases. A growing field in genetic research is aimed at the identification of genes involved in multifactorial diseases of the old and in longevity. Important issues in these studies include the definition of phenotype, which maximally highlights the genetic contribution, whether earlier and later onset phenotypes have loci in common, and how to rank or reject the many candidate disease loci found in different studies. These issues will be illustrated from research on cardiovascular disease and osteoarthritis.

INTRODUCTION

Aging in higher species is associated with a gradual accumulation of a diverse spectrum of pathological conditions. This process seems to be partly under the control of genetic factors. The maximum life span (age at death of the single last survivor), among others, is a species-specific characteristic. This implicates the presence of species-specific genes that influence a basic aging rate and longevity. Evolutionary theories predict that such genes contribute to aging because of germline mutations that affect the organism only late in life (after the period of maximal reproductivity,[1] or because of germline mutations that have a positive effect early in life and a deleterious effect late in life (antagonistic pleiotropy[2]). The disposable soma theory[3] predicts that sets of such genes act in a broad network of somatic maintenance functions and that the energy reserved for these functions is in evolutionarily defined balance with energy reserved for reproduction. According to this concept, aging rates are determined by the interplay between accumulation of damaged (mac-

[a]Address for correspondence: P. Eline Slagboom, TNO-PG, Department of Vascular and Connective Tissue Research, Leiden, The Netherlands.

p.slagboom@pg.tno.nl

ro) molecules in somatic tissue and maintenance/repair functions evolved to restrict such accumulation. Because energy resources allocated to reproduction and somatic maintenance vary among species, the effectiveness of maintenance functions to control somatic damage varies among species. This is reflected, for example, in how error-prone DNA repair enzymes determining the accumulation rate of somatic mutations in the genome are.

The genome should be considered both determinant and target of aging.[4] Here, we focus on the determinant role of the genome in aging, the nature of the genes expected to contribute to aging, and the methods used to find such genes in humans.

EVOLUTIONARY CONCEPTS AND HUMAN AGING

Genes contributing to species differences in basic aging rate may also determine differences in aging rate between individuals of one species by variations in such genes. Many different genes may be expected to contribute to the variance between humans in physical condition, occurrence of disease, age of onset of disease, age at death, cause of death, and so forth. The evolutionary concepts given above help us to imagine how gene variants contributing to aging may become distributed in the human population. Neutral germline mutations, slightly altering gene products and/or expression levels affecting health only late in life, may become distributed slowly throughout human populations. Such mutations escape natural selection if they have no impact on reproductive success. The apolipoprotein E (ApoE) ε4 allele, for example, is carried by 15% of Caucasians, 20% of African Americans, and 10% of Japanese. This allele is associated with the risk of dementia[5–7] and cardiovascular disease.[8] The even more common 4G allele in the promoter of the plasminogen activator inhibitor–1 (PAI-1) gene,[9] associated with risk of cardiovascular disease,[10] is present in 45–50% of Caucasians. Such frequent variants are expected to have arisen 100,000–150,000 years ago and may contribute to aging phenomena in all human populations. Pleiotropic mutations may, when advantageous early in life, become dispersed throughout the population more rapidly. Such mutations may be those that would result in an increased inflammatory response, protecting against infection early in life and contributing (by chronic inflammation) to cardiovascular disease or dementia late in life. Gene variants that alter the balance of energy involved in somatic maintenance and reproduction may also become actively distributed throughout the population. Recently, a study on the relation between age at death and the number of progeny in a historical dataset from the British aristocracy revealed that either the concept of the disposable soma (trade-offs between fertility and lifespan) or that of antagonistic pleiotropy may indeed contribute to aging in humans.[11] A larger study on progeny and age at death in the Icelandic population, however, did not reveal such associations between fertility and longevity.[12]

The completion of the map and sequence (2000–2003) of the human genome, combined with the development of statistical methods for genetic studies of complex traits, has greatly stimulated research programs aimed at the localization and identification of genes involved in longevity and common diseases of the old.

HUMAN DISEASE

Common late-onset diseases in humans may arise through a combination of basic processes of tissue aging (shared by all humans but at different rates) and the sum of genetic and environmental risk factors to which the individual is exposed. Together, these determine the individual disease risk. The process of atherosclerosis, for example, is a universal aspect of aging vasculature in humans contributing to cardiovascular disease. Only a part of the population, however, develops a myocardial infarction (MI) due to a number of genetic and environmental risk factors and events such as rupture of the atherosclerotic plaque that leads to MI. Many studies are currently being focused on the dissection of genetic influences on common complex diseases such as cardiovascular disease, osteoporosis, osteoarthritis, type II diabetes, cancer, and dementia. The genetic make-up of individuals contributes considerably to the risk of disease within the context of age, various systemic risk factors, comorbidity, lifestyle, and environment. For most of the diseases mentioned, systematic and elaborate genome searches are being performed aimed at the localization of the major gene loci that contribute to the disease. Such studies use a tool provided by nature itself: neutral variations between individuals in DNA sequence organization (polymorphisms). These polymorphic markers are present at high density in the human genome and can be detected quite easily in the laboratory. The approach of a genome search using markers evenly spaced over the entire genome allows for the unbiased identification of yet-unknown genes with a major effect on the disease.[13,14] In addition to the approach of genome searching/scanning to find disease genes, there is the candidate gene approach, in which case known genes are being investigated as potential disease susceptibility loci. The hypothesis that a specific (candidate) gene is involved in a disease is usually based on the function of the gene in a process assumed to contribute to the pathophysiology of the disease. Alternatively, the gene may be identified as the causal gene in an early-onset form of the disease, with a relatively simple mode of inheritance, or in an animal model.

In contrast to the study of shorter lived individuals, extremely long-lived humans and animal strains may be studied. Identification of factors contributing to the "risk" of becoming extremely long-lived may reveal pathways of protection against basic aspects of aging and disease. It is an intriguing issue whether pathways can be identified that modulate a basic aging rate of tissues. It may be expected that some of these basic pathways will contribute to aging in many species.

Genetic research of human aging summarized in the next sections deals with the following questions. Which genes determine early-onset disease and how do we investigate their relevance to diseases of the old? How are causal genes specific for common diseases identified in elderly patients? Which of these genes contribute to the exponential increase in mortality of the population that occurs late in life? How do disease loci relate to longevity; how could longevity loci be found and be related to disease?

EARLY-ONSET DISEASE GENES

The most feasible strategy for identifying loci that contribute to human aging is by studying animal models or early-onset human diseases that resemble common diseases of the old. Such early-onset diseases may be caused by a relatively small number of genes (mono- or oligogenic). Affected subjects can more accurately be distinguished from unaffected subjects than is often the case for common late-onset diseases. Usually, DNA is collected from families in which the disease is transmitted in two or more generations. Linkage studies can be performed using polymorphic DNA markers with known locations in the genome. It is tested whether a given marker and the disease phenotype are transmitted together from parents to offspring. The extent to which marker and disease phenotype segregate together in a pedigree (linkage) indicates how close the marker and disease gene must be located on a single chromosome. Because the location of the markers is known, the location of the disease gene, close to the marker that shows statistically significant evidence for linkage, can be estimated. Positive findings of linkage are eventually followed by an extensive search for mutations in the gene with the shortest physical distance to the marker. Mutations have, for example, been identified in linkage studies of families with early-onset cardiovascular disease (hypercholesterolemia caused by mutations in the low-density lipoprotein receptor gene,[15,16] osteoarthritis (generalized osteoarthritis with mild dysplasia caused by mutations in the collagen type II gene[17,18]), dementia (familial Alzheimer's dementia caused by mutations in the presenilline 1 and 2 genes[19]), and progeroid/accelerated aging showing multiple features of aging (the Werner syndrome caused by mutations in a gene of the recQ helicase family[20,21]). The early-onset disease gene may contribute to more common forms of disease when additional mutations of the gene reside in the population with a mild effect, resulting in late-onset symptoms. Mild mutations are, for example, sequence variations in the promoter region of genes (promoter variants) leading to slightly altered levels of gene expression. Even if the early-onset disease gene itself is not relevant for the more common disease, the study of early-onset disease may reveal pathophysiologic pathways that contribute to more common forms of disease in older patients.

LATE-ONSET DISEASE GENES

The relevance of an early-onset disease gene to a more common form of the disease can be investigated by genetic association studies comparing gene frequencies in DNA from patient and control groups. Because functional (mild) variations are usually not yet identified in a candidate gene, variants (alleles) of a polymorphic marker within or close to the candidate gene are compared between unrelated subjects with and without the disease (cases and controls, respectively). A significantly increased frequency of an allele in patients compared to controls (positive association) may indicate that the marker segregates in the population together (in linkage disequilibrium) with a nearby mutation that increases the risk for disease. This implies that the gene(s) close to the marker may harbor mutations with a late deleterious effect. Common alleles of polymorphisms in the Werner gene have been associated with the risk of myocardial infarction in the Japanese population[22]; common alleles of the collagen type II (COL2A1) gene have been associated with the risk of osteoarthritis in Caucasians.[23] The presence of gene defects underlying these disease associations remains to be identified.

It can be expected that many of the genes causing early-onset disease are not the major contributors to the common disease phenotype. A huge number of candidate genes are usually hypothesized to play an important role in the disease. Association studies are very sensitive detectors of both major and minor disease loci in the population. However, since the function of only a minority (about 6000) of the 100,000–150,000 genes in the human genome is known, the odds are high that the major disease loci will go undetected by the candidate gene approach. This is why genome scanning is frequently applied. First, the site of a major disease gene in the genome is localized in steps of increasing accuracy (initial scan and fine mapping). Then evidence is obtained to show which of the genes at a specific genome location is the actual disease gene (identification of the gene by physical mapping and mutation analysis), followed by an in-depth study of the function of the gene and its defects in different patients.

To localize and identify major causal genes directly in elderly patients is much more complex then to perform linkage studies in families with a monogenic disease. Common diseases of the old mostly aggregate in families (rather than showing a clear pattern of inheritance), and multiple generations cannot be collected. Parents of patients are usually not alive, and children are too young to reveal whether they will become affected or not. A typical genome scan for a late-onset disease is therefore done by genotyping marker loci at regular distances (10–20 cM) over the entire genome in populations of sibling pairs both affected by the disease. For each marker, the number of siblings sharing alleles identical by descent is tested. For markers linked to the disease locus, significantly more sibling pairs affected by the trait are assumed to share (0, 1, 2) alleles than would be expected by chance (25%, 50%, 25%, respectively). Today, many variations of such genome searches are being used (including nonaffected siblings).[24,25] This is a very elaborate approach. Localization of the genes requires 10 to hundreds of thousands of genotypings in the initial genome searches indicating genome areas containing 10 to hundreds of genes. By fine mapping, physical mapping, and mutation analysis, the evidence must be provided to tell which of the genes in a genomic fragment is the disease gene.

Much attention lately is focused on the prospects for performing genome scans in populations of unrelated individuals (instead of sibling pairs) by using dense maps of single nucleotide polymorphisms (SNPs). A complete genome search in such association (linkage disequilibrium) studies would require the typing of 500,000 SNP markers,[26] which can only be performed using advanced and expensive automated DNA chip technology. Such studies will also be performed for fine mapping of genome regions that have shown positive linkage in sibling-pair studies. Because siblings share much larger pieces of chromosome than unrelated subjects, the sibling-pair approach does not allow fine mapping of gene loci beyond 2–4 cM, whereas linkage diseqilibrium mapping in unrelated individuals with SNPs does allow this.

Although many diseases are importantly influenced by genetic factors, the genes involved may be hard to find. Many different genes may interactively cause the disease in different patients (genetic heterogeneity), and the clinical expression of a disease gene may vary widely among patients (clinical heterogeneity). Measurable endpoints other than presence or absence of disease may allow the identification of major loci contributing to disease more easily. Therefore, in addition to genome searches for disease in patients and their relatives, genome searches are being per-

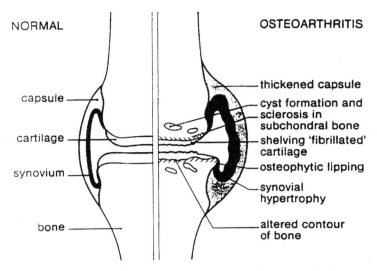

FIGURE 1. Schematic representation of a normal and osteoarthritic joint.

formed in healthy subjects for loci determining (quantitative) risk factors of disease. Examples are the searches that are being performed for quantitative trait loci (QTLs) determining blood pressure, cholesterol (and other lipid) levels,[27] weight, and scores for personality traits such as anxiety or depression.

Genetic research on osteoarthritis is now described to illustrate the different aspects of genetic research into a complex common disease: linkage studies in early-onset disease; establishing the genetic component of osteoarthritis in the general population, and subsequent candidate gene studies and genome searches.

THE GENETICS OF OSTEOARTHRITIS

Osteoarthritis (OA) is a prevalent chronic disease of the joints that causes considerable pain and mobility problems in the elderly. Radiological characteristics of OA are joint space narrowing (representing degeneration of articular cartilage) and formation of osteophytes (representing the formation of new subchondral bone) (see Fig. 1). Cartilage degradation, as observed by radiographic photography, is a basic aspect of human aging in all humans, but only a part of the population develops the clinical symptoms of OA. Because radiological characteristics of OA (ROA) can much better be measured and quantified than clinical ones such as pain, most genetic studies have used ROA as a measure for OA. The role of genetic factors in increasing susceptibility to OA is being investigated both in early-onset families (age of onset 20–50 years) and at later ages of onset in the population at large.

ROA records for Dutch families expressing early-onset, generalized OA (GOA occurring in multiple joints before the age of 50 years) were collected.[28] In these families the disease was transmitted in a dominant Mendelian fashion, probably caused by a single gene. Linkage studies were initially performed for a large number of candidate genes encoding components of the extracellular matrix in cartilage and proteins involved in cartilage metabolism and repair.[28] Because most of these genes were excluded as causal factors, a genome-wide scan was performed in seven early-onset Dutch GOA families. The results of this genome-wide scan indicated the presence of a novel disease gene on chromosome 2q. Studies are now being pursued to identify this gene and to find the mutation(s) in the various OA families. Because OA is such a common disease, the question is raised whether the gene at 2q plays a role in common OA. But is OA in the population at large influenced by genetic factors at all?

This was investigated in a Dutch prospective, population-based study, called the Rotterdam study.[29] Radiographic characteristics of OA were assessed in hip, knee, hand, and spine in a random sample of 1600 subjects of the ages between 55 and 70 years. For 118 of these subjects (probands), brothers and sisters were included in the study (257 siblings in total). By comparing the occurrence of ROA in these siblings with the prevalence of ROA in the population sample, heritabilities of ROA could be estimated. This revealed that essentially three definitions of ROA were influenced significantly by genetic factors[29]:

(1) ROA at multiple joints with a heritability of 0.78, 95% CIL 0.34–0.76 (meaning that 78% of the variation in the population is due to genetic variation between individuals);

(2) ROA in hands (0.56, 95% CIL 0.34–0.76);

(3) Disk degeneration of the spine, which is clinically not considered to be OA but still represents degradation of cartilage (heritability 0.75, 95% CIL 0.30–1.00).

The first definition is relatively rare and was present in only 14% of 55- to 65-year-old subjects. Hand OA and disk degeneration of the spine are very frequent in the population. As yet, it is not clear whether these definitions represent different patient subgroups in which the disease in caused by different sets of genes or different clinical expressions of one set of OA genes. Other groups also demonstrated considerable genetic influences for different definitions of OA.[30–32]

Candidate gene studies for OA revealed associations for a number of genes, among which are the collagen type II gene,[23] the insulin-like growth factor-1 gene,[33] and two genes encoding components of the extracellular matrix, the cartilage matrix protein (CRTM) gene[34] and the aggrecan gene.[35] The influence of these loci, however, explains only a small fraction of the genetic influences on OA in the population at large.

Genome searches of OA demonstrated different areas of positive linkage depending on OA definitions that were used to collect sibling pair populations (presence of OA in hands or in hip, for example).[36–38] Linkage was found at chromosomes 11q12 (LOD score 2.40), 9q33-34 (LOD score 2.23), 4q26-q27 (LOD score 2.02), Xp11.3 (LOD score 1.65), and 7p15-p21 (LOD score 1.29). Three groups reported linkage in a broad region on human chromosome 2q. Summarized, these studies indicate linkage at a 5-cM interval at 2q12-q14 containing the interleukin-1 (IL-1) gene cluster and linkages at 2q13-q32 and 2q32-q35. The last location on chromosome 2q, found in patients with hip OA, overlaps with the region detected in Dutch families with early-onset generalized OA. If these results point to a single disease gene, the gene may harbor different mutations in the population, leading to milder and more severe symptoms of the disease. Fine mapping and mutation analysis will eventually lead to identification of OA susceptibility genes.

CARDIOVASCULAR DISEASE GENES AND MORTALITY

Between the ages of 65 and 95, the mortality rate increases exponentially. The heritability of age at death in this phase as established in a Danish twin study is 20–30%.[39] Over the last few years, a large number of common functional variants have been identified in genes that may contribute to major age-related pathologic conditions. More than 30 gene loci were shown to be associated with the risk of cardiovascular disease. The gene products of these loci have a function in lipid metabolism, fibrinolysis, coagulation, blood pressure, methionine/homocysteine metabolism, extracellular matrix metabolism, inflammation, and so forth. A great variety of populations and clinical endpoints was selected in different studies. In general therefore, it has not yet been possible to rank or reject these loci for their role in cardiovascular disease. As the number of polymorphisms associated with disease is growing, the need for distinguishing major ones for in-depth studies increases. The relation between a number of functional variants and polymorphisms previously associated with cardiovascular disease was examined in a population-based study among subjects aged 85 years and over (Leiden 85-plus Study[40]). Associations of these gene variants with mortality before the age of 85 years was studied cross-sectionally by comparing gene frequencies in the elderly with a control group of young subjects with families from the same geographic region. In a 10-year follow-up period, the relation of the gene variants to all-cause and cause-specific mortality of the Leiden 85-plus population was studied prospectively. Functional variants in the paraoxonase,[41] factor V,[42] angiotensin-converting enzyme (ACE),[43] and tumor necrosis factor-α (TNF-α) genes did not reveal any major associations with population mortality. Such associations were, however, observed for the apolipoprotein (Apo) E gene with mortality before 85 years, for the PAI-1 gene with the risk of ischemic heart disease after the age of 85 years,[43] and for the methylene tetrahydrofolate reductase (*MTHFR*) gene.[44]

A common *ala*-to-*val* mutation (677C→T) in the *MTHFR* gene is considered a factor that could contribute to mortality. The mutation is associated with a disturbed methionine/homocysteine metabolism (FIG. 2) and with increased plasma homocysteine levels,[45] which are associated with cardiovascular disease[46] and death from coronary artery disease.[47] The frequency of the mutation was significantly lower in the subset of 365 elderly subjects who were born in Leiden, The Netherlands, than in 250 young subjects whose families originated from the same geographic region. This difference was only present in men. The estimated mortality risk up to 85 years in men homozygous for the mutation was 3.7 (95% CI, 1.3–10.9). The complete cohort of 666 elderly subjects was followed over a period of 10 years for all-cause and cause-specific mortality. Over the age of 85 years, mortality in men homozygous for the mutation was increased 2.0-fold (95% CI, 1.1–3.9) (FIG. 3), and gene environment interactions with smoking habits were observed. Among women aged 85 years and over, no deleterious effect of the *MTHFR* mutation was detected. Replication of

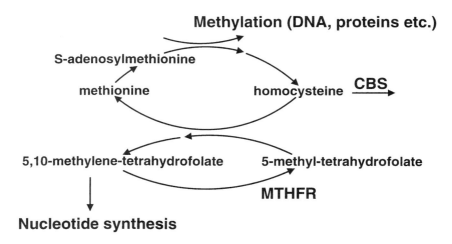

FIGURE 2. Schematic representation of the function of the *MTHFR* gene product in the methionine–homocysteine pathway.

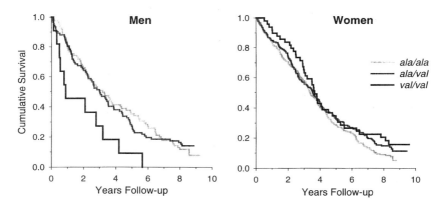

FIGURE 3. Kaplan-Meier estimate of 10-year cumulative survival according to *MTHFR* genotype for men and women aged 85 years and over.

our findings in a 65- to 85-year-old Dutch population sample is currently being tested. The findings obtained in our studies may be universal but could also reflect selection bottlenecks specific for the Dutch population or this cohort. Our findings for the *MTHFR* mutation were supported by reduction in the frequency of the homozygous mutated genotype among French centenarians,[48] but not by two other cross-sectional studies of elderly populations.[49,50]

Surprisingly, cancer rather than cardiovascular diseases contributed to the increased mortality in men homozygous for the mutation. The *MTHFR* mutation has previously been associated with colon cancer,[51,52] and a recent meta-analysis has shown that the mutation was not associated with vascular disease.[53] Genes may contribute to multiple pathologic conditions simultaneously (such as ApoE variants contribute to both dementia and cardiovascular disease). This may be a common feature of gene variants contributing to mortality at older ages.

LONGEVITY

Extreme longevity aggregates in families,[54] showing a tendency for a maternal component of inheritance.[55] The quest for identification of longevity loci is very appealing. A growing number of groups is investigating candidate loci in centenarians.[56–59] Analogous to other genetic association studies, the design of a proper control group is very critical in these studies. Many of the loci investigated thus far are those that have been associated with increased disease risk in younger cohorts. Also, drug metabolism loci[60] and polymorphisms in the mitochondrial genome have been investigated. The results of genetic association studies in centenarians need a very careful interpretation. Toupance *et al.*[61] have shown that risk alleles contributing to the period of exponential mortality increase can be expected to occur at higher frequencies in centenarians than in younger cohorts. To test whether disease genes contribute to mortality may therefore be better investigated in cohorts of increasing ages between 65 and 95 than in centenarians. Centenarian studies have revealed, however, that gene variants or other factors associated with disease risk in the population at large may have beneficial (or neutral) effects at extremely high ages.[62,63]

The longevity studies performed thus far do not yet suggest which pathways are critical to achieving an extreme old age. In view of the disposable soma theory of aging, general candidate longevity loci may be considered to be those involved in somatic maintenance (DNA replication and repair; systemic and cellular response to endo/exogenic exposure, etc.) linked in interaction or in other ways related to networks involved in fecundity. Because of the limited knowledge that exists so far on the rate-limiting pathways of human aging, genome-scanning approaches seem to be the best option for finding major longevity loci.

The phenotype in longevity that is expected to be a genetically influenced trait has not yet been thoroughly described. Does the genetic component in longevity promote a long life by simultaneously providing protection from all major diseases or from specific diseases or just from death due to such diseases? Longevity may, for example, be compatible with atherosclerosis that does not lead to myocardial infarction; with joint space narrowing that does not lead to OA; with insulin resistance that does not lead to diabetes and cardiovascular disease. Families with a high proportion of extremely long-lived members in each generation are being identified worldwide. A description of the long-lived phenotype in such families, including age at death, specific cause of death, medical history of diseases, age of onset of such diseases, clinical assays, lifestyle factors, and so forth, will be imperative in the search for longevity loci. Investigating extreme old age at death as the trait may not reveal major longevity loci in a genome scan, because of the etiological heterogeneity underlying this characteristic. Other (quantitative) phenotypes representing basic aging may have higher heritabilities and may help reveal major longevity loci in a search. Longevity studies in animal models and the concept of the disposable soma theory offer interesting phenotypic characteristics and candidate pathways to be studied in relation to longevity, such as parameters of metabolic control (insulin signaling), stress resistance, and genetic instability. Studies in *C. elegans* indicated the presence of interactions between pathways of insulin signaling and antioxidant defense promoting longevity.[61]

FUTURE PERSPECTIVES

One of the challenges of genetic research in the next millennium will be to identify genetic variations that have a major role in affecting the basic aging rate (of all humans), the variation in susceptibility to diseases of the old, and in the potential to become extremely long-lived. This requires an understanding of the human genome in terms of gene sequence and function, epigenetic mechanisms of gene regulation, gene–gene and gene–environment interactions. Such research will eventually reveal pathophysiologic pathways underlying the clinical heterogeneity in complex diseases. Once a disease gene or pathway is identified, a search can be performed for environmental factors modulating its action, potentially leading to new strategies and targets for therapeutic intervention or prevention. For some true polygenic diseases, however, it may not become feasible to dissect the genetic component. In the near future SNP mapping will become available for the localization of common disease genes. The study of extremely long-lived subjects and their families, on the other hand, may uncover shortcuts to disease-protecting mechanisms.

REFERENCES

1. MEDAWAR, P.B. 1952. An Unsolved Problem of Biology. Lewis. London.
2. WILLIAMS, G.C. 1999. Pleiotropy, natural selection, and the evolution of ageing. Evolution **11:** 398–411.
3. KIRKWOOD, T.B. 1977. Evolution of ageing. Nature **270:** 301–304.
4. SLAGBOOM, P.E. 1990. The aging genome: determinant or target? Report of the EURAGE meeting on "Genomic Instability and Aging," Nerja (Spain). Mutat. Res. **237:** 183–187.
5. CORDER, E.H., A.M. SAUNDERS, W.J. STRITTMATTER, *et al.* 1993. Gene dose of apolipoprotein E type 4 allele and the risk of Alzheimer's disease in late onset families. Science **261:** 921–923.
6. STRITTMATTER, W.J., A.M. SAUNDERS, D. SCHMECHEL, *et al.* 1993. Apolipoprotein E: high-avidity binding to beta-amyloid and increased frequency of type 4 allele in late-onset familial Alzheimer disease. Proc. Natl. Acad. Sci. USA **90:** 1977–1981.
7. FARRER, L.A., L.A. CUPPLES, J.L. HAINES, *et al.* 1997. Effects of age, sex, and ethnicity on the association between apolipoprotein E genotype and Alzheimer disease. A meta-analysis. APOE and Alzheimer Disease Meta Analysis Consortium [see comments]. JAMA **278 :** 1349–1356.
8. VAN DER CAMMEN, T.J., C.J. VERSCHOOR, C.P. VAN LOON, *et al.* 1998. Risk of left ventricular dysfunction in patients with probable Alzheimer's disease with APOE*4 allele. J. Am. Geriatr. Soc. **46:** 962–967.
9. DAWSON, S.J., B. WINNAN, A. HAMSTEN, *et al.* 1993. The two allele sequences of a common polymorphism in the promoter of the plasminogen activator inhibitor-1 (PAI-1) gene respond differently to interleukin-1 in HEPG2 cells. J. Biol. Chem. **268:** 10739–10745.
10. ERIKSSON, P., B. KALLIN, H. VAN 'T, *et al.* 1995. Allele-specific increase in basal transcription of the plasminogen-activator inhibitor 1 gene is associated with myocardial infarction. Proc. Natl. Acad. Sci. USA **92:** 1851–1855.

11. WESTENDORP, R.G. & T.B. KIRKWOOD. 1998. Human longevity at the cost of repro-ductive success. Nature **396:** 743–746.
12. FRIGGLE, M.L., D.F. GUDBJARTSSON, H. GUDMUNDSSON & K. STEFANSSON. 1999. Lon-gevity in Iceland: fertility and genetics. Am. J. Hum. Genet. **65:** 1118.
13. LANDER, E.S. & D. BOTSTEIN. 1986. Strategies for studying heterogeneous genetic traits in humans by using a linkage map of restriction fragment length polymor-phisms. Proc. Natl. Acad. Sci. USA **83:** 7353–7357.
14. LANDER, E.S. & D. BOTSTEIN. 1986. Mapping complex genetic traits in humans: new methods using a complete RFLP linkage map. Cold Spring Harbor Symp. Quant. Biol. **51:** 49–62.
15. WIDHALM, K., C. IRO, A. LINDEMAYR, *et al.* 1999. Heterozygous familial hypercholes-terolemia: a new point-mutation (1372del2) in the LDL-receptor gene which causes severe hypercholesterolemia. Hum. Mutat. **14:** 357.
16. OSE, L. 1999. An update on familial hypercholesterolaemia. Ann. Med. **31**(Suppl. 1): 13–18.
17. BLEASEL, J.F., D. HOLDERBAUM, V. BRANCOLINI, *et al.* 1998. Five families with argi-nine 519-cysteine mutation in COL2A1: evidence for three distinct founders. Hum. Mutat. **12:** 172–176.
18. PROCKOP, D.J., L. ALA-KOKKO, D.A. MCLAIN & C. WILLIAMS. 1997. Can mutated genes cause common osteoarthritis? Br. J. Rheumatol. **36:** 827–829.
19. LENDON, C.L., F. ASHALL & A.M. GOATE. 1997. Exploring the etiology of Alzheimer disease using molecular genetics. JAMA **277:** 825–831.
20. YU, C.E., J. OSHIMA, Y.H. FU, *et al.* 1996. Positional cloning of the Werner's syn-drome gene. Science **272:** 258–262.
21. PENNISI, E. 1996. Premature aging gene discovered [news; comment]. Science **272:** 193–194.
22. YE, L., T. MIKI, J. NAKURA, *et al.* 1997. Association of a polymorphic variant of the Werner helicase gene with myocardial infarction in a Japanese population [pub-lished erratum appears in Am. J. Med. Genet. 1997. **70**(1): 103]. Am. J. Med. Genet. **68:** 494–498.
23. MEULENBELT, I., C. BIJKERK, S.C.M. DE WILDT, *et al.* 1999. Haplotype analysis of three polymorphisms of the COL2A1 gene and associations with generalised radio-logical osteoarthritis. Ann. Hum. Genet. **63:** 393–400.
24. RISCH, N. & K. MERIKANGAS. 1996. The future of genetic studies of complex human diseases. Science **273:** 1516–1517.
25. RISCH, N. & H. ZHANG. 1995. Extreme discordant sib pairs for mapping quantitative trait loci in humans. Science **268:** 1584–1589.
26. KRUGLYAK, L. 1999. Prospects for whole-genome linkage disequilibrium mapping of common disease genes. Nat. Genet. **22:** 139–144.
27. VOGLER, G.P., G.E. MCCLEARN, H. SNIEDER, *et al.* 1997. Genetics and behavioral medicine: risk factors for cardiovascular disease. Behav. Med **22:** 141–149.
28. MEULENBELT, I., C. BIJKERK, F.C. BREEDVELD & P.E. SLAGBOOM. 1997. Genetic link-age analysis of 14 candidate gene loci in a family with autosomal dominant osteoarthritis without dysplasia. J. Med. Genet. **34:** 1024–1027.
29. BIJKERK, C., J.J. HOUWING-DUISTERMAAT, H.A. VALKENBURG, *et al.* 1999. Heritabili-ties of radiologic osteoarthritis in peripheral joints and of disc degeneration of the spine. Arthritis Rheum. **42:** 1729–1735.
30. FELSON, D.T., N.N. COUROPMITREE, C.E. CHAISSON, *et al.* 1998. Evidence for a Men-delian gene in a segregation analysis of generalized radiographic osteoarthritis: the Framingham Study. Arthritis Rheum. **41:** 1064–1071.
31. HIRSCH, R., M. LETHBRIDGE-CEJKU, R. HANSON, *et al.* 1998. Familial aggregation of osteoarthritis. Arthritis Rheum. **41:** 1227–1232.
32. WRIGHT, G.D., M. REGAN, C.M. DEIGHTON, *et al.* 1998. Evidence for genetic anticipa-tion in nodal osteoarthritis. Ann. Rheum. Dis. **57:** 524–526.
33. MEULENBELT, I., C. BIJKERK, H.S. MIEDEMA, *et al.* 1998. A genetic association study of the IGF-1 gene and radiological osteoarthritis in a population-based cohort study (the Rotterdam Study). Ann. Rheum. Dis. **57:** 371–374.

34. MEULENBELT, I., C. BIJKERK, S.C. DE WILDT, et al. 1997. Investigation of the association of the CRTM and CRTL1 genes with radiographically evident osteoarthritis in subjects from the Rotterdam study. Arthritis Rheum. **40:** 1760–1765.

35. HORTON, W.E.J., M. LETHBRIDGE-CEJKU, M.C. HOCHBERG, et al. 1998. An association between an aggrecan polymorphic allele and bilateral hand osteoarthritis in elderly white men: data from the Baltimore Longitudinal Study of Aging (BLSA). Osteoarth. Cartil. **6:** 245–251.

36. CHAPMAN, K., Z. MUSTAFA, C. IRVEN, et al. 1999. Osteoarthritis-susceptibility locus on chromosome 11q, detected by linkage. Am. J. Hum. Genet. **65:** 167–174.

37. LEPPAVUORI, J., U. KUJALA, J. KINNUNEN, et al. 1999. Genome Scan for Predisposing Loci for Distal Interphalangeal Joint Osteoarthritis: evidence for a locus on 2q. Am. J. Hum. Genet. **65:** 1060–1067.

38. WRIGHT, G.D., A.E. HUGHES, M. REGAN & M. DOHERTY. 1996. Association of two loci on chromosome 2q with nodal osteoarthritis. Ann. Rheum. Dis. **55:** 317–319.

39. LJUNGQUIST, B., S. BERG, J. LANKE, et al. 1998. The effect of genetic factors for longevity: a comparison of identical and fraternal twins in the Swedish Twin Registry. J. Gerontol. A Biol. Sci. Med. Sci. **53:** M441–M446.

40. LAGAAY, A.M., J. D'AMARO, G.J. LIGTHART, et al. 1991. Longevity and heredity in humans. Association with the human leucocyte antigen phenotype. Ann. N.Y. Acad. Sci. **621:** 78–89.

41. HEIJMANS, B.T., R.G.J. WESTENDORP, A.M. LAGAAY, et al. 1999. Common paraoxonase gene variants, mortality risk and fatal cardiovascular events in elderly subjects. Atherosclerosis, in press.

42. HEIJMANS, B.T., R.G. WESTENDORP, D.L. KNOOK, et al. 1998. The risk of mortality and the factor V Leiden mutation in a population-based cohort. Thromb. Haemost. **80:** 607–609.

43. HEIJMANS, B.T., R.G.J. WESTENDORP, D.L. KNOOK, et al. 1999. Angiotensin I–converting enzyme and plasminogen activator inhibitor-1 gene variants: the risk of mortality and fatal cardiovascular disease in an elderly population-based cohort. J. Am. Coll. Cardiol. **34:** 1176–1183.

44. HEIJMANS, B.T., J. GUSSEKLOO, C. KLUFT, et al. 1999. Mortality risk in men is associated with a common mutation in the methylenetetrahydrofolate reductase gene (MTHFR). Eur. J. Hum. Genet. **7:** 197–204.

45. FROSST, P., H.J. BLOM, R. MILOS, et al. 1995. A candidate genetic risk factor for vascular disease: a common mutation in methylenetetrahydrofolate reductase. Nat. Genet. **10:** 111–113.

46. BOUSHEY, C.J., S.A. BERESFORD, G.S. OMENN & A.G. MOTULSKY. 1995. A quantitative assessment of plasma homocysteine as a risk factor for vascular disease. Probable benefits of increasing folic acid intakes. JAMA **274:** 1049–1057.

47. NYGARD, O., J.E. NORDREHAUG, H. REFSUM, et al. 1997. Plasma homocysteine levels and mortality in patients with coronary artery disease. N. Engl. J. Med. **337:** 230–236.

48. FAURE-DELANEF, L., I. QUERE, J.F. CHASSE, et al. 1997. Methylenetetrahydrofolate reductase thermolabile variant and human longevity. Am. J. Hum. Genet. **60:** 999–1001.

49. GALINSKY, D., C. TYSOE, C.E. BRAYNE, et al. 1997. Analysis of the apo E/apo C-I, angiotensin converting enzyme and methylenetetrahydrofolate reductase genes as candidates affecting human longevity. Atherosclerosis **129:** 177–183.

50. HARMON, D.L., D. MCMASTER, D.C. SHIELDS, et al. 1997. MTHFR thermolabile genotype frequencies and longevity in Northern Ireland. Atherosclerosis **131:** 137–138.

51. MA, J., M.J. STAMPFER, E. GIOVANNUCCI, et al. 1997. Methylenetetrahydrofolate reductase polymorphism, dietary interactions, and risk of colorectal cancer. Cancer Res. **57:** 1098–1102.

52. CHEN, J., E. GIOVANNUCCI, K. KELSEY, et al. 1996. A methylenetetrahydrofolate reductase polymorphism and the risk of colorectal cancer. Cancer Res. **56:** 4862–4864.

53. BRATTSTROM, L., D.E. WILCKEN, J. OHRVIK & L. BRUDIN. 1998. Common methylenetetrahydrofolate reductase gene mutation leads to hyperhomocysteinemia but not to vascular disease: the result of a meta-analysis. Circulation **98:** 2520–2526.

54. PERLS, T.T., E. BUBRICK, C.G. WAGER, *et al.* 1998. Siblings of centenarians live longer. Lancet **351:** 1560.
55. KORPELAINEN, H. 1999. Genetic maternal effects on human life span through the inheritance of mitochondrial DNA. Hum. Hered. **49:** 183–185.
56. DE BENEDICTIS, G., L. CAROTENUTO, G. CARRIERI, *et al.* 1998. Gene/longevity association studies at four autosomal loci (REN, THO, PARP, SOD2). Eur. J. Hum. Genet. **6:** 534–541.
57. DE BENEDICTIS, G., G. ROSE, G. CARRIERI, *et al.* 1999. Mitochondrial DNA inherited variants are associated with successful aging and longevity in humans. FASEB J. **13:** 1532–1536.
58. IVANOVA, R., V. LEPAGE, D. CHARRON & F. SCHACHTER. 1998. Mitochondrial genotype associated with French Caucasian centenarians. Gerontology **44:** 349.
59. IVANOVA, R., N. HENON, V. LEPAGE, *et al.* 1998. HLA-DR alleles display sex-dependent effects on survival and discriminate between individual and familial longevity. Hum. Mol. Genet. **7:** 187–194.
60. MUIRAS, M.L., P. VERASDONCK, F. COTTET & F. SCHACHTER. 1998. Lack of association between human longevity and genetic polymorphisms in drug-metabolizing enzymes at the NAT2, GSTM1 and CYP2D6 loci. Hum. Genet. **102:** 526–532.
61. TOUPANCE, B., B. GODELLE, P.H. GOUYON & F. SCHACHTER. 1998. A model for antagonistic pleiotropic gene action for mortality and advanced age. Am. J. Hum. Genet. **62:** 1525–1534.
62. SCHACHTER, F., L. FAURE-DELANEF, F. GUENOT, *et al.* 1994. Genetic associations with human longevity at the APOE and ACE loci. Nat. Genet. **6:** 29–32.
63. LUFT, F.C. 1999. Bad genes, good people, association, linkage, longevity and the prevention of cardiovascular disease. Clin. Exp. Pharmacol. Physiol. **26:** 576–579.
64. HONDA, Y. & S. HONDA. 1999. The *daf-2* gene network for longevity regulates oxidative stress resistance and Mn-superoxide dismutase gene expression in *Caenorhabditis elegans*. FASEB J. **13:** 1385–1393.

Genetics of Survival

FRANÇOIS SCHÄCHTER

CESTI-ISMCM, Université Léonard de Vinci, 92916 Paris-la-Défense, France

ABSTRACT: The fields of gerontology and genetics have merged, spawning novel lines of investigation and generating a wealth of new results in recent years. However, the lack of clarity and consistency in the basic definitions upon which the science of gerontology must rest has fostered a certain amount of enduring confusion. Among the unclear issues are the genetic components of life span and the distinction between "normal" and "pathologic" aging. At a time of massive world population aging, such issues have, beyond their scientific importance, a momentous social and economic impact. A simple axiomatic framework, consisting of three definitions and five axioms, is proposed that clarifies the aforementioned issues and reconciles disparate data in gerontology. Based on this framework, a new classification of genes involved in survival is proposed. Within the Compensatory Adaptation Theory of aging, apparent paradoxes are solved and problems in gerontology may be formulated anew.

INTRODUCTION

"Genetics of longevity" is a misnomer that may have led astray other scientists interested in human gerontology. The reason is simple: longevity is the outcome of a lifetime of manifold events, integrating all types of intrinsic and extrinsic influences on viability — or vulnerability, a term that conveys the downward trend of an aging organism. What makes it so misleading is that the apparent simplicity of a simple scalar value — longevity — hides the diversity of its contributing components. One might as well try to understand a complex three-dimensional landscape through its projection on a straight line. A statement closer to reality can be outlined in four stages: (1) mortality is age-dependent, (2) age-dependent mortality is partially under genetic control, (3) the adjective "partially" makes room for environmental contributions, and (4) the genetic components of age-dependent mortality are themselves age-dependent. Survival, as a function of age, of which mortality is the first derivative with a minus sign, lends itself to the necessary distinctions. Therefrom results a new framework calling for new questions: What are the major genetic components of age-dependent mortality? Are these components interrelated? In what ways? Can a new classification of genes involved in aging be derived? What is to be learned from the genetics of age-dependent mortality about aging processes, and vice-versa? What are the present frontiers of survival and what paths of exploration are opened? In what follows we start addressing these questions.

THE COMPENSATORY ADAPTATION THEORY OF AGING

The tenets of this theory have been described elsewhere.[1,2] I shall present the theory in the condensed form of a set of definitions and axioms, propose some predictions to be checked in future investigations, and finally offer one biologic illustration.

Definition 1: Aging is the composite outcome of irreversible alterations and partially reversible compensations. It always corresponds to a decrease in viability or an increase in vulnerability.

Definition 2: Primary aging is the core component of aging, resulting from irreversible alterations undergone by mere functioning.

Definition 3: Compensatory adaptation (CA) refers to reorganizations of metabolic pathways and coordinated functions consecutive to primary aging or to external stresses. It is the second component of aging.

Axiom 1: The genetic design of the organism is optimized for reproductive fitness.

Axiom 2: The mere functioning of an organism brings about irreversible alterations in its molecular, cellular, and tissular components.

Axiom 3: The genome itself is the target of some irreversible and cumulative damage with chronologic time, which entails a progressive loss or alteration of genetic information.

Axiom 4: In an effort to survive, the organism reorganizes its physiology and metabolism to compensate for its deficiencies.

Axiom 5: Both components of aging are under partial genetic control.

Now let us briefly comment upon these definitions and axioms:

Definition 1: This is a biologic definition of aging at the individual level. It ties up with the formal statistical definition at the population level in terms of an increase in mortality rates. It stems from this definition that aging is not synonymous with the passage of time, because some compensations may be reversed and even some damage may be repaired, thereby reducing the irreversible kernel of aging. This definition offers the advantage that it directly points to processes and mechanisms, opening the way for action.

Definition 2: It is noteworthy that primary aging is not synonymous with intrinsic or endogenous aging. Indeed, under the best environmental conditions, minimizing damage from external sources, some molecular damage will still accumulate due to free radical reactions, etc. This damage, resulting from basic combustion reactions, cannot be attributed exclusively to either external or internal sources that are nearly impossible to separate in an open system. Therefore, there is little sense in talking about endogenous aging. Intrinsic aging includes some of the consequences of primary aging under conditions of minimal external stress.

Definition 3: CA derives from both primary aging and supplementary stresses. One might say that the CA component of aging contains (intrinsic minus primary) aging plus the lasting effects of other stresses. The latter covers pathologic aging.

The definitions could be summarized in the following equation:

Total Aging = Primary Aging + CA = Intrinsic Aging + Pathological Aging

Axiom 1: The genetic design is understandable in light of the whole life history, from conception through development and reproduction, with their respective time scales. This is Darwinian evolution in a nutshell. Implicit in this axiom is the dwindling evolutionary pressure on the postreproductive period of an organism's life span and its consequences in terms of the evolutionary appearance of aging.[3]

Axiom 2: In fact, this axiom has a much wider generality as it applies to any physical system. Biologic systems have the unique property of being able to repair and renew themselves to some extent. Their maintenance systems are not sufficient, however, to fend off cumulative damage, because of axiom 1. Therefore, the rate of alteration is the resultant of two opposite processes: damage and repair.

Axiom 3: This axiom is the basis of the old somatic mutation theory of aging. Although we know this to be true, the information is yet very scant about the sequence- and tissue-specificity of such damage. The age-dependence also remains to be plotted out. The key question at this point is the extent that genomic damage contributes to aging.

Axiom 4: Individual adaptation stands in clear distinction with adaptation in the evolutionary sense. The latter has been the molding force of the germline-transmitted genome; it contains the blueprint of every situation that was faced in the past, with solutions that were devised to cope and survive; all pathways in their intricate coordination can function perfectly in the internal milieu of a pre-reproductive organism with most of its genetic information intact. The former, however, is a solution put together with defective components so that it is by necessity suboptimal.

Axiom 5: Both components of aging may display environmental influences and gene × environment interactions, although these are probably far more important for the second component.

Here are some predictions of the theory: Modulation of aging rate and therefore of maximum life span could be accomplished by modulating one of the two terms of the damage–repair equation; the common denominator of aging manifestations in a given species is the compound of primary aging and minimal resulting CA; given the characteristic common patterns of aging, the distribution of damage in time and space may exhibit a much greater specificity than is currently thought; by identifying the targets of primary aging, it may become feasible to compensate directly for the incurred damage and therefore minimize the CA component of aging.

A biologic illustration of CA is cellular senescence. Ever since the seminal work of Hayflick and Moorhead,[4] laboratories have been fascinated with the "Hayflick limit," researching the mechanisms behind the cells' limited division capacity. Telomere shortening is one of the more studied genomic alterations that correlates well with cellular senescence,[5] in the sense that the doubling potential may be modulated to some extent by switching on or off the telomerase activity. Other genomic alterations, such as reorganization of chromatin before entry into the senescent stage, have received less attention although they are probably at least as important. Innumerable papers have fueled the never-ending debate as to whether cellular senescence is a cause or a consequence of aging. The alternative to cellular senescence is the division of cells containing altered genetic information, leading to cancer and other clonal disorders.[6] Pre-senescent cells may be forced to continue dividing until a tiny proportion of the culture will finally yield immortalized cells, while the remainder of the culture undergoes a massive death crisis. These data argue clearly in favor of the compensatory adaptive nature of cellular senescence as it occurs during aging. Indeed, senescent cells are associated with pathologic rather than normal aging.[7] The progressive emergence of senescent cell populations is a common but not universal feature of aging, as seen for example in centenarians.[8]

A NOVEL CLASSIFICATION OF GENES INVOLVED IN SURVIVAL

This classification is naturally derived from the CA theory of aging. It consists in separating the genetic contributions to the primary and CA components of aging.

Genes Involved in Primary Aging

These genes modulate one term of the damage–repair equation. Intervening in the rate of damage are genes of the endocrine pathways regulating basal metabolism, genes that provide the first line of defense against reactive by-products of the metabolism, encompassing the common antioxidant enzymes. Intervening in maintenance and repair are DNA repair genes, which have been extensively characterized in several organisms, but also genes involved in protein and lipid repair and turnover.

In this category of genes, a change in activity would affect many other pathways, necessitating a coordinate regulation. Therefore, little genetic variation is expected. If a variant increased life span, it would do so in diverse environments. Because these genes depend on basal energy metabolism in their activity, putative longevity-extending variants could display antagonistic pleiotropy with fitness.[9] Interactions are expected with the environment whenever it affects rates of damage or maintenance. Such is the case with nutritional intake or exposure to chemical pollutants or radiation. It is noteworthy that caloric restriction falls into that category by increasing the turnover rates of proteins and lipids and maybe even of DNA.

Genes Involved in Compensatory Adaptation

Many examples of CA can be envisioned at all levels of biologic organization. They are obviously interrelated, which in itself is sufficient to account for the compensation effect on mortality and the negative correlations between the incidences of major age-related pathologies.[10] Among these, one may cite the increase in cardiac cell volume at the cellular level, the increase in left ventricular volume at the organ level, and the hypercoagulation phenotype found in centenarians.[2]

Because these compensations occur at the expense of one another, a fair amount of genetic variation is expected. The effects should be largely dependent on the environment and individual life history. Antagonistic pleiotropy between survival at different periods of the lifetime may take place.[11]

To this day, the so-called human longevity genes have fallen into the second category. Therefore, the "paradoxes of longevity" whereby the risk status of certain alleles or genotypes in adulthood turns into a protective status at very old ages should come as no surprise.

In the nematode *Caenorhabditis elegans,* however, genes of the first category have been found. It is greatly interesting that one of them bears homology to an insulin receptor gene and another to a mitochondrial component of the respiratory chain.[12] They appear to plug right into the energy metabolism.

THE "RATE-OF-LIVING" THEORY OF AGING REVISITED

This is perhaps the most enduring theory of aging, formulated by Raymond Pearl in the early portion of our century. Like the sea serpent, this theory reemerges periodically under a different guise. It is based on the observation that life span is, by and large, inversely proportional to the basal metabolic rate.

I want to show that this theory is contained in the CA theory of aging herein presented. Primary aging is but one term in the equation of aging. It is already the resultant of two categories of processes: damage and alterations, on the one hand; maintenance, replacement, and repair, on the other. Let us label these terms symbolically by: *DA* for the global rate of damage and alteration processes; *MR* for the global rate of maintenance, replacement, and repair processes. *DA* depends on the biochemical and biophysical parameters of a living system: temperature, pH, and redox potential. It has been estimated for various types of DNA damage.[13] There is a good case for it being proportional to the rate of oxygen consumption, because free radicals generated during respiration are the main culprits of biochemical damage. What remains to be assessed though is the extent to which other biochemical pollutants pervading our present environment and entering the body through the atmosphere and the food – the fuel components – may increase this term, or, stated otherwise, what is the minimal value and the range of variation of *DA*. We know also that *DA* may be increased by exposure to certain radiations, and the possibility that it might be reduced by specific radiations cannot be discarded, although there are no data about this interesting hypothesis for the time being.

MR on the other hand is an evolved response to *DA* encoded in the genome. The whole machinery of *MR* can be defined in terms of networks of genes. There is no biochemical or biophysical reason *a priori* why *MR* should not be able to completely counter *DA*, except that it makes no evolutionary sense for any species to have developed such a system.[9] Whereas *DA* is largely conserved across species, there exists an endless variety of combinations to modulate *MR*. This explains why the inverse correlation between metabolic rate and life span holds only across related species. For example, flying birds have a much higher life span energy potential (total amount of energy available in a life time, equal to the product of the maximum life span potential and the specific metabolic rate) than do mammals of similar size.[14] This is associated with higher antioxidant activity, better mitochondrial coupling resulting in less leakage of free radicals, and greater genome stability.

GENOME DYNAMICS ON TWO TIME SCALES

A fundamental polarity lies at the heart of evolution and aging: the balance between plasticity and stability of the genome. Evolution would not be possible without the appearance of new genetic diversity in the germ line at each generation. Yet such diversity includes the pool of genetic variants that are responsible for inherited diseases and the pool of slightly deleterious mutations that remain more covert but make up the genetic load.[15] Providing an estimate of the deleterious mutation rate is an open and actively investigated problem.[16]

The progressive and unavoidable contamination of species' genomes by late-acting deleterious mutations is regarded as the major evolutionary mechanism accounting for the universal occurrence of aging across animal phyla.[17,18] The same basic phenomena are at work in the soma. They are summarized in the interplay between *DA* processes and *MR* systems. Sexual recombination may provide a means of eliminating deleterious mutations in bunches, thereby escaping the genetic decay that could threaten our mutation-prone race.[19,20] But sexual reproduction may also impose a limit to longevity, because genomes from male gametes that have undergone too many divisions bear a certain genetic load.[21] We may be led to discover hitherto unsuspected mechanisms of "averaging selection" that would prevent mutants of increased longevity from taking over, such as negative correlations between genetically determined early and late mortality. Whereas these correlations have emerged as properties of a simple mathematical model,[11] their biologic substrate is far from elucidated.

EVOLUTIONARY PRESSURE FROM WITHIN

The level of environmental hazard is an evolutionary determinant of a species' life span: that is a fundamental tenet of evolutionary theories of aging, such as the disposable soma or antagonistic pleiotropy theories. Danger is thought to come from macroscopic predators scattered in the external environment. Left aside are inner microscopic predators gnawing at the tissues, cells, and genome. Such predators, belonging to the category of parasites, exist in many shapes and sizes. Metazoans, protozoans, bacteria, and viruses thrive in the body. They are exquisitely sensitive and responsive to their host's stresses, which they are apt to exploit to their own ends. They are certainly active in compensatory adaptation and may even have a share in primary aging. The complex ecosystem of parasites influences not only the species' patterns of aging as a driving evolutionary force, bringing about the HLA associations with longevity and disease, but also individual aging. The same versatile agents are at work in macroevolution and microevolution.

The human race may have acquired mastery over its external environment and overcome predators of times past. Its control over inner predators is much less advanced, as evidenced by recent outbreaks in infectious epidemics. In fact, we may not even be aware of the range and extent of parasitism in developed countries. Exploring this hidden frontier will be a major challenge ahead in gerontology.

REFERENCES

1. SCHÄCHTER, F. 1997. Genetics of Aging. *In* Longevity: To the Limits and Beyond. J.M. Robine, J.W. Vaupel, B. Jeune & M. Allard, Eds. :131–138. Springer. Heidelberg.
2. SCHÄCHTER, F. 1998. Causes, effects and constraints in the genetics of human longevity. Am. J. Hum. Genet. **62:** 1008–1014.
3. WILLIAMS, G.C. 1957. Pleiotropy, natural selection, and the evolution of senescence. Evolution **11:** 398–411.
4. HAYFLICK, L. 1965. The limited in vitro lifetime of human diploid cell strains. Exp. Cell Res. **37:** 614–636.
5. HARLEY, C.B., A.B. FUTCHER & C.W. GREIDER. 1990. Telomeres shorten during ageing of human fibroblasts. Nature **345:** 458–460.
6. SING, C.F. & S.L. REILLY. 1993. Genetics of common diseases that aggregate, but do not segregate, in families. *In* Genetics of Cellular, Individual, Family, and Population Variability. C.F. Sing & C.L. Hanis, Eds. : 140–161. Oxford University Press. Cambridge.
7. MACIEIRA-COELHO, A. 1995. The implications of the "Hayflick limit" for aging of the organism have been misunderstood by many gerontologists. Gerontology **41:** 94–97.

8. BOUCHER, N., T. DUFEU-DUCHESNE, E. VICAUT et al. 1998. CD28 expression in T cell aging and human longevity. Exp. Gerontol. **33:** 267–282.
9. Kirkwood, T.B.L. & M.R. Rose. 1991. Evolution of senescence: late survival sacrificed for reproduction. Phil. Trans. R. Soc. Lond. B **332:** 15–24.
10. GAVRILOV, L.A. & N.S. GAVRILOVA. 1991. The Biology of Lifespan: A Quantitative Approach. Harwood Academic. Chur, Switzerland.
11. TOUPANCE, B., B. GODELLE, P.-H. GOUYON & F. SCHÄCHTER. 1998. A model for antagonistic pleiotropic gene action for mortality and advanced age. Am. J. Hum. Genet. **62:** 1525–1534.
12. WOOD, W.B. 1998. Aging of *C. elegans*: mosaics and mechanisms. Cell **95:** 147–150.
13. LINDAHL, T. 1993. Instability and decay of the primary structure of DNA. Nature **362:** 709–715.
14. PEREZ-CAMPO, R., M. LOPEZ-TORRES, S. CADENAS et al. 1998. The rate of free radical production as a determinant of the rate of aging: evidence from the comparative approach. J. Comp. Physiol. B **168:**149–158.
15. KONDRASHOV, A.S. 1995. Contamination of the genome by very slightly deleterious mutations: why have we not died 100 times over? J. Theor. Biol. **175:** 583–594.
16. EYRE-WALKER, A. & P.D. KEIGHTLEY. 1998. High genomic deleterious mutation rates in hominids. Nature **397:** 344–347.
17. PATRIDGE, L. & N. PROWSE. 1994. Mutation, variation and the evolution of ageing. Curr. Biol. **4:** 430–432.
18. PARTRIDGE, L. & N.H. BARTON. 1993. Optimality, mutation and the evolution of ageing. Nature **362:** 305–311.
19. CROW, J.F. 1997. The high spontaneous mutation **rate:** is it a health risk? Proc. Natl. Acad. Sci. USA **94:** 8230–8386.
20. CROW, J.F..1999. The odds of losing at the genetic roulette. Nature **397:** 293–294.
21. GAVRILOV, L.A., N.S. GAVRILOVA, V.N. KROUTKO et al. 1997. Mutation load and human longevity. Mutat. Res. **377:** 61–62.

Melanin Accumulation Accelerates Melanocyte Senescence by a Mechanism Involving p16^{INK4a}/CDK4/pRB and E2F1

DEBDUTTA BANDYOPADHYAY[a,b] AND ESTELA E. MEDRANO[a-d]

[a]Huffington Center on Aging and [b]Departments of Molecular and Cellular Biology and [c]Department of Dermatology, Baylor College of Medicine, One Baylor Plaza M320 and VAMC, Houston, Texas 77030, USA

ABSTRACT: Cellular and molecular evidence suggests that senescence is a powerful tumor-suppressor mechanism that prevents most higher eukaryotic cells from dividing indefinitely *in vivo*. Recent work has demonstrated that α-melanocyte stimulating hormone (α-MSH) or cholera toxin (CT) can activate a cAMP pathway that elicits proliferative arrest and senescence in normal human pigmented melanocytes. In these cells, senescence is associated with increased binding of p16^{INK4a} to CDK4 and loss of E2F-binding activity. Because senescence may provide defense against malignant transformation of melanocytes, and because pigmentation is a strong defense against melanoma, we examined the ability of melanocytes derived from light and dark skin to respond to CT. Here we demonstrate that in melanocytes derived from dark-skinned individuals, CT-induced melanogenesis is associated with accumulation of the tumor suppressor p16^{INK4a}, underphosphorylated retinoblastoma protein (pRb), downregulation of cyclin E, decreased expression of E2F1, and loss of E2F-regulated S-phase gene expression. In contrast to other senescent cell types, melanocytes have reduced or absent levels of the cyclin-dependent kinase inhibitors p27^{Kip1} and p21^{Waf-1}. Importantly, melanocytes derived from light-skinned individuals accumulated smaller amounts of melanin than did those from dark-skinned individuals under the same conditions, and they continued to proliferate for several more division cycles. This delayed senescence may result from reduced association of p16 with CDK4, reduced levels of underphosphorylated pRb, and steady levels of cyclin E and E2F1. Because cyclin E-CDK2 inhibition is required for p16-mediated growth suppression,[1] upregulation of p16 and downregulation of cyclin E appear essential for maintenance of terminal growth and senescence. Given the rising incidence of melanoma, identification of major growth regulatory proteins involved in senescence should shed light on the biology of this genetically mysterious tumor.

INTRODUCTION

Normal human cells have a finite proliferative potential *in vitro* (50–100 population doublings), after which they exit the cell cycle in an irreversible manner.[2] This process has been termed replicative senescence, and it could be implicated in organ-

[d]To whom correspondence should be addressed at V A Medical Center, Research Building 110, 2002 Holcombe Boulevard, Houston, Texas 77030. Phone: 713-791-1414, ext. 4174; fax: 713-794-7978.

medrano@bcm.tmc.edu

ismal aging, tumor suppression, and terminal differentiation. In contrast to apoptosis, it represents a stable, protracted state in which cells remain viable and metabolically active for extended periods of time. Irreversible growth arrest and senescence can be induced by oxidative damage or oncogenic stimuli.[3–5] Cellular senescence is phenotypically characterized by a flat, rounded cell morphology, by activation of senescence-associated β-galactosidase (SA-β-gal),[6] and by reduction in telomere length (mostly in replicatively senescent cells).[7]

Progression through the cell cycle is controlled by cell-cycle regulatory proteins known as cyclins, cyclin-dependent kinases (CDKs), CDK inhibitors (CDK-Is), retinoblastoma (Rb) protein family members, and E2F transcription factors.[8,9]

The E2F family of transcription factors contains six members named E2F1 through E2F6. Among these, E2F1, 2, 3, 4, and 5 take part in regulating cell-cycle progression by influencing DNA synthesis and expression of proteins required for G1 to S-phase transition.[10] E2F activity is regulated by formation of E2F/DP heterodimers and by association with cyclins and members of the pRb family.[11–13] In general, binding of free E2F/DP heterodimers to E2F sites activates transcription, whereas complex formation between E2F and pRb silences transcription of target genes.[14,15] On the other hand, E2F-6 has been reported to act as a transcriptional repressor by countering the activity of other E2F complexes via a pRb-, p107-, or p130-independent mechanism.[16] Rb protein family members remain in the underphosphorylated state and bind E2F proteins, thereby inhibiting E2F's transcriptional activity. As cells progress towards S phase, pRB proteins become hyperphosphorylated by cyclin/CDKs and can no longer bind E2Fs, leading to E2F activation. Activated E2Fs augment transcription of genes such as thymidylate synthetase (TS)[17] and thymidylate kinase (TK),[18] which are required for G1 to S-phase progression. A decrease in E2F activity is a prerequisite for induction of quiescence and terminal differentiation in many cell types, whereas mitogenic stimulation of quiescent cells is accompanied by an increase in E2F-dependent transcription.[19] During *in vitro* myogenic differentiation, muscle cells irreversibly exit the cell cycle,[20] but this pathway can be prevented by overexpression of E2F1.[11] In differentiating myocytes, cyclin/CDK activity is suppressed, and pRb is activated by dephosphorylation.[11]

The growth-inhibitory and E2F-suppressive properties of the Rb protein family are regulated by cyclin/CDK-dependent phosphorylation during G1 to S-phase progression.[21] This phosphorylation is dependent on CDK activity and can be positively regulated by cyclins or negatively regulated by CDK-inhibitors (p16^{INK4a}, p21^{Waf-1}, p27^{Kip1} etc.).[8,22] The p27/p21 family was initially thought to interfere with cyclin D-, E-, and A-dependent kinase activity. More recent work has altered this view, revealing that although p27/p21 proteins are potent inhibitors of cyclin E- and A-dependent CDK2, they act as assembly factors for cyclin D/CDK4 complexes.[23] On the other hand, p16 is a specific inhibitor of CDK4 and CDK6.[24] Mounting evidence suggests that p16 is a tumor suppressor. It maps physically to chromosome 9p21, a locus rearranged in many human cancers, including familial melanomas.[25] p16 is commonly deleted, mutated, or hypermethylated and transcriptionally silenced in tumors that retain wild-type Rb, and its ectopic expression results in G1 arrest.[21,26,27] The precise mechanism by which p16 functions as a tumor suppressor is not completely understood. One possibility is that p16 inactivation is required for entry into S phase. However, many normal cells express p16 throughout the G1 phase and are still able to proliferate. An alternative mechanism involves the link between p16 expression and cellular senescence.[28,29] p16 levels increase with age in fibroblast, epithelial, and melanocyte cultures. Importantly, loss of p16^{INK4a} expression might be required for cells to escape senescence during tumor progression. In addition, DNA double strand-breaking agents, such as bleomycin and oncogenic stimuli, can cause p16 accumulation.[31,32]

Melanocytes are specialized cells that are located in the basal layer of the epidermis and that synthesize and transfer melanin pigments to surrounding keratinocytes, leading to uniform skin pigmentation. Melanin pigments play a key protective role against the carcinogenic effects of solar ultraviolet light *in vivo*.[33,34] Dark-skinned individuals have substantially lower risks of melanoma than do light-skinned populations. In melanocytes isolated from dark skin, sustained activation of the cAMP pathway by chronic treatment with the pharmacologic agent cholera toxin (CT) results in synthesis and accumulation of melanin, followed by withdrawal from the cell cycle and expression of a senescent phenotype.[35,36]

In this communication, we compared the response to CT of melanocytes from light and black skin tissues. We propose that (1) premature senescence is accelerated in melanocytes that induce high levels of pigmentation and (2) induction of p16^{INK4a} and downregulation of cyclin E and E2F are necessary for this process.

MATERIALS AND METHODS

Cell Culture. Human neonatal melanocytes were cultured as described previously.[37] The melanocyte cell line NHM-B was derived from a dark-skinned individual (Black), whereas line NHM-C was from a light-skinned individual (Caucasian). Cells were grown in human melanocyte-proliferating medium which consisted of MCDB-153 medium supplemented with 8 nM phorbol 12-myristate 13-acetate (PMA), 0.3 ng/ml human recombinant fibroblast growth factor, 5 µg/ml transferrin, 4% fetal bovine serum (Sigma Co., MO), and 30 µg/ml pituitary extract (Clonetics, CA). The differentiation medium was the same but lacked PMA and was supplemented with 10 nM CT plus 0.1 mM 3-isobutyl-methylxantine (IBMX).[30,35] Melanocytes were considered senescent when they did not show an increase in cell number over a 2-week period.

Antibodies and Other Reagents. Anti-pRb, anti-p16^{INK4a}, and anti-E2F1 mouse monoclonal antibodies were purchased from Labvision Inc. (Fremont, CA) and anti-p27^{Kip1} monoclonal antibodies from Transduction Laboratories, Inc. (Lexington, KY) (1:200). Mouse anti-p21$^{Waf-1/SDI-1/Cip-1}$ antibodies (OP64) were acquired from Oncogene Research Products (MA) and all others from Santa Cruz Biotechnology (CA). Antibodies were used at dilutions recommended by the manufacturers. The BrdU detection kit was purchased from Boehringer-Mannheim. Histone H1 and GST-Rb were acquired from Gibco-BRL and Santacruz Biotechnology (CA), respectively.

BrdU Labeling and Immunocytochemistry. Subconfluent melanocyte cultures were seeded on 12-mm pre-washed glass coverslips in growth factor-free MCDB-153 medium. After 24 hours, the medium was replaced with differentiating medium. At appropriate times, 20 mM BrdU was added, and cells were incubated at 37°C, 5% CO$_2$ for 4 hours. Melanocytes were then fixed with 70% ethanol in 50 mM glycine, pH 2.0, for 20 minutes at −20°C. Cells were washed and incubated with mouse anti-

BrdU antibodies for 30 minutes at 37°C. After washing with phosphate-buffered saline (PBS), cells were incubated with anti-mouse alkaline phosphatase-conjugated secondary antibodies as described by the manufacturer (Boehringer-Mannheim, IN). Coverslips were washed and mounted on glass slides using fluoromount-G. Cells were counted and photographed using an Olympus light microscope.

Preparation of Cell Extracts. Whole cell extracts were prepared from melanocytes by adding twice each cell pellet's volume of lysis buffer (50 mM Tris-Cl, pH 7.5, 0.4% NP-40, 120 mM NaCl, 1.5 mM $MgCl_2$, 2 mM PMSF, 80 µg/ml Leupeptin, 3 mM NaF, 1 mM DTT) and incubating on ice for 30 minutes with occasional vortexing. Lysate was collected after centrifuging at 14,000 g for 10 minutes. The resulting supernatant was used for immunoblotting, immunoprecipitation, and CDK kinase assays. Pellets were used to determine melanin content (to be described).

Immunoblotting and Immunoprecipitation. Whole cell lysates (50 µg) were electrophoresed in SDS-polyacrylamide gel slabs and transferred to nitrocellulose membranes. Blots were washed with T-TBS (20 mM Tris, 150 mM NaCl, 0.2% Tween-20) and blocked with 5% milk in T-TBS for 1 hour at room temperature. Membranes were incubated overnight at 4°C in primary antibodies (using the manufacturer's recommended dilution), followed by three 10-minute washes in T-TBS buffer. Membranes were then incubated with HRP-conjugated donkey anti-rabbit or sheep anti-mouse immunoglobulin G antibodies in T-TBS (1:3,000) for 1 hour at room temperature. After washing, target proteins were detected using the ECL immunoblotting detection kit (Amersham, IL). Actin was used as a control for protein loading. For immunoprecipitations, 200 µg of total cell extract was routinely used in a 200 µl volume. Recommended concentrations of antibodies were added, followed by incubation at 4°C for 3 hours. Protein G- (for IgG1) or protein A- (for IgG2a or polyclonal antibodies) sepharose beads were used for immobilizing the antibody complexes. Sepharose-bound immunocomplexes were washed three times with 10 volumes of lysis buffer and analyzed by SDS-PAGE.

Melanin Content. Melanin content was determined as described previously.[38] Briefly, cell pellets were solubilized overnight at 60°C in 0.2 M NaOH. Absorbance of aliquots was read at 475 nm against a standard curve of synthetic melanin (Sigma).

CDK Activity Assay. Total cell extracts were diluted to a concentration of 1 µg/µl and immunoprecipitated with an anti-CDK2 or anti-CDK4 polyclonal antibody (Santa Cruz, CA). CDK kinase activity associated with immunocomplexes was analyzed by incubating the immunocomplex-bound beads in H1-kinase buffer (25 mM β-glycerophosphate, pH 7.3, 10 mM $MgCl_2$, 1 mM DTT, 25 mM HEPES, pH 7.3, 10 mM EGTA, 0.1 mg/ml Histone H1, 0.5 µCi [32P]-γ-ATP) at 37°C for 30 minutes. Finally, phosphorylated H1 was analyzed by SDS-PAGE and autoradiography. H1 was replaced with GST-Rb for CDK4 kinase assays.

RESULTS

Activation of Melanogenesis Correlates with Premature Senescence in Human Melanocytes: Role of Melanin Pigments

Alpha-melanocyte stimulating hormone (α-MSH) is one of the most potent activators of melanogenesis.[39] α-MSH binds to a G protein-coupled receptor and leads to Gαs protein activation and an increase in intracellular cAMP. In cultured melanocytes, the effect of α-MSH can be mimicked by other cAMP-inducing agents, such as CT, forskolin, and the phosphodiesterase inhibitor isobutylmethylxanthine (IBMX).[40–42] CT has been shown to induce melanogenesis and withdrawal from the cell cycle in melanocytes isolated from dark-skinned individuals.[35,36]

We were interested in determining whether melanocytes isolated from light-skinned individuals could also enter senescence after prolonged exposure to CT. For this study, we chose two melanocyte strains at opposite ends of the pigmentation spectrum, derived from very light (Caucasian) and dark (Black) neonatal skin tissues (designated NHM-C and NHM-B, respectively). Melanocytes were cultured in a medium supplemented with CT and IBMX for up to 6 (chronic treatment) or 12 weeks. Initially, both cell strains showed comparable percentages of cells in S phase (measured by BrdU incorporation): 29.3% for NHM-C and 32.2% for NHM-B. However, 22.1% of NHM-C cells were still proliferative after 6 weeks, compared to only 5.2% of NHM-B cells (FIG. 1A). Eventually, exposure to CT-containing medium for up to 12 weeks also resulted in senescence of the lightly pigmented NHM-C melanocytes (FIG. 1A). It was previously demonstrated that exposure to cAMP-inducing agents results in progressive morphologic changes in pigmented melanocytes.[43,44] Melanocytes become multidendritic within 1 week and by 6 weeks acquire an enlarged and flat, round-shaped morphology, characteristic of senescent cells. In our study, exposure to CT resulted in a dramatic increase in melanin content in NHM-B cells but only a negligible increase in NHM-C melanocytes (FIG. 1B). Identical results were obtained in several other melanocyte strains isolated from dark and light skin tissue and from a tyrosinase-negative albino sample (data not shown). However, melanocytes with light-medium pigmentation accumulated variable (although low) melanin levels in response to CT. Correspondingly, BrdU staining showed data intermediate between NHM-C and NHM-B cells (data not shown). Senescent melanocytes with varying degrees of pigmentation survive in culture for many months. Heavily melanized melanocytes (containing numerous stage III and IV melanosomes) can be found in Black skin, indicating that melanin accumulation as such does not impair human melanocyte viability. Altogether our results indicate that terminal differentiation and senescence is accelerated in melanocytes able to accumulate substantially high levels of melanin.

Characterization of Cell Cycle Changes in Senescent Melanocytes: pRb Phosphorylation

Phosphorylation, dephosphorylation, and protein levels of pRb family members play critical roles in cell-cycle progression.[8,9,45] In quiescent, Go cells or those in early G1, pRb is present in non- to underphosphorylated forms. During progression from G1 to S phase, pRb becomes increasingly hyperphosphorylated. We were interested in determining the phosphorylation status of pRb in CT-treated melanocytes. Protein lysates from NHM-C and NHM-B cells exposed to CT for several weeks were analyzed by SDS-PAGE, blotted, and probed with an anti-pRb antibody. FIGURE 2A (*upper panel*) shows that pRb in NHM-C melanocytes was present in both hypo-and hyperphosphorylated forms through 3 weeks (*lanes 1 through 3*), whereas it was predominantly in the hypophosphorylated form at 6 weeks (*lane 4*). In contrast, pRb in NHM-B melanocytes was mostly in the nonphosphorylated form

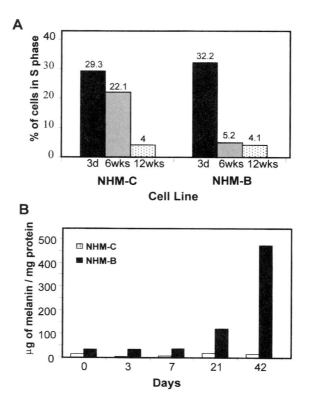

FIGURE 1. (**A**) BrdU incorporation in melanocytes. NHM-C or NHM-B cells were treated with cholera toxin (CT) for 3 days, 6 weeks, or 12 weeks. After incubation, cells were tested for BrdU incorporation. Percentage of BrdU-incorporated cells after CT treatment is shown above bars at different time periods. (**B**) Melanin accumulation in NHM-C or NMH-B cells treated with CT for up to 6 weeks. The *vertical axis* represents the amount of melanin in μg/mg total protein.

by 3 weeks of CT treatment. Intriguingly, overall pRb levels were lower at any given time in NHM-B as compared to NHM-C cells (*lanes 5 through 8*).

Changes in CDK2 and CDK4 Activity

Studies from different laboratories support a model in which pRb is initially phosphorylated by D-type cyclins complexed to CDK4. As cells progress through the cell cycle, cyclin E/CDK2 further increases pRb phosphorylation levels.[46] We therefore investigated whether exposure to CT would induce changes in CDK activity in NHM-C and NHM-B melanocytes. Extracts from these cells were immunoprecipitated with either CDK4 or CDK2 polyclonal antibodies. Kinase activity was assayed in precipitated complexes using GST-Rb (for CDK4) or histone H1 (for

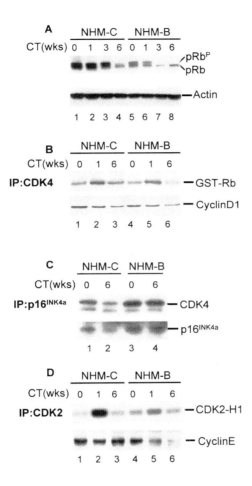

FIGURE 2. (**A**) Changes in pRb phosphorylation after CT treatment. Extracts from CT-treated cells were analyzed by immunoblotting with an anti-pRb antibody (*upper panel*). pRb^P indicates the phosphorylated forms of pRb. The *lower panel* shows an actin blot of the same nitrocellulose, confirming equal protein loading. (**B**) Extracts from CT-treated melanocytes were immunoprecipitated with an anti-CDK4 antibody. Kinase activity assays were performed on immunocomplexes using GST-pRb as a substrate (*upper panel*). The CDK4 immunoprecipitates were analyzed by SDS-PAGE and immunoblotted with an anti-cyclin D1 antibody to assess CDK4-cyclin D1 association under indicated conditions (*lower panel*). (**C**) The p16^{INK4a} immunoprecipitates were analyzed with an anti-CDK4 antibody (*upper panel*) or with an anti-p16^{INK4a} antibody (*lower panel*) to assess CDK4- p16^{INK4a} association. (**D**) Anti-CDK2 immunoprecipitates from CT-treated melanocytes were assayed for CDK-kinase activity using Histone H-1 as substrate (*upper panel*). The immunoprecipitates were analyzed by probing with an anti-cyclin E antibody to assess CDK2-cyclin E association (*lower panel*).

CDK2) as substrates. At 6 weeks, CDK4-GST-Rb kinase activity was notably reduced in NHM-B as compared to NHM-C cells (FIG. 2B, upper panel). CDK activity is regulated by association to the cyclin regulatory subunit and to inhibitory proteins (CDK-Is). To determine the reason(s) of low cyclin D/CDK4 activity in pigmented melanocytes (FIG. 2B, lane 6), we used coimmunoprecipitation to analyze levels of cyclin D1 bound to CDK4. FIGURE 2B (lower panel) shows a more pronounced decrease in cyclin D1 levels in CDK4 complexes from NHM-B melanocytes. p16^{INK4a} has been proposed to play a critical role in regulating senescence and terminal differentiation in several cell types.[31,47,48] To determine the role of p16 in melanocyte senescence, we analyzed its association with CDK4 by coimmunoprecipitation with an anti-p16^{INK4a} monoclonal antibody. FIGURE 2C (upper panel) shows that over a 6-week period, levels of CDK4 bound to p16 decreased dramatically in NHM-C melanocytes, whereas it remained almost unchanged in NHM-B cells. A decrease in the p16-CDK4 association resulted from reduced levels of p16 in immunoprecipitates (FIG. 2C, lower panel, lane 2) and in total cell extracts (FIG. 3C, lower panel). Thus, higher cyclin D1 levels and reduced association of p16 with CDK4 resulted in high kinase activity in NHM-C cells (FIG. 2B, lane 3). In turn, CDK2-H1 kinase activity decreased similarly in both cell strains at 6 weeks (FIG. 2D, upper panel), despite a dramatic decrease in levels of cyclin E associated with CDK2 in NHM-B cells (lower panel). This can be explained by the fact that p27 levels increased in NHM-C cells with CT treatment, whereas they decreased in NMH-B cells. Together, these results indicate that melanin accumulation accelerates melanocyte senescence by a p16/CDK4/pRb-regulated pathway.

Dissimilar Regulation of Cyclin E and CDK-Is

Changes in association levels of p16 and cyclin D1 with CDK4 and cyclin E with CDK2 prompted us to analyze total protein levels by immunoblotting. CDK4, cyclin D1, and most dramatically, cyclin E levels decreased in NHM-B cells over a 6-week period, whereas CDK2 remained relatively constant over time (FIG. 3A and 3B). Interestingly, both p21 and p27 were decreased notably in NHM-B as compared to NHM-C cells, whereas p16 decreased in NHM-C cells only (FIG. 3C). Because cyclin D-CDK complex assembly and activation is facilitated by interaction with p21 and p27, reduced levels of both CDK-Is would result in poor complex assembly and activity in NHM-B cells. Furthermore, p16 would not have to compete with p27/p21 proteins for CDK4,[23] resulting in a greater chance to bind CDK4. Considering that high levels of p21 are associated with senescence in fibroblasts,[49] decreased levels of p21 in melanocytes strongly suggest that cells from different tissues use cell-specific pathways to reach senescence.

Downregulation of E2F1 and Thymidylate Synthetase

E2F mRNA levels increase in mid-G1 and reach their highest levels in S phase of the cell cycle.[50,51] E2F1 activity is positively regulated by the protein's association with DP1/2 and negatively regulated by its association with hypophosphorylated pRb.[45] Since E2F1 and E2F2 proteins are downregulated in pigmented melanocytes undergoing senescence,[44] we were interested in determining whether E2F1 levels were differentially affected in lightly versus darkly pigmented melanocytes. FIGURE

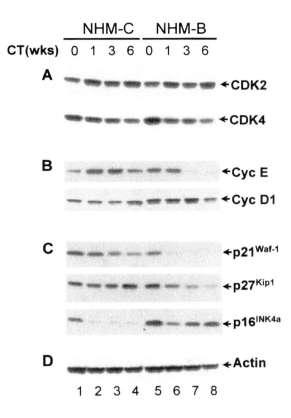

FIGURE 3. (**A-C**) Total extracts from CT-treated NHM-C or NHM-B cells were analyzed for CDK2 (**A**, *upper panel*), CDK4 (**A**, *lower panel*), cyclin E (3, upper panel), cyclin D1 (**B**, lower panel), p21^{Waf-1} (**C**, *top panel*), p27^{Kip1} (**C**, *middle panel*), or p16^{INK4a} (**C**, *bottom panel*). (**D**) Actin blot confirming equal protein loading.

4A shows that in NHM-C melanocytes, E2F1 levels did not substantially change over time with exposure to CT. However, E2F1 levels in pigmented melanocytes were downregulated by 6 weeks of treatment, as demonstrated previously.[44] Because E2F factors are required for transcription of S-phase genes, low levels of E2F may result in poor transcriptional activity and consequent inability of melanocytes to enter S phase. To test this hypothesis, we analyzed thymidylate synthetase (TS) protein levels by immunoblotting in total cell extracts. After 1 week of CT treatment, TS protein levels in NHM-C cells increased approximately twofold and were retained throughout the experiment. As predicted herein, TS was dramatically downregulated over time in NHM-B melanocytes (FIG. 4B).

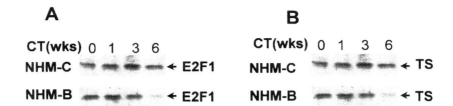

FIGURE 4. (**A**) NHM-C (*upper panel*) or NHM-B melanocytes (*lower panel*) were treated with CT for indicated time periods, and extracts were analyzed for E2F1 protein. (**B**) NHM-C (*upper panel*) or NHM-B (*lower panel*) extracts were analyzed for thymidylate synthetase (TS) protein under indicated conditions.

DISCUSSION

Multiple mechanisms may contribute to a senescent phenotype, and different cell types may achieve this phenotype by activation of similar, but not identical pathways. Although senescence, either induced by replicative senescence, oncogenic stimuli, or DNA damage, has been extensively characterized in fibroblasts,[3–5] little is known about this process in other cell types. In a medium supplemented with the phorbol ester TPA (tetradecanoyl-phorbol-myristate acetate), neonatal melanocytes reach replicative senescence after 48–80 population doublings (reviewed in ref. 43). However, removal of TPA and subsequent addition of CT dramatically shorten the replicative capacity of the melanocytes, although they still have potential to undergo additional rounds of cell division otherwise.[44] Interestingly, melanocytes (derived from both Caucasian and Black skin) have substantially lower melanin levels when cultured with TPA as opposed to CT (FIG. 1). Thus, it is likely that TPA artificially contributes to increasing the melanocyte life span by inhibition of both tyrosinase activity[67,68] and transcription of TRP-2, an important enzyme in the melanin synthesis pathway (Schwahn and Medrano, unpublished results).

We have shown here that the ability to elicit a potent melanogenic response correlates with accelerated senescence. Similar results have been observed in B16 mouse melanoma cells after inducing hyperpigmentation with millimolar concentrations of L-tyrosine, a precursor of the melanin synthesis pathway.[52] Furthermore, the addition of forskolin to these cells results in melanin accumulation and growth arrest by a mechanism involving PI-3 and S6 kinase inhibition.[53]

In our study, prolonged exposure to CT resulted in extensive dephosphorylation of pRb in the heavily pigmented NHM-B cells. Because pRb is the only known physiologic substrate of cyclin D/CDK4, the CDK4 kinase appears to be an early sensor in the melanocyte senescence cascade. CDK4 activity can be inhibited by decreasing cyclin D1 levels, by increasing p16's association with CDK4, and by decreasing CDK4 levels. We found reduced levels of cyclin D1 and CDK4 in NHM-B cells treated with CT. This, together with p16-CDK4 association, resulted in low pRb kinase activity. On the contrary, p16-CDK4 association decreased in NHM-C cells, resulting in active cyclin D/CDK4 complexes (FIG. 2B, lane 3 vs lane 6). These results highlight the importance of the p16/pRb pathway in regulating melanocyte senescence. p16 is involved in G1 checkpoints in response to DNA damage. Furthermore, DNA double strand-damaging agents can induce p16 and premature senescence in fibroblasts.[31] Reactive oxygen species (ROS) are also critical signals for mediating senescence and can be induced by forced expression of oncogenic ras proteins.[56] In melanocytes, ROS are generated by oxidation of tyrosine and dihydroxyphenylalanine (DOPA)[54] during melanin synthesis. In the Bcl-2[-/-] mouse, ROS may generate oxidant injury during melanogenesis, resulting in melanocyte death.[57] It is likely that they promote a DNA-damage response that leads to p16 induction in human melanocytes. Oxidative damage increases with age and may also be a major causal factor for senescence *in vivo*.[55]

Importantly, NHM-C melanocytes downregulate p16 protein levels upon CT treatment. Loss of p16 expression could result from increased protein instability or decreased transcription through increased methylation and/or acetylation, resulting in chromatin condensation at the p16 locus.[58] Equally intriguing is the dramatic and rapid reduction in p27 and p21 levels in NHM-B melanocytes. p21 can be transactivated by E2F-containing complexes.[64] Therefore, loss of p21 in NHM-B melanocytes might be a direct consequence of reduced E2F-1 protein levels and activity. p21 also decreases in differentiating keratinocytes, and its forced expression inhibits expression of terminal differentiation markers.[59] These observations suggest that downregulation of p21 may be required for terminal differentiation and senescence in some cells.

We have shown that in NHM-C cells, sustained E2F1 protein levels correlate with thymidilate synthetase expression and proliferation ability, whereas reduced or absent E2F-1 correlates with accelerated senescence.[44] These results support previous observations of reduced proliferation in melanocytes derived from E2F-1[-/-] animals[65] and suggest that E2F-1 may be a major regulator of melanocyte proliferation. In normal cells, E2F-1 controls cyclin E expression by binding to E2F-responsive elements in the cyclin E promoter.[61] Interestingly, studies in Drosophila embryos have demonstrated that cyclin E expression induces E2F-dependent transcription.[60] Loss of cyclin E in NHM-B cells might be the consequence of reduced E2F-1 levels. Dependence of one gene upon the other can be reversed in different tissues or stages, depending on which gene product becomes rate limiting.[62] Because downregulation of E2F-1 correlates with melanocyte senescence, sustained E2F-1 levels could generate abnormal cell cycles in melanoma tumors. The presence of abnormally high cyclin E levels in primary invasive melanomas[66] supports this prediction.

There appears to be strong selection for cells with increased proliferative potential during tumor development. Immortality, or an extended replicative life span, greatly increases the ability of cells to progress toward malignancy, because it permits the extensive cell division that is needed to acquire multiple, successive mutations.[63] Because senescence-related growth arrest is essentially irreversible, a timely senescence response *in vivo* would protect an organism from melanoma. It is evident from our studies that the ability to trigger a potent melanogenic response in normal melanocytes could impart such a protective mechanism.

REFERENCES

1. JIANG, H., H.S. CHOU & L. ZHU. 1998. Requirement of cyclin E-Cdk2 inhibition in p16(INK4a)-mediated growth suppression. Mol. Cell Biol. **18:** 5284–5290.
2. HAYFLICK, L. 1965. The limited in vitro lifespan of human diploid cell strains. Exp. Cell Res. **25:** 585–621.
3. CRISTOFALO, V.J. & R.J. PIGNOLO. 1993. Replicative senescence of human fibroblast-like cells in culture. Physiol.Rev. **73:** 617–638.
4. GOLDSTEIN, S. 1990. Replicative senescence: the human fibroblast comes of age. Science **249:** 1129–1133.
5. SMITH, J.R. & D.W. LINCOLN. 1984. Aging of cells in culture. Int. Rev. Cytol. **89:** 151–177.
6. DIMRI, G.P., X. LEE, G. BASILE et al. 1995. A biomarker that identifies senescent human cells in culture and in aging skin in vivo. Proc. Natl. Acad. Sci. USA **92:** 9363–9367.
7. HARLEY, C.B., A.B. FUTCHER & C.W. GREIDER. 1990. Telomeres shorten during ageing of human fibroblasts. Nature **345:** 458–460.
8. MORGAN, D.O. 1995. Principles of CDK regulation. Nature **374:** 131–134.
9. GRANA, X., J. GARRIGA & X. MAYOL. 1998. Role of the retinoblastoma protein family, pRB, p107 and p130 in the negative control of cell growth. Oncogene **17:** 3365–3383.
10. LAM, E.W. & T.N. LA. 1994. DP and E2F proteins: coordinating transcription with cell cycle progression. Curr. Opin. Cell Biol. **6:** 859–866.
11. WANG, J., K. HELIN, P. JIN & B. NADAL-GINARD. 1995. Inhibition of in vitro myogenic differentiation by cellular transcription factor E2F1. Cell Growth Differ. **6:** 1299–1306.
12. WEINTRAUB, S.J., C.A. PRATER & D.C. DEAN. 1992. Retinoblastoma protein switches the E2F site from positive to negative element. Nature **358:** 259–261.
13. BANDARA, L.R., V.M. BUCK, M. ZAMANIAN et al. 1993. Functional synergy between DP-1 and E2F-1 in the cell cycle-regulating transcription factor DRTF1/E2F. EMBO J. **12:** 4317–4324.
14. HELIN, K., C.L. WU, A.R. FATTAEY et al. 1993. Heterodimerization of the transcription factors E2F-1 and DP-1 leads to cooperative trans-activation. Genes Dev. **7:** 1850–1861.
15. ZAMANIAN, M. & T.N. LA. 1993. Transcriptional repression by the Rb-related protein p107. Mol. Biol. Cell **4:** 389–396.
16. TRIMARCHI, J.M., B. FAIRCHILD, R. VERONA et al. 1998. E2F-6, a member of the E2F family that can behave as a transcriptional repressor. Proc. Natl. Acad. Sci. USA **95:** 2850–2855.
17. JOHNSON, L.F. 1992. G1 events and the regulation of genes for S-phase enzymes. Curr. Opin. Cell Biol. **4:** 149–154.
18. DOU, Q.P., P.J. MARKELL & A.B. PARDEE. 1992. Thymidine kinase transcription is regulated at G1/S phase by a complex that contains retinoblastoma-like protein and a cdc2 kinase. Proc. Natl. Acad. Sci. USA **89:** 3256–3260.
19. CHITTENDEN, T., D.M. LIVINGSTON & J.A. DECAPRIO. 1993. Cell cycle analysis of E2F in primary human T cells reveals novel E2F complexes and biochemically distinct forms of free E2F. Mol. Cell Biol. **13:** 3975–3983.
20. NADAL-GINARD, B. 1978. Commitment, fusion and biochemical differentiation of a myogenic cell line in the absence of DNA synthesis. Cell **15:** 855–864.
21. MIHARA, K., X.R. CAO, A. YEN et al. 1989. Cell cycle-dependent regulation of phosphorylation of the human retinoblastoma gene product. Science **246:** 1300–1303.
22. MITTNACHT, S. 1998. Control of pRB phosphorylation. Curr. Opin. Genet. Dev. **8:** 21–27.
23. SHERR, C.J. & J.M. ROBERTS. 1999. CDK inhibitors: positive and negative regulators of G1-phase progression. Genes Dev. **13:** 1501–1512.
24. SERRANO, M., G.J. HANNON & D. BEACH. 1993. A new regulatory motif in cell-cycle control causing specific inhibition of cyclin D/CDK4 [see comments]. Nature **366:** 704–707.
25. KAMB, A., N.A. GRUIS, J. WEAVER-FELDHAUS et al. 1994. A cell cycle regulator potentially involved in genesis of many tumor types [see comments]. Science **264:** 436–440.
26. LIN, B.T., S. GRUENWALD, A.O. MORLA et al. 1991. Retinoblastoma cancer suppressor gene product is a substrate of the cell cycle regulator cdc2 kinase. EMBO J. **10:** 857–864.

27. ZACKSENHAUS, E., Z. JIANG, , R.A. PHILLIPS & B.L. GALLIE. 1996. Dual mechanisms of repression of E2F1 activity by the retinoblastoma gene product. EMBO J. **15:** 5917–5927.

28. ALCORTA, D.A., Y. XIONG, D. PHELPS *et al.* 1996. Involvement of the cyclin-dependent kinase inhibitor p16 (INK4a) in replicative senescence of normal human fibroblasts. Proc. Natl. Acad. Sci. USA **93:** 13742–13747.

29. SERRANO, M., H. LEE, L. CHIN *et al.* 1996. Role of the INK4a locus in tumor suppression and cell mortality. Cell **85:** 27–37.

30. HADDAD, M.M., W. XU & E.E. MEDRANO. 1998. Aging in epidermal melanocytes: cell cycle genes and melanins. J. Invest. Dermatol. Symp. Proc. **3:** 36–40.

31. ROBLES, S.J. & G.R. ADAMI. 1998. Agents that cause DNA double strand breaks lead to p16INK4a enrichment and the premature senescence of normal fibroblasts. Oncogene **16:** 1113–11233.

32. SERRANO, M., A.W. LIN, M.E. MCCURRACH *et al.* 1997. Oncogenic ras provokes premature cell senescence associated with accumulation of p53 and p16INK4a. Cell **88:** 593–602.

33. FRIEDMANN, P.S. & B.A. GILCHREST. 1987. Ultraviolet radiation directly induces pigment production by cultured human melanocytes. J. Cell Physiol. **133:** 88–94.

34. ABERDAM, E., C. ROMERO & J.P. ORTONNE.. 1993. Repeated UVB irradiations do not have the same potential to promote stimulation of melanogenesis in cultured normal human melanocytes. J.Cell Sci. **106:** 1015–1022.

35. MEDRANO, E.E., F. YANG, R. BOISSY *et al.* 1994. Terminal differentiation and senescence in the human melanocyte: repression of tyrosine-phosphorylation of the extracellular signal- regulated kinase 2 selectively defines the two phenotypes. Mol. Biol. Cell **5:** 497–509.

36. ABDEL-MALEK, Z.A., V.B. SWOPE, J.J. NORDLUND & E.E. MEDRANO. 1994. Proliferation and propagation of human melanocytes in vitro are affected by donor age and anatomical site. Pigment Cell Res. **7:** 116–122.

37. MEDRANO, E.E. & J.J. NORDLUND. 1990. Successful culture of adult human melanocytes obtained from normal and vitiligo donors. J. Invest. Dermatol. **95:** 441–445.

38. ABDEL-MALEK, Z., V. SWOPE, D. SMALARA *et al.* 1994. Analysis of the UV-induced melanogenesis and growth arrest of human melanocytes. Pigment Cell Res. **7:** 326–332.

39. HUNT, G., C. TODD, J.E. CRESSWELL & A.J. THODY. 1994. Alpha-melanocyte stimulating hormone and its analogue Nle4DPhe7 alpha- MSH affect morphology, tyrosinase activity and melanogenesis in cultured human melanocytes. J. Cell Sci. **107:** 205–211.

40. O'KEEFE, E. & P. CUATRECASAS. 1974. Cholera toxin mimics melanocyte stimulating hormone in inducing differentiation in melanoma cells. Proc. Natl. Acad. Sci. USA **71:** 2500–2504.

41. HALABAN, R., S.H. POMERANTZ, S. MARSHALL *et al.* 1983. Regulation of tyrosinase in human melanocytes grown in culture. J. Cell Biol. **97:** 480–488.

42. ENGLARO, W., R. REZZONICO, M. DURAND-CLEMENT *et al.* 1995. Mitogen-activated protein kinase pathway and AP-1 are activated during cAMP-induced melanogenesis in B-16 melanoma cells. J. Biol. Chem. **270:** 24315–24320.

43. MEDRANO, E.E. 1998. Aging, replicative senescence and the differentiated function of the melanocyte. *In* The Pigmentary System: Physiology and Pathophysiology. J. Nordlund, R. Boissy, V.J. Hearing, King R.A. & J.P. Ortonne, Eds.: 151–158. Oxford University Press. New York.

44. HADDAD, M.M., W. XU, D.J. SCHWAHN *et al.* 1999. Activation of a cAMP pathway and induction of melanogenesis correlate with association of p16INK4 and p27kip-1 to CDKs, loss of E2F binding activity and premature senescence of human melanocytes. Submitted.

45. SIDLE, A., C. PALATY, P. DIRKS *et al.* 1996. Activity of the retinoblastoma family proteins, pRB, p107, and p130, during cellular proliferation and differentiation. Crit. Rev. Biochem.Mol.Biol. **31:** 237–271.

46. MATSUSHIME, H., M.E. EWEN, D.K. STROM *et al.* 1992. Identification and properties of an atypical catalytic subunit (p34PSK- J3/cdk4) for mammalian D type G1 cyclins. Cell **71:** 323–334.

47. HUSCHTSCHA, L.I. & R.R. REDDEL. 1999. p16(INK4a) and the control of cellular proliferative life span. Carcinogenesis **20:** 921–926.

48. SHAPIRO, G.I., C.D. EDWARDS, M.E. EWEN & B.J. ROLLINS. 1998. p16INK4A participates in a G1 arrest checkpoint in response to DNA damage. Mol. Cell Biol. **18:** 378–387.
49. NODA, A., Y. NING, S.F. VENABLE *et al.* 1994. Cloning of senescent cell-derived inhibitors of DNA synthesis using an expression screen. Exp. Cell Res. **211:** 90–98.
50. SLANSKY, J.E., Y. LI, W.G. KAELIN & P.J FARNHAM. 1993. A protein synthesis-dependent increase in E2F1 mRNA correlates with growth regulation of the dihydrofolate reductase promoter [published erratum appears in Mol. Cell Biol. 1993. **13:** 7201]. Mol. Cell Biol. **13:** 1610–1618.
51. KAELIN, W.G.J., W. KREK, W.R. SELLERS *et al.* 1992. Expression cloning of a cDNA encoding a retinoblastoma-binding protein with E2F-like properties. Cell **70:** 351–364.
52. BENNETT, D.C. 1983. Differentiation in mouse melanoma cells: initial reversibility and an on-off stochastic model. Cell **34:** 445–453.
53. BUSCA, R., C. BERTOLOTTO, J.P. ORTONNE & R. BALLOTTI. 1996. Inhibition of the phosphatidylinositol 3-kinase/p70(S6)-kinase pathway induces B16 melanoma cell differentiation. J.Biol. Chem. **271:** 31824–31830.
54. JOSHI, P.C., C. CARRARO & M.A. PATHAK. 1987. Involvement of reactive oxygen species in the oxidation of tyrosine and dopa to melanin and in skin tanning. Biochem. Biophys. Res. Commun. **142:** 265–274.
55. JOHNSON, F.B., D.A. SINCLAIR & L. GUARENTE. 1999. Molecular biology of aging. Cell **96:** 291–302.
56. LEE, A.C., B.E. FENSTER, H. ITO *et al.* 1999. Ras proteins induce senescence by altering the intracellular levels of reactive oxygen species. J. Biol. Chem. **274:** 7936–7940.
57. VEIS, D.J., C.M. SORENSON, J.R. SHUTTER & S.J. KORSMEYER. 1993. Bcl-2-deficient mice demonstrate fulminant lymphoid apoptosis, polycystic kidneys, and hypopigmented hair. Cell **75:** 229–240.
58. COSTELLO, J.F., M.S. BERGER, H.S. HUANG & W.K. CAVENEE. 1996. Silencing of p16/CDKN2 expression in human gliomas by methylation and chromatin condensation. Cancer Res. **56:** 2405–2410.
59. DI, C.F., G. TOPLEY, E. CALAUTTI *et al.* 1998. Inhibitory function of p21Cip1/WAF1 in differentiation of primary mouse keratinocytes independent of cell cycle control [see comments]. Science **280:** 1069–1072.
60. DURONIO, R.J. & P.H. O'FARRELL. 1995. Developmental control of the G1 to S transition in Drosophila: cyclin Eis a limiting downstream target of E2F. Genes Dev. **9:** 1456–1468.
61. OHTANI, K., J. DEGREGORI & J.R. NEVINS. 1995. Regulation of the cyclin E gene by transcription factor E2F1. Proc. Natl. Acad. Sci. USA **92:** 12146–12150.
62. ALEVIZOPOULOS, K., J. VLACH, S. HENNECKE & B. AMATI. 1997. Cyclin E and c-Myc promote cell proliferation in the presence of p16INK4a and of hypophosphorylated retinoblastoma family proteins. EMBO J. **16:** 5322–5333.
63. CAMPISI, J. 1999. Cellular senescence, aging and cancer. *In* Molecular Biology of Aging. V.A. Bohr, B.F.C. Clarke & T. Stevnsner, Eds.: 112–113. Munksgaard. Copenhagen.
64. HIYAMA, H., A. IAVARONE & S.A. REEVES. 1998. Regulation of the cdk inhibitor p21 during cell cycle progression is under the control of the transcription factor E2F. Oncogene **16:**1513–1523.
65. HALABAN, R., E. CHENG, Y. ZHANG *et al.* 1998. Release of cell cycle constraints in mouse melanocytes by overexpressed mutant E2F1E132, but not by deletion of p16INK4A or p21WAF1/CIP-1. Oncogene **16:** 2489–2501.
66. BALES, E.S., C. DIETRICH, D. BANDYOPADHYAY *et al.* 1999. High levels of expression of p27^{KIP1} and cyclin E in invasive primary malignant melanomas. J. Invest. Dermatol. In press.
67. MEDRANO, E.E, J. FAROOQUI, R.E. BOISSY *et al.* 1993. Chronic growth-stimulation of human adult melanocytes by inflammatory mediators *in vitro*: implications for nevi formation and initial steps in melanocyte oncogenesis. Proc. Natl. Acad. Sci. USA **90:** 1790–1794.
68. BERTOLOTTO, C., K. BILLE., J.P. ORTONNE. & R. BALLOTTI. 1998. In B16 melanoma cells, the inhibition of melanogenesis by TPA results from PKC activation and diminution of microphthalmia binding to the M-box of the tyrosinase promoter. Oncogene **16:** 1665–1670.

Stress-Induced Premature Senescence

Essence of Life, Evolution, Stress, and Aging

OLIVIER TOUSSAINT,[b] PATRICK DUMONT, JEAN-FRANÇOIS DIERICK,
THIERRY PASCAL, CHRISTOPHE FRIPPIAT, FLORENCE CHAINIAUX,
FRANCIS SLUSE,[a] FRANÇOIS ELIAERS, AND JOSÉ REMACLE

*Department of Biology, Unit of Cellular Biochemistry and Biology,
University of Namur (FUNDP), B- 5000 Namur, Belgium*

[a]Laboratory of Bioenergetics, University of Liège, B-4000 Liège (Sart-Tilman), Belgium

INTRODUCTION. WHAT IS STRESS?

The stress syndrome was discovered accidentally by Hans Selye while searching for new hormones in the placenta.[1] After injecting rats with crude preparations, Selye found adrenal enlargements and involution of thymus and lymph nodes, which he thought were specific for a particular hormone. It occurred to Selye that these symptoms might represent a nonspecific response to noxious agents. Indeed, this was found to be the case when he injected rats with diverse agents. Selye defined the stress response as the "general adaptation syndrome."[2,3] According to this theory, the initial reaction to stress is shock, it is followed by a countershock phase, and gradually resistance develops to the stressor. This resistance may turn into exhaustion, however, if the stressor persists, and death may ensue. Both specific and nonspecific resistance develops during stress.[4] In his last scientific book, Selye defined biologic stress as "the non-specific response of the body to any demand made upon it."[5] Beside the transfer of the word "stress" from physics to biology, Selye also coined the words corticosteroids, glucocorticoids, and mineralocorticoids.[6] Nowadays, the concept of stress has invaded most fields of the biologic, medical, and social sciences. Cellular and molecular biology has become interested in the study of the stress response of human, animal, and plant cells, the consensus being that "any environmental factor potentially unfavourable to living organism" is stress.[7] It is also generally agreed that "if the limits of tolerance are exceeded and the adaptive capacity is over-worked, the result may be permanent damage or even death."[8] Three phases of the stress response have been defined based on experimental observations: (1) the response phase of alarm reaction with deviation of functional norm, decline of vitality, and excess of catabolic processes over anabolism, (2) the restitution phase or stage of resistance with adaptation processes and repair processes, and (3) either the end phase, that stage of exhaustion or long-term response when stress intensity is too high, leading to overcharge of the adaptation capacity, damage, chronic dis-

[b]Corresponding author: Olivier Toussaint, Department of Biology, Unit of Cellular Biochemistry & Biology, University of Namur (FUNDP), Rue de Bruxelles, 61, B-5000 Namur, Belgium.
Phone: +32 81 72 41 32; fax: +32 81 72 41 35.
    olivier.toussaint@fundp.ac.be

ease, senescence, or even death, or the regeneration phase, full restoration of physiologic function, when the stressor is removed.[9]

STRESS AND LIFE SPAN

It has been proposed that a subset of genes, termed "public," may be of particular importance in how most individuals of a species age, whereas "private" mechanisms of aging would control the rate of aging of particular individuals.[10] It has also been proposed that genes involved in the stress response, such as genes coding for DNA repair enzymes and antioxidant enzymes, could be among these genes regulating aging.

A theory analogically derived from the darwinian theory of evolution proposed that organisms (or cells from organisms) that have the longest life span should secure their longevity through investment in a more durable soma and including enhanced cellular resistance to stress, arguing that long life span means evolution-driven better adaptation to stress. To investigate whether cells from long-lived species have better mechanisms to cope with oxidative and nonoxidative stress, the survival of primary skin fibroblasts and lymphocytes from eight mammalian species with a range of life spans was estimated following stresses induced by oxidative and alkaline stress. Cellular resistance to the stresses tested was positively correlated with mammalian longevity.[11]

The evolutionary theory of aging suggests that senescence occurs because the force of natural selection declines with age and because longevity is only acquired at some metabolic cost. Organisms may trade late survival for enhanced reproductive investments in earlier life.[12,13] It has been proposed in the "network theory of aging" that "an important corollary of this hypothesis is that multiple mechanisms of aging operate in parallel,"[14] thereby connecting this theory to other theories of aging. For instance, the theory of catastrophic error proposed by Orgel[15] argued that aging is caused by the accumulation of many kinds of defects. The "global concept of failure of maintenance," proposed by Holliday,[16] and the concept of "critical threshold"[17,18] were the basis for the network theory of aging. Indeed, these theories proposed that aging could be a consequence of the accumulation of defects, thereby including all "wear and tear" theories such as the theory of free radicals and the mitochondrial theory of aging. In those theories, it was also postulated that this accumulation of defects may be due not only to an increase in damage generation but also to a decrease in the efficiency of elimination and repair systems such as the antioxidant system, the system responsible for the degradation of altered proteins, DNA repair systems, etc. Theoretical developments based on a model of free radical production and elimination also confirmed the possibility of a threshold beyond which the cellular system is unstable.[18]

The fundamental question is why do cells, during aging, undergo internal modifications either spontaneously due to molecular decay or due to interactions with "any environmental factor potentially unfavourable to living organisms," that is to say, stress (oxidative, mechanical, heat, etc), reaching finally a threshold of error leading to irreversible degeneration and death. The answer to this question is probably linked to why and how cells exist and evolve. Therefore, to propose an answer to this question, it was necessary to use a theoretical background, considering the very essence of living matter, in order to avoid analogical extrapolations about the theory of aging. This question was approached by using the thermodynamics of open systems, which allows a global systemic description of the cell's behavior and, in this way, transcends the genetic or stochastic considerations of the aging process as well as evolutionary questions.

THE ESSENCE OF LIFE AND AGING

The first law of thermodynamics states that energy cannot be created or destroyed and that the total energy within a closed system remains unchanged. The second law of thermodynamics requires that any process underway in a system irreversibly degrades the quality of the energy in that system in a direction that results in an entropy increase.

Let us focus our attention towards the class of thermodynamic systems that are open, moved away from equilibrium by the fluxes of material and energy across their boundary, and maintain their form or structure by continuous exchanges. This is done at a cost of increasing the entropy of the larger "global" system which contains such systems, because, following the second law, that overall entropy in the global sense must increase. Dissipative systems such as living systems, from cells to ecosystems, and nonliving organized systems, such as convection cells, tornadoes and lasers, are dependent on outside energy fluxes to maintain their organization in a locally reduced entropy state. The entropy relationships in dissipative systems were put forward by Denbeigh[19] and Prigogine[20] in the following relationship: $dS = dS_i + dS_e$, where dS is the total entropy change in a system and dS_i is the internal entropy in the system. According to the second law, dS_i is always greater than or equal to zero. dS_e is the entropy exchange with the environment, which may be positive, negative, or zero. For the system to maintain itself in a nonequilibrium steady state, dS_e must be negative and larger than the entropy produced by internal processes, dS_i, that is, metabolism in the case of living cells.

Implicit in the observation that a system will tend towards its unique equilibrium state is that a system will also resist being removed from the equilibrium state. The degree to which a system has been moved from equilibrium can be measured by the gradients imposed on the system. As systems are moved away from equilibrium, they will utilize all avenues available to counter the applied gradients. Attractors can emerge so that the system organizes in a way that reduces or degrades the gradient. As the applied gradients increase, so does the system's ability to oppose further movement from equilibrium. If dynamic and/or kinetic conditions permit, self-organization processes are to be expected. The building of organizational structure and associated processes degrades the imposed gradient more effectively than if the dynamic and kinetic pathways for those structures were unavailable.[21,22]

This corollary of the second law holds for all living systems from cells to ecosystems and describes their expected behavior. Prigogine and others have shown that dissipative structures self-organize through fluctuations and instabilities that lead to irreversible bifurcations and new stable system states. Dissipative structures are stable over a finite range of conditions and are sensitive to fluxes and flows from outside the system. Glansdorff and Prigogine[23] have shown that these thermodynamic relationships can be represented by coupled nonlinear relationships, that is, autocatalytic positive feedback cycles, many of which lead to stable macroscopic structures that exist away from the equilibrium state. Convection cells, hurricanes, autocatalytic chemical reactions, and living systems are all examples of nonequilibrium dissipative structures that exhibit coherent behavior. Living cells are dissipative structures exhibiting processes of informed, self-replicating, autocatalytic cycles. This paradigm provides for a thermodynamically consistent explanation of why there is life, including the origin of life, biological growth, development of ecosystems, patterns of biological evolution observed in the fossil record, stress response, and aging.[21,22,24,25]

Maintenance of life, ontogenic (growth, development, and differentiation) and phylogenic evolution of living processes, from cells to ecosystems, is the response to the thermodynamic imperative of dissipating gradients. Biological growth occurs when the system adds more of the same types of pathways for degrading imposed gradients. Biological development occurs when new types of pathways emerge in the system for degrading imposed gradients. In biological terms, growth means an increase in the amount of substrates entering preexisting metabolic pathways, whereas a new step of development means the appearance of supplementary pathways. This can explain, for instance, the phylogenic appearance of more and more complex multicellular organisms with more and more complex metabolic pathways using and degrading the energy present in their environment, thus optimizing the production and use of free energy in function of the available substrate.[21,22,24,25]

At the present stage of evolution, mechanisms exist, such as predation, parasitism, infectious diseases, aging, and age-linked risks of getting sick, being caught by a predator, or starving, that lead to a decrease in the number of individuals of a species. We have seen that a possible criterion of growth, differentiation, development, and evolution, is the capacity of cells and organisms to degrade energy gradients from the environment, which is the capacity to produce more entropy per unit of *time* and *moles* of molecules *entering* the cellular system (i.e., input fluxes) and per unit of *volume* or *weight* of the system considered. A direct corollary is that this capacity decreases with aging, which is experimentally fulfilled. From this corollary, it can be predicted that evolution, which arrow of time depends on thermodynamics as we have seen, has selected mechanisms that are triggered when the capacity of the organisms to degrade energy starts to decrease and leads to the death of the organisms.[25,26]

STRESS AND GENOMES

A modification of the balance between the capacities of repairing the damage caused by stresses and the accumulation of damage due to stress might be a factor responsible for this decrease in the capacity of gradient degradation.

In the thermodynamic model of aging, stresses are sorted into three different types depending on the ratio between the damage they generate and the capacity of stress response of the biological system considered. The thermodynamic development has been given in reference 26. First, let us consider the constant, unavoidable, ubiquitous mild stresses such as steady state concentrations of reactive oxygen species in normal conditions. Thermodynamically these mild stresses do not immediately alter the stability of the system. However, the possibility remains for internal modifications to accumulate without immediately altering the cellular stability as long as they can be counterbalanced by other cellular mechanisms. This is the only possibility that allows the prediction that cells shift to a new steady state when the level of damage has reached a critical threshold where compensation mechanisms become momentarily overwhelmed. This new steady state is characterized by a higher level of damage and lower global metabolic activity.[26] This is normal aging.

The second type of stress is chronic stress or (repeated) stress of sublethal intensity resulting from abnormal situations such as inflammation, anoxia-reoxygenation, and exposure to radioactivity, UV irradiation, and toxic chemicals. During such stress, the level of damage may rise so much that the capacity for a specific stress response is momentarily overwhelmed and damage spreads to various cell components, requiring other defense systems. If these systems can eliminate/repair all the damage, then the cell remains stable. When defense systems cannot eliminate/repair quickly all the damage, this may lead to a transient increase in the level of damage and a transient decrease in the capacity to produce (ATP, redox potential, etc.) and use (biochemical reactions) free energy. Such conditions represent the thermodynamic conditions for destabilization of the steady state of far from equilibrium systems and straightforwardly of living cells. Thermodynamically speaking, if the cell can reorganize at new steady state, there will be two major irreversible differences between this new steady state and the previous one, that is, a higher level of internal damage and lower global metabolic activity. These conditions lead to stress-induced premature senescence (SIPS). In this perspective, normal and accelerated aging would be ways for decreasing the metabolic activity of those cells that have a given level of damage, possibly to avoid cell transformation and premature decrease in the capacity of the organism to degrade energy gradients. It is also possible that depending on the cell type, the increase in the level of damage will trigger programmed mechanisms of self-destruction rather than accelerated senescence, that is, apoptosis.[27]

In the third type of stress, the injury is so harmful that cells cannot find a new steady state: the level of intracellular damage rises and global metabolic activity decreases so much that the cells cannot run their housekeeping pathways, and they die by necrosis.[26]

HUMAN CELLS AND NONCYTOTOXIC STRESS

In vivo, human cells are constantly exposed to various types and degrees of noncytotoxic stress of various nature and intensity. These types of stress include, for instance, UV irradiation of skin keratinocytes, fibroblasts, and melanocytes, effects of oxidized low density lipoproteins (LDL) on endothelial cells and smooth muscle cells, effects of urban pollution or tobacco smoke on lung pneumocytes and endothelial cells, effects of food oxidants on enterocytes, effects of repeated exposures of hepatocytes to ethanol, reactive oxygen species in skeletal myocytes during intense exercise or in inflammation processes as well as pathobiological hypoxia/reoxygenation processes.

Two hypotheses have been put forward for the role of senescent cells in tissues: either senescent cells could accumulate *in vivo* in tissues, as shown for fibroblasts,[28,29] using two different biomarkers, or the senescent cells are removed by immune cells or undergo apoptosis, thereby participating in the aging of tissues through the installation of a microinflammatory state or/and activation of neighboring cells.[30]

The accumulation of such senescent cells might be very slow *in vivo* under normal physiologic conditions. Nevertheless, exposure of cells to repeated or chronic nonlethal stress may participate in an increase in the accumulation of stress-induced senescent cells, thereby accelerating tissue aging. It is also suggested that local accumulation of such stress-induced senescent cells may be responsible for the alteration in tissue function, the differentiation status of neighboring cells, and may be a focus responsible for the start of pathological processes linked with aging.[31]

IMMEDIATE AND LONG-TERM STRESS RESPONSES

The immediate response to stimulations by growth factors, cytokines, and the immediate response to stresses, such as oxidative stress, ionizing radiation, osmotic stress, mechanical stress, hypoxia, heavy metals, and heat shock, have received much attention from many investigators. The long-term cellular response to stresses, from 48 hours to several weeks after stress or after stimulation by cytokines, has received much less attention. The development of proper experimental noncytotoxic models allowed the start of research programs on the long-term cellular response to noncytotoxic stresses or stimulations.

LONG-TERM STRESS RESPONSES AND
STRESS-INDUCED PREMATURE SENESCENCE

Exposure of human skin fibroblasts to repeated noncytotoxic oxidative or UV stress triggers the nonreversible appearance of the biomarkers of cellular senescence, beginning 48–72 hours after the end of the stress.

First, morphologic studies showed that human diploid fibroblasts exposed to repeated sublethal stress under tert-butylhydroperoxide (t-BHP)[32] or single H_2O_2 stress[33] display the phenotype of SIPS. The first studies were designed to follow cellular morphology after stress. This work was based on the description of the successive fibroblast morphotypes observed during *in vitro* aging (three successive mitotic fibroblasts followed by three successive postmitotic fibroblasts and a degenerative state of short existence). Various interpretations have been given about these seven successive morphotypes. Whatever this interpretation, many morphologists have been able to determine reproducibly the proportions of the various morphotypes in culture of fibroblasts at very low density, allowing spreading of cells.[32,34–37] It is possible to discriminate human diploid fibroblasts into classes according to their morphology: a software able to sort fibroblast morphologies from CCD camera pictures could be developed (unpublished data). Using this type of classification, it was possible to show that after sublethal stress under UV light, t-BHP, ethanol, mitomycin C,

A. Percentage of morphotypes (1) and S-A β-gal positive cells in each morphotype (2)

Morpho-types	Fibroblasts at early CPDs		Fibroblasts at late CPDs		Fibroblasts at early CPDs + 5 stresses under 30 μM t-BHP	
	(1)	(2)	(1)	(2)	(1)	(2)
I	2.0 ± 0.2	**0.0 ± 0.0**	2.3 ± 0.7	**0.0 ± 0.0**	0.6 ± 1.0	**0.0 ± 0.0**
II	86.2 ± 1.3	**2.8 ± 0.3**	41.3 ± 4.1	**20.6 ± 7.8**	19.9 ± 3.2	**35.3 ± 13.8**
III	10.1 ± 1.2	**41.3 ± 8.2**	46.5 ± 5.6	**61.1 ± 15.8**	54.5 ± 6.1	**54.3 ± 10.1**
IV	1.3 ± 0.5	**88.7 ± 19.6**	7.9 ± 2.1	**75.3 ± 6.3**	21.2 ± 3.7	**79.0 ± 12.1**
V	0.5 ± 0.2	**83.3 ± 18.9**	1.8 ± 1.7	**94.4 ± 9.6**	2.0 ± 1.5	**93.3 ± 11.6**

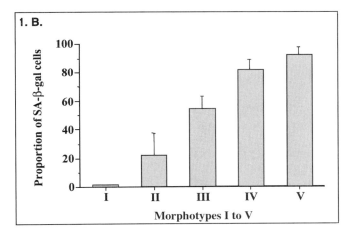

FIGURE 1. (**A**) Relationship between the fibroblast morphotypes and the senescence-associated (S-A) β-gal activity in WI-38 fibroblasts at early CPDs (50% of proliferative life span), at late CPDs (>95 % of proliferative life span), or at early CPDs after exposure to 5 stresses under 30 μM t-BHP performed as described earlier (1 hr/day for 5 days followed by a 48-hr recovery).[39] *Morphotype proportions:* At 24 hours after seeding cells at a density of 700 cells per cm^2, the proportions of the different morphotypes were recorded by two different examiners as previously described[32] on 400 cells/dish. Results are mean values ± SD of 4 different countings. Results are expressed as percentages of the total number of cells counted. *Senescence-associated β-galactosidase activity (S-A β-gal):* At 24 hours after seeding cells at a density of 700 cells per cm^2, S-A β-gal was detected as described in reference 29. The proportions of cells positive for S-A β-gal are given as percentages of the number of cells of each morphotype and as mean values ± SD in 4 different experiments (400 cells/experiment). S-A β-gal detection was never performed on confluent cells to avoid effects of confluency. (**B**) Average proportions of S-A β-gal positive cells for morphotype I to V in WI-38 fibroblasts. The average proportion of S-A β-gal positive cells is presented as the mean value obtained for each morphotype in fibroblasts at early CPDs, at late CPDs, and at early CPDs after 5 stresses under 30 μM t-BHP. Morphotypes VI and VII were not presented because they represented less than 0.025% of the total population in these experimental conditions.

and the like, the treated fibroblasts acquired the morphologic features of fibroblasts at late cumulative population doublings (CPDs).[32,38] (For a review see ref. 31.) More recently, it was shown that the proportion of fibroblasts positive for senescence-associated β-galactosidase activity was increased after successive sublethal stresses under t-BHP[39] after at least 2 days of recovery after the stresses.

CORRELATION BETWEEN FIBROBLAST MORPHOTYPES AND SENESCENCE-ASSOCIATED β-GAL ACTIVITY IN SIPS

The proportions of fibroblasts positive for senescence-associated β-galactosidase activity (S-A β-gal) were determined for each morphotype in cultures at early or late CPDs and after five successive exposures to 30 μM t-BHP at early CPDs (FIG. 1) performed as described earlier (1 stress/day for 5 days + 2 days of recovery).[32,39] In all three situations, there were very few S-A β-gal positive cells in morphotypes I and II. Most morphotypes IV and V were positive for the S-A β-gal. Morphotypes III could be seen as intermediate morphotypes with a mean value of 52% positive cells among the three situations.

In all three situations, each morphotype displayed similar proportions of cells positive for S-A β-gal, whatever the relative proportion of the morphotype in the cultures. The only significant variation concerned morphotype II at late CPDs where a few more cells were positive for S-A β-gal than in young cultures.

OTHER BIOMARKERS OF REPLICATIVE SENESCENCE IN SIPS

After sublethal stresses under 30 μM t-BHP, the cells developed other biomarkers of replicative senescence such as loss of replicative potential, elevation of cyclin-dependent kinase inhibitors p21[Waf-1/SDI-1/Cip1], and failure to hyperphosphorylate the retinoblastoma protein.[39,40] The common 4,977-bp mitochondrial DNA deletion was detected in WI-38 HDFs at late CPDs and at early CPDs after 30 μM t-BHP stresses.[39]

The amounts of mRNAs transcribed from at least 15 genes specifically over- or underexpressed in normal senescence changed similarly in SIPS.[39,41] Human fibroblasts stressed under 40% hyperoxia displayed accelerated senescence as judged by accelerated shortening of their telomeric sequences, growth arrest, lipofuscin accumulation, and mitochondrial respiration. Telomere shortening can trigger a senescence-like growth arrest via liberation of G-rich single-stranded DNA and activation of a DNA damage-response pathway dependent on p53 and its downstream effector p21[waf-1 42]. Retrovirally mediated stable transfection of mutants showed that the retinoblastoma protein is involved in the control of the appearance of the senescent phenotype and is responsible for the relocalization of focal adhesion proteins 72 hours after noncytotoxic stress under H_2O_2. De novo protein synthesis is necessary for the appearance of the senescent phenotype.[43]

Human normal melanocytes exposed to repeated sublethal doses of UV also express biomarkers of cellular senescence, namely, modifications in the regulation of the cell cycle as, for example, the prolonged stable overexpression of p21[waf-1 44] as seen in fibroblasts undergoing SIPS.

Taken together, these data strongly suggest that exposure of human proliferative cell types to repeated noncytotoxic stresses leads to SIPS provided the stress is intense enough in terms of concentration, duration, dose, or chronicity.

SENESCENCE-ASSOCIATED β–GAL ACTIVITY AND MORPHOTYPE TRANSITION AFTER STIMULATION WITH IL-1α AND TNF-α

Stimulations of fibroblasts with TNF-α or IL-1α transiently increase the production of reactive oxygen species in human fibroblasts. It was shown that repeated stimulations of WI-38 fibroblasts with TNF-α or IL-1α accelerate the transition from early to late fibroblast morphotypes. Such stimulations also increase the proportion of fibroblasts positive for the S-A β-gal. The involvement of reactive oxygen species was suggested by experiments in which the stimulations of fibroblasts with TNF-α or IL-1α were performed in the presence of *N*-acetylcysteine, which increases the intracellular antioxidant potential.[37,45] *N*-acetylcysteine protected against stimulation-induced changes in the proportions of morphotypes and in the proportions of cells positive for S-A β-gal.

ROLE OF ENERGETICS IN THE RESISTANCE TO STRESS-INDUCED PREMATURE SENESCENCE

The thermodynamic theoretical model of aging described herein predicts that sublethal stress may induce SIPS. It also predicts that the level of global metabolic activity may protect the cells against SIPS and against cell death caused by lethal stress. Review and data in favor of this prediction as far as cell death is concerned have been presented.[31,46] Preliminary results suggesting that substrates of the energy metabolism protect against SIPS were already published in references 34 and 35. In this paper, we provide more experimental arguments in favor on this hypothesis: two biomarkers of replicative senescence, that is, S-A β-gal and morphotype proportions, were analyzed after five successive sublethal stresses under 25 μM t-BHP or 4% v/v ethanol diluted in PBS buffer (1 hour of stress/day for 5 days + 48 hours of recovery) in the presence of D-glucose or vitamin E. D-glucose (5 mM) was added in PBS buffer for each stress (1 hour), whereas vitamin E (1 mM) was present throughout the experiment beginning 24 hours before the first stress.

At 48 hours after t-BHP treatment, it was found that D-glucose and vitamin E protected the cells against the stress-induced decrease in the proportions of morphotype II and increase in morphotypes III to VI (FIG. 2A). After ethanol stresses, D-glucose similarly protected the cells against changes in the morphotype proportions, whereas vitamin E did not (FIG. 2B).

Similar experiments were carried out in which the proportions of cells positive for the S-A β-gal were determined at day 2 after the stresses. D-glucose had a protective effect against the increase in the proportion of positive cells observed after stress under ethanol and t-BHP. Vitamin E protected only against the stress under t-BHP (FIG. 3).

From these experiments, it is suggested that D-glucose protects the cells against SIPS induced by several types of stress, whereas specific defense mechanisms, in

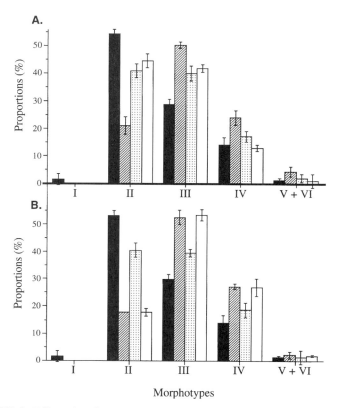

FIGURE 2. Effect of D-glucose and vitamin E on the proportions of morphotypes obtained after 5 successive stresses under 25 µM t-BHP (**A**) or 4% (v/v) ethanol (**B**). Cells were exposed to the stressor for 1 hour for 5 consecutive days and were thereafter given 2 days of recovery after the stresses, as described earlier.[39] Cells were plated at a density of 700 cells/cm² for 24 hours before morphotype determination performed as described earlier.[32]

this case antioxidant protection, protect only against specific stress, in this case oxidative stress. In fibroblasts, ethanol has been described as exerting fluidification effects on biological membranes. Ethanol also generates osmotic stress.[47] These experimental arguments are in favor of the theoretical model, which predicts that SIPS is favored when the cellular energetic potential is decreased.

CONCLUSION

Corollaries of the second law of thermodynamics allow the proposal that all living systems have common energetic traits. In all of them, stress and/or aging results in lower energy flow and lower specific entropy production (weight-normalized entropy production), meaning lower global metabolic activity.

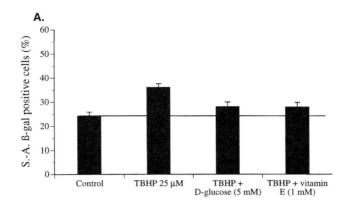

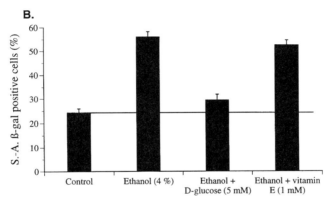

FIGURE 3. Effect of D-glucose and vitamin E on the proportions of cells positive for the S-A β-gal obtained after 5 successive stresses under 25 μM t-BHP (**A**) or 4 % (v/v) ethanol (**B**). Cells were exposed to the stressor for 1 hour for 5 consecutive days and were thereafter given 2 days of recovery after the stresses, as described earlier.[39] Cells were plated at a density of 700 cells/cm^2 for 24 hours before histochemical detection of S-A β-gal performed as described earlier.[29]

Evolution resulted in cells that optimize their production and use of free energy in given ecologic conditions. Instabilities, however, can occur that can destabilize the biologic systems.[21,22,26] As far as aging is concerned, cells can shift through a succession of steady states characterized by a higher level of damage and lower global metabolic activity, until a critical threshold is reached where the cells can no longer survive.

ACKNOWLEDGMENTS

C. Frippiat, J.-F. Dierick, F. Chainiaux, and P. Dumont have a fellowship of the FRIA, Belgium, and O. Toussaint from the FNRS, Belgium. The European Union

Biomed and Health Research Programme, Concerted Action Programme 'Molgeron' (BMH1 CT94), and shared-cost action 'Genage' (BMH2 CT98), the Fulbright program, and the Belgian FRFC and FRSM allowed us to carry out this work.

REFERENCES

1. SELYE, H. 1952. The Story of Adaptation Syndrome. Acta Inc. Medical Publishers. Montreal.
2. SELYE, H. 1936. A syndrome produced by diverse noxious agents. Nature **138:** 32.
3. SELYE, H. 1946. The general adaptation syndrome and the diseases of adaptation. J. Clin. Endocrinol. **6:** 117–230.
4. BERCZI, I. 1998. The stress concept and neuroimmunoregulation in modern biology. Ann. N.Y. Acad. Sci. **851:** 3–12.
5. SELYE, H. 1976. Stress in Health and Disease. Butterworth. Boston.
6. CZABO, S. 1998. Hans Selye and the development of the stress concept. Ann. N.Y. Acad. Sci. **851:** 19–27.
7. LEVITT, J. 1980. Responses of Plants to Environmental Stress. Vol 1. Academic Press. New York.
8. LARCHER, W. 1987. Streß bei Pflantzen. Naturwissenschaften **74:** 158–167.
9. LICHTENHALER, H.K. 1998. The stress concept in plants: an introduction. Ann. N.Y. Acad. Sci. **851:** 187–198.
10. MARTIN, G.M. 1997. Genetics and the pathobiology of ageing. Philos. Trans. R. Soc. Lond. B. Biol. Sci. **352:** 1773–1780.
11. KAPAHI, P., M.E. BOULTON & T.B. KIRKWOOD. 1999. Positive correlation between mammalian life span and cellular resistance to stress. Free Rad. Biol. Med. **26:** 495–500.
12. KIRKWOOD, T.B. 1997. The origins of human ageing. Philos. Trans. R. Soc. Lond. B. Biol. Sci. **352:** 1765–1772.
13. WESTENDORP, R.G. & T.B. KIRKWOOD. 1998. Human longevity at the cost of reproductive success. Nature **396:** 743–746.
14. KIRKWOOD, T.B. & A. KOWALD. 1997. Network theory of aging. Exp. Gerontol. **32:** 395–399.
15. ORGEL, L.E. 1973. Ageing of clones of mammalian cells. Nature **243:** 441–445.
16. HOLLIDAY, R. 1988. Towards a biological understanding of the ageing process. Perspect. Biol. Med. **32:** 109–123.
17. MICHIELS, C., M. RAES, E. PIGEOLET *et al.* 1990. Importance of a threshold for error accumulation in cell degenerative processes. Modulation of the threshold in a model of free radical-induced cell degeneration. Mech. Ageing Dev. **51:** 41–54.
18. REMACLE, J., D. LAMBERT, E. PIGEOLET *et al.* 1992. Importance of the various antioxidant enzymes for the stability of cells. Confrontation between theoretical and experimental data. Biochem. J. **286:** 41–46.
19. DENBEIGH, K.G. 1951. The Thermodynamics of the Steady State. Methuen Ltd. London.
20. PRIGOGINE, I. 1955. Thermodynamics of Irreversible Processes. John Wiley. New York.
21. SCHNEIDER, E.D. & J.J. KAY. 1994. Life as a manifestation of the second law of thermodynamics. Mathl. Comput. Modelling **19:** 25–48.
22. SCHNEIDER, E.D. & J.J. KAY. 1995. Order from disorder: the thermodynamics of complexity in biology. *In* What Is Life: The Next Fifty Years. M. P. Murphy & L.A. O'Neil, Eds.: 85–99. Cambridge University Press. Cambridge.
23. GLANSDORFF, P. & I. PRIGOGINE. 1971. Thermodynamics of Structure, Stability, and Fluctuations. Wiley Interscience. New York.
24. SCHNEIDER, E.D. 1988. Thermodynamics, ecological succession and natural selection: a common thread. *In* Entropy, Information, and Evolution: New Perspectives on Physical and Biological Evolution. B.H. Weber, D.J. Deprew & J.D. Smith, Eds.: 107–138. MIT Press. Boston.
25. TOUSSAINT, O. & E.D. SCHNEIDER. 1998. The thermodynamics and evolution of complexity in biological systems. J. Comp. Physiol. Biochem. Part A **120A:** 3–9.

26. TOUSSAINT, O., M. RAES & J. REMACLE. 1991. Aging as a multi-step process character-ized by a lowering of entropy production leading the cell to a sequence of defined stage. Mech. Ageing Dev. **64:** 45–64.

27. REMACLE, J., M. RAES, O. TOUSSAINT & G. RAO. 1995. Low levels of reactive oxygen species as modulators of cell function. Mutation Res. **316:** 103–122.

28. BAYREUTHER, K., P.I. FRANCZ & H.-G. MEINRATH. 1991. The molecular biology of the terminal differentiation of dermel fibroblasts in the fibroblast stem cell system *in vivo* and *in vitro*. *In* Cell Culture Methods in Dermatological Research. Springer-Ver-lag. Heidelberg.

29. DIMRI, G.P., X. LEE, G. BASILE *et al.* 1995. A biomarker that identifies senescent cells in culture and in aging skin *in vivo*. Proc. Natl. Acad. Sci. USA **92:** 9363–9367.

30. GIACOMONI, P. & P. D'ALESSIO. 1996. Skin ageing: the relevance of antioxidants. *In* Molecular Gerontology. S. Rattan & O. Toussaint, Eds.: 177–192. Plenum Press. New York.

31. TOUSSAINT, O. & J. REMACLE. 1996. Role of the cellular energetic metabolism in age-related processes. *In* Molecular Gerontology. S. Rattan & O. Toussaint, Eds.: 87–109. Plenum Press. New York.

32. TOUSSAINT, O., A. HOUBION & J. REMACLE. 1992. Aging as a multi-step process charac-terized by a lowering of entropy production leading the cell to a sequence of defined stage II. Experimental results with cultivated cells. Mech. Ageing Dev. **65:** 65–83.

33. CHEN, Q. & B. N. AMES. 1994. Senescence-like growth arrest induced by hydrogen peroxide in human diploid fibroblast F65 cells. Proc. Natl. Acad. Sci. USA **91:** 4130–4134.

34. TOUSSAINT, O., C. MICHIELS, M. RAES & J. REMACLE. 1995. Importance of the ener-getic factors in cellular ageing. Exp. Gerontol. **30:** 1–22.

35. TOUSSAINT, O., F. ELIAERS, A. HOUBION *et al.* 1995. Protective effect of EGb 761 and bilobalide on mortality and accelerated cellular ageing in stressful conditions. *In* Advances in Ginkgo Biloba Extract Research $\underline{4}$. Effect of Ginkgo Biloba Extracts (EGb 761) on Aging and Age-related Disorders. Y. Christen, Y. Courtois & M.-T. Droy-Lefaix, Eds.: 1–16. Elsevier Publications. Paris.

36. HAMELS, S., O. TOUSSAINT, L. LE *et al.* 1995. AGO4432 fibroblasts also shift through seven successive morphotypes during in vitro ageing: comparison with WI-38 human foetal lung fibroblasts and HH-8 human skin fibroblasts. Arch. Int. Physiol. Bio-chem. Biophys. **103:** B16.

37. RODEMANN, H.P., K. BAYREUTHER, F. DITTMANN *et al.* 1989. Selective enrichment and biochemical characterization of seven human skin fibroblasts cell types *in vitro*. Exp. Cell Res. **180:** 84–93.

38. TOUSSAINT, O., M. RAES, C. MICHIELS & J. REMACLE. 1998. Cellular response to stress: relationship with ageing and pathology. Medecine/Science. **14:** 622–635.

39. DUMONT, P., Q.M. CHEN, M. BURTON *et al.* 2000. Induction of replicative senescence biomarkers by sublethal oxidative stresses in normal human fibroblasts. Free Rad. Biol. Med. In press.

40. CHEN, Q.M., J.C. BARTHOLOMEW, J. CAMPISI *et al.* 1998. Molecular analysis of H_2O_2-induced senescent-like growth arrest in normal human fibroblasts: p53 and Rb con-trol G1 arrest but not cell replication. Biochem. J. **332:** 43–50.

41. SARETZKI, G., J. FENG, T. VON ZGLINICKI & B. VILLEPONTEAU. 1998. Similar gene expression pattern in senescent and hyperoxic treated fibroblasts. J. Gerontol. 53A: B438–B442.

42. VON ZGLINICKI, T., G. SARENTZKI, W. DÖCKE & C. LOTZE. 1995. Mild hyperoxia short-ens telomeres and inhibits proliferation of fibroblasts, a model for senescence. Exp. Cell Res. **220:** 186–213.

43. CHEN, Q.M., S. TU, V. CATANIA *et al.* Induction of senescent morphology by H_2O_2 involves de novo synthesis, functional Rb Protein and redistribution of focal adhe-sion proteins. Submitted for publication.

44. MEDRANO, E.E., S. IM, F. YANG & Z.A. ABDEL-MALEK. 1995. Ultraviolet B induces G1 arrest in human melanocytes by prolonged inhibition of retinoblastoma protein phos-phorylation associated with long-term expression of the p21Waf-1/SDI-1 protein. Cancer Res. **55:** 4047–4052.

45. DUMONT, P., L. BALBEUR, J. REMACLE & O. TOUSSAINT. 2000. Appearance of biomarkers of *in vitro* ageing after successive stimulations of WI-38 fibroblasts with IL-1 α and TNF-α: senescence associated ß-galactosidase activity and morphotype transition. J. Anat. In press.
46. TOUSSAINT, O., A. HOUBION & J. REMACLE. 1994. Effects of modulations of the energetical metabolism on the mortality and ageing of cultured cells. Biochim. Biophys. Acta **1186:** 209–220.
47. BARNES, Y., S. HOUSER & F.A. BARILE. 1990. Temporal effects of ethanol on growth, thymidine uptake, protein and collagen production in human foetal lung fibroblasts. Toxicol. in Vitro **4:** 1–7.

Role of Oxidative Stress in Telomere Length Regulation and Replicative Senescence

THOMAS VON ZGLINICKI[a]

Institute of Pathology, Charité, Humboldt University Berlin, Germany

ABSTRACT: Replicative senescence is tied into organismal aging processes in more than one respect, and telomeres appear to be the major trigger of replicative senescence under many conditions *in vitro* and *in vivo*. However, the structure-function relationships in telomeres, the mechanisms of telomere shortening with advancing replicative age, and the regulation of senescence by telomeres are far from understood. Combining recent data on telomere structure, function of telomere-binding proteins, and sensitivity of telomeres to oxidative damage, an integrative model of telomere shortening and signaling is developed. The model suggests that t-loop formation hinders access of repair proteins to telomeres, leading to accumulation of abasic sites and single-strand breaks. These might contribute to accelerated telomere shortening by transient stalling of replication as well as, if present in high concentrations, to a relief of torsional tension which might destabilize the telomeric loop structure. As a result, the single-stranded G-rich overhang, which is present at the very ends of telomeres but is normally protected at the base of the telomeric loop, will be exposed to the nucleoplasm. Free G-rich telomeric single strands are a strong inductor of the p53 pathway, and exposure of the overhangs seems to be the first step in the signal transduction cascade to replicative senescence.

TELOMERE MAINTENANCE AND SENESCENCE OF HUMAN CELL CULTURES

Human somatic cells in culture are mortal, that is, their division potential becomes exhausted after a certain number of population doublings. This effect is called replicative senescence, and there is now ample evidence that replicative senescence is an irreversible cell cycle block in G0/G1 triggered via a concerted activation of the p53/p21/p19 and p16/pRb pathways.[1] The discovery that telomeres shorten with each cell division in somatic human cells,[2] but not in immortal tumor cell cultures, led to the suggestion that telomeres might act as a mitotic clock, counting the number of divisions a clone has gone through and eventually activating replicative senescence as an ultimate DNA damage checkpoint. It was shown that telomeres shorten with each division in mortal human somatic cells by 30–200 bp, and this shortening was thought to be due to the so-called end-replication problem, the inability of DNA polymerases to replicate the 3′-end of a linear DNA molecule to its very end.[3] On the other hand, telomere shortening was not found in mortal cells that were not al-

[a]Address for correspondence: Dr. Thomas von Zglinicki, Department of Gerontology, University of Newcastle, 1HE, Newcastle General Hospital, Newcastle upon Tyne, NE4 6BE, UK. Phone: +44191 256 3310; fax: +44191 219 5074.

t.vonzglinicki@ncl.ac.uk

lowed to proliferate,[4] and it was not found in immortalized and tumor cells that maintain their telomeres for an unlimited number of generations by activation of either the telomere-elongating enzyme, telomerase, or an alternative recombination-dependent pathway.

The suggestion of telomeres being instrumental in replicative senescence gained considerable credence when it was demonstrated that ectopic expression of telomerase would lead to elongation of telomeres followed by immortalization of some human cell clones.[5,6] Specific and successful inhibition of telomerase in tumor cells resulted in telomere shortening and growth arrest with the occurrence of a senescent phenotype.[7,8] In conclusion, it appears that the integrity of the telomere complex is checked by the p53/pRb system before passage through S phase is granted. The property that is actually checked seems to be closely related to telomere length. However, it is probably not telomere length itself (see below).

A number of stress conditions can induce a senescence-like arrest prematurely. These conditions include radiation,[9] treatment with hydrogen peroxide,[10] treatment with organic hydroperoxides,[11] culture under chronic mild hyperoxia,[12] or transfection with oncogenes.[13] In each case, activation of at least one of the p53 and p16/pRb pathways was shown. Different markers of the senescent phenotype as, for instance, senescence-associated β-gal activity,[10] mitochondrial dehydration,[14] lipofuscin accumulation,[15] G0/G1 specificity of cell cycle arrest,[12] or senescence-associated gene expression[11,16] confirmed the similarity of these irreversible, premature cell cycle blocks and replicative senescence. Telomeres might not always be involved in the signaling pathway. On the other hand, chronic mild oxidative stress accelerates telomere shortening.[12,17] Irradiation or acute oxidative stress might cause DNA damage anywhere in the genome and give rise to a DNA damage response. However, telomeres are preferential targets for acute oxidative damage[18,19] and are deficient in repair of oxidatively generated single-strand breaks.[20] Together, the data indicate strongly that telomere damage and/or shortening is a major mechanism in triggering replicative senescence.

It is currently unknown how this trigger functions. It recently became clear that different mechanisms, especially telomeric single-strand damage, contribute greatly to telomere shortening.[20–22] Again, the mechanism is not clear. Recent data have shown that the structure of human telomeres might be more complicated than originally thought.[23] Combining structural and functional data, a hypothesis of how telomere shortening occurs and how it eventually signals replicative senescence will be developed.

THE STRUCTURE OF HUMAN TELOMERES

Human telomeres are simple repeats of the sequence TTAGGG. Their length is highly heterogeneous, even within a single cell.[24] On average, the longest telomeres are found in the germline (mean telomere length about 16 kb), whereas mean telomere length in peripheral blood lymphocytes of old people or in senescent fibroblast cultures is about 5–6 kb. The 3′-end of telomeres (the G-rich strand) forms an overhang of about 100–200 nt in length. This overhang is present in both somatic and tumor cells, and it appears not to be cell cycle dependent.[25] The overhang length does not change following DNA damaging treatments.[22] Vertebrate telomeres are

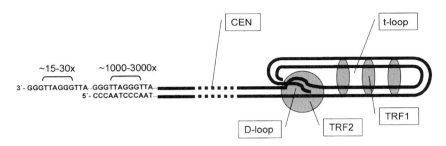

FIGURE 1. Telomere structure of a chromosome. CEN: Whole subtelomeric part of the chromosome. (*Left*) Linear structure of the telomere, showing the 3' G-rich single-stranded overhang. (*Right*) Telomeric loop structure. The telomeric DNA loops back onto itself (t-loop) with the G-rich overhang invading the dsDNA (D-loop). TRF2 stabilizes the D-loop, while TRF1 might contribute to t-loop stability. Adapted from Griffith *et al.*[23]

constructed of closely spaced nucleosomes[26] that, in addition, bind a number of telomere-specific proteins. The best characterized of these are called TRF1 and TRF2.[27]

Telomeres have classically been regarded as a simple linear structure, possibly capped by specific proteins. However, this simple structural view was challenged recently[23] (FIG. 1). It was shown by electron microscopy that artificial telomeres, consisting of some kilobases of double-stranded telomeric repeats plus G-rich single- stranded overhang, would form lasso-like loops (called t- or telomere-loop) if brought together with TRF2. The single-stranded overhang invades a partial unwinding (a D-loop) at the base of the t-loop, forming essentially the first step in a Holliday junction. TRF2 was localized at the t-loop base, suggesting that it might help to stabilize the D-loop and/or the invasion of the overhang. After preparation of psoralen-crosslinked DNA highly enriched for telomeres, the same structures were found in at least 40% of putative *ex vivo* telomeres from human and mouse liver cells. Given the complexities of the preparation, it appears very likely that t-loops might exist at the ends of most, if not all, human telomeres. The size of the t-loops was highly variable, but tended to increase with telomere length. TRF1 might contribute to the formation of t-loops because it was shown that it catalyzes telomeric synapsis, resulting in a coiled telomeric structure.[28]

These results are in line with functional data showing that both TRF1[29] and TRF2[23] are negative regulators of telomere length in telomerase-positive cells. One could easily imagine that overproduction of TRF1 and/or TRF2 would result in stabilization of the t-loop and, therefore, would not allow access of telomerase to the G-rich telomeric end. On the other hand, it was shown that inhibition of TRF2 function by overexpression of a dominant-negative TRF2 mutant resulted in degradation of the G-rich overhang and in telomeric end-to-end fusions.[30] Together, these data suggest the t-loop as an elegant possibility for protecting the single-stranded chromosomal ends. They also suggest that t-loops must be very dynamic structures. They have to be opened during DNA replication in order to allow passage for the DNA polymerase machinery. During or after completion of telomere DNA replication,

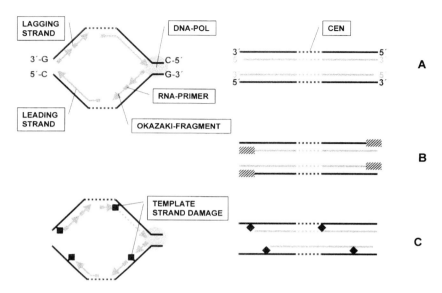

FIGURE 2. Telomere shortening models. (*Left*) Replication forks move towards the chromosomal ends. The polymerase holoenzyme has just finished replicating the left telomere and is very close to the right end. *Grey arrows*: Newly synthesized strands (leading strand or Okazaki fragments). *Grey blocks*: RNA primers. (*Right*) Ends of the two daughter chromosomes. (**A**) End replication model. The 5'ends of the newly synthesized strands are recessed by the length of the most distal RNA primer. Telomeres shorten on average by one quarter of the primer length per round of replication.[3] (**B**) Generation of 3'overhangs on all chromosomal ends by the action of a putative 5'- specific exonuclease (*hatched areas*). Telomeres shorten by half the overhang length per replication cycle.[25] (**C**) Single-strand damage contributes to telomere shortening.[12,20] Damage in the template strand stalls replication temporarily. Depending on the position of the damaged site, this may lead to loss of telomeric sequences (see text).

new overhangs have to be generated before t-loops can be reformed. This could happen by either elongation of the G-rich strand by telomerase, by limited degradation of the C-rich strand by a specific exonuclease, or by the presence of a lesion in the parental overhang, which would not allow the formation of an Okazaki fragment at this site (see below). It is expected that the formation of overhangs must be fast, because exposure of free single-stranded overhangs leads to either degradation and end-to-end fusions[30] or to activation of a p53-dependent DNA damage response.[31]

In vitro, TRF2 is sufficient to catalyze the formation of t-loops, indicating that TRF2 may be sufficient to both partially unwind a region of dsDNA (D-loop formation) and to catalyze overhang invasion.[23] In *Escherichia coli*, the formation of D-loops by partial unwinding of dsDNA and recognition of homology to the ssDNA are performed by the RecA protein.[32] RecA remains bound to the partially unwound DNA.[33] It interacts with topoisomerase I.[34] hRad51 is the structural and functional human homologue of RecA.[35] It is not clear, however, if hRad51 and topoisomerases may be required to efficiently form D-loops–t-loops in human cells *in vivo*.

OXIDATIVE STRESS, TELOMERE DAMAGE, AND
TELOMERE SHORTENING

Three different mechanisms are currently thought to contribute to telomere short-ening. The first one is the so-called end replication problem[3] (FIG. 2A). While syn-thesis of the leading strand (the new 3' G-rich end of the telomere) goes straight through to the very end of the template, synthesis of the lagging strand occurs via a multitude of Okazaki fragments primed by short RNAs (about 8–12 nt long). The gap between two Okazaki fragments is filled in and ligated later, but the gap left by removal of the most distal primer remains, resulting in a G-rich overhang on one of the chromosome ends and finally in a shortening of the telomere by at least one quar-ter of the primer length. This effect is rather small compared to the observed telom-ere shortening rates of human cells. When it was found that human telomeres contain G-rich overhangs about 100–200 nt in length, the action of a C-strand–specific exo-nuclease was suggested.[25] The action of this exonuclease would shorten each telom-ere by half the overhang length per round of replication (FIG. 2B). It is currently not yet clear if overhangs of that size exist on one or both chromosomal ends. There are experimental arguments for both observations.[25,36] Therefore, it is not clear if t-loops are the only end structure of human chromosomes. However, it is hard to imag-ine that principally different structural and functional rules should be valid on the two ends of a chromosome.

Both the end replication problem and the C-strand degradation model of telomere shortening do not take into account the possibility that the shortening rate of telom-eres depends on external influences, especially oxidative stress-dependent DNA damage. In fact, the telomere shortening rate in human fibroblasts in culture varies by more than one order of magnitude in dependence on oxidative stress[12,17,22] (FIG. 3). The telomeric G triplet is especially sensitive to cleavage by oxidative dam-age.[18,19] Moreover, it was shown that oxidative stress increases the frequency of S1 nuclease-sensitive sites, especially in telomeres.[12,20,21] In strong contrast to all the rest of the genome, S1-sensitive sites induced in telomeres of human fibroblasts by an acute dose of hydrogen peroxide are never completely repaired if the cells are not allowed to proliferate.[20] On the other hand, a quantitative correlation exists between the generation of S1-sensitive sites in telomeres of quiescent cells and the disappear-ance of these sites concomitant to accelerated telomere shortening after release from a cell cycle block[21] (FIG. 4).

S1 nuclease detects a number of different lesions including abasic sites (because the assay is performed at low pH), single-strand breaks, ssDNA loops, gaps, and overhangs. The first question was, therefore, whether the length of the G-rich over-hangs would change under conditions that influence the S1 sensitivity of telomeres and the telomere shortening rate. However, when the overhang length was measured by quantitative nondenaturing in-gel hybdridization, no change whatsoever could be detected following different DNA-damaging treatments.[22] Moreover, using native/alkaline two-dimensional electrophoresis/Southern blotting, a homogeneous distri-bution of strand breaks along the whole telomere following alkylating damage was demonstrated.[22]

It is not clear if D-loops are sensitive to S1 nuclease and if they are located at ran-dom along the telomere. In the experiments just described, S1 probing was done af-

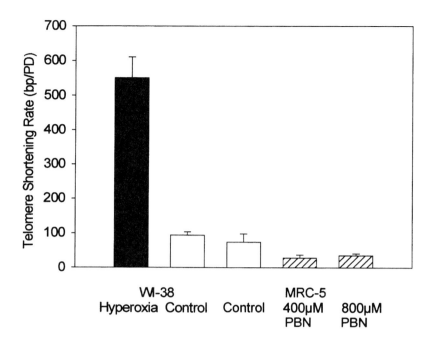

FIGURE 3. Telomere shortening rates in human fibroblasts depend on oxidative stress. Telomere shortening (in bp/population doubling) was measured in human WI-38 fibroblasts cultured under 40% ambient oxygen partial pressure[12] or in MRC-5 fibroblasts treated with the free radical scavenger α-phenyl-*t*-butyl-nitrone (PBN) in the indicated concentrations.[22]

ter deproteination of the DNA. This means that t-loops were opened and single-stranded overhangs would be degraded by S1 nuclease. Because the overhang length is constant under all tested conditions, overhang degradation would simply result in a constant background signal. D-loops, on the other hand, represent local under-winding of dsDNA stabilized by proteins. If torsional stress is not relieved by a single-strand break in proximity of the loop, D-looped DNA might in fact turn back to a native double-stranded confirmation by removal of proteins. In other words, D-loops might not be recognized by the S1 probe except in conjunction with a single-strand break.

The high sensitivity of telomeres to oxidative stress-mediated damage, the constancy of the length of the G-rich overhangs, and the homogeneous distribution of single-strand breaks along the telomere after alkylation all suggest that telomeres accumulate the same type of single-strand damage that is readily repaired all over the bulk of the genome, namely, randomly distributed abasic sites and single-strand breaks.[22] The t-loop structure suggests a possible explanation for the repair deficiency of telomeres: D-loops might block the access of repair complexes to all the DNA within the t-loop. Only repair coupled to replication could be functional within the t-loop.

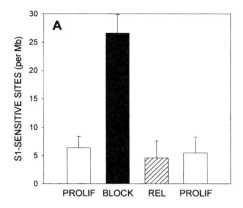

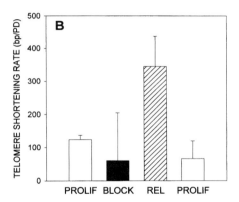

FIGURE 4. Telomeric single-strand damage accumulates in nonproliferating human fibroblasts and is transformed into telomere shortening during replication. Telomeric single-strand damage (as frequency of S1 nuclease-sensitive sites in telomeres, **A**) and telomere shortening rate (in bp/population doubling, **B**) were measured in human MRC-5 fibroblasts during normal proliferation (PROLIF), at the end of a block to replication by confluency for 60–90 days (BLOCK), at less than 2 PD after release from the block (REL), and at more than 5 PD after release from the block (PROLIF). Adapted from Sitte *et al.*[21]

How can unrepaired single-strand damage contribute to telomere shortening? A simple abasic site stalls polymerase III in *E. coli* transiently. Replication finally proceeds either by reinitiation further downstream of the damaged base or by insertion of a (potentially incorrect) base opposite the damaged one under control of the inducible SOS repair, consisting of RecA, UmuC, and UmuD.[37] In mammals, replication bypassing the prototypical UV lesion – a thymidine dimer – has been studied most extensively. Again, replication is stalled transiently but ultimately proceeds, with polymerase h (the enzyme mutated in xeroderma pigmentosum variant) being responsible for correct base insertion.[38] The effects of a stalling lesion will be different depending on the position of the lesion. If the lesion is in the template to the lagging strand, both leading and lagging strand synthesis will proceed with only a temporary interruption of the lagging strand by one Okazaki fragment (about 100–150 nt).[39] If, on the other hand, the lesion is on the template to the leading strand, lagging strand synthesis may proceed for more than 1,000 nt before translesion synthesis of the leading strand occurs. It is important to note that the replication fork moves onward all the time, leaving the leading strand template as a single-stranded, unreplicated stretch behind.[40] With respect to telomeres (FIG. 2C), these data suggest that a lesion in the G-rich strand (the template for lagging strand synthesis) should be without effect on telomere length in the daughter strand, except that it is within the most distal 150 nt. An abasic site or a bulky lesion in the G-rich overhang would prohibit the formation of the most distal Okazaki fragment. This is one possible way to reproduce the G-rich overhang. A telomeric lesion in the C-rich strand (the template to the leading strand) could have a much stronger effect. If leading strand synthesis would be stalled until the replication fork reaches the distal end of the telomere, the polymerase complex might fall off and translesional synthesis might not occur at all, leaving the newly synthesized 3'-end greatly recessed (FIG. 2C, lower left). Because single-stranded C-rich telomeric DNA has never been found,[22,25] the protruding 5'end must be degraded by an exonuclease.

In conclusion, the model predicts that the persistence of lesions in the C-rich telomeric strand should have a much larger effect on telomere shortening than do lesions in the G-rich strand. This prediction should be true for lesions able to block DNA replication transiently, such as bulky lesions and abasic sites. Whether the same holds true for single-strand breaks in telomeres is not clear at present.

HOW DOES TELOMERE SHORTENING TRIGGER REPLICATIVE SENESCENCE?

Recent data show that telomere shortening triggers a p53-dependent DNA damage response not only in human cells, but also in mouse cells and tissues.[41] Late-generation telomerase-knockout tissues and fibroblasts display short telomeres and show apoptosis or cell cycle arrest, respectively. Both phenotypes are rescued by a telomerase/p53 double knockout.

Originally it was thought that shortening of telomeres to a certain length would directly trigger replicative senescence. A number of cell clones[42] as well as parallel cultures under different conditions of oxidative stress[12] senesce at about the same average telomere length of about 5 kb. It was assumed that at this length at least one telomere per cell should be short enough to trigger a DNA damage response either directly or via formation of a fusion–breakage–fusion cycle.[4] The frequency of chromosome fusions was repeatedly found to increase in late-passage human or mouse fibroblast cultures.[41] However, measurement of telomere length in single chromosomes revealed that the shortest telomeres are already present many cell divisions before senescence.[24] Moreover, antioxidative treatment of human fibroblasts with α-phenyl-*t*-butyl-nitrone, a free radical scavenger, slows telomere shortening much more than it elongates the replicative lifespan. This means that parallel cultures of MRC-5 fibroblasts can senesce with significantly different telomere length.[22]

We tested the ability of single-stranded DNA fragments to trigger a cell cycle block.[43] It was found that G-rich telomeric DNA, but not C-rich DNA, inhibited proliferation of human cells in a p53-dependent fashion. G-rich scrambled DNA was less efficient than telomeric DNA. Two interpretations are possible. G-rich single-stranded DNA could either trigger the p53 checkpoint directly or it could compete with telomeric overhangs for t-loop formation by invasion in the D-loop. Both interpretations point to exposure of the G-rich overhang as the *in vivo* trigger of replicative senescence. This is in line with data showing that overexpression of a dominant-negative TRF2 mutant, which exposes the overhang, also elicits a p53-dependent DNA damage response.[23,31]

Is there a causal relationship between telomere shortening and exposure of the G-rich overhang? A very short telomere length might impose a torsional constraint on t-loop formation. However, loops can be formed by rather short stretches of DNA. Together with the data just cited it seems rather improbable that telomere length by itself prohibits t-loop formation. Another possibility is raised by the fact that high concentrations of S1-sensitive sites are found in telomeres from senescent cells both at the "normal" Hayflick limit and at premature senescence under hyperoxic culture.[12] Single- strand breaks relieve tension locally and might therefore interfere with the formation and/or maintenance of D-loops. It can be expected that certain torsional stress is necessary to maintain the D-loop structure with the integrated G-rich overhang. If the concentration of single-strand breaks in telomeres becomes too high, D-loops may no longer be stable, and the overhangs might be released.

This suggestion is in accord with the fact that all DNA damaging treatments known to trigger premature senescence induce much more single- than double-strand breaks.[9–12] Telomeres accumulate single-strand breaks due to their inherent repair deficiency.[20] The suggestion is also in accord with data showing that topoisomerase inhibitors (which allow cleavage of the DNA but block the religation step) induce a senescent phenotype in human fibroblasts.[44] DNA topoisomerase II cleaves telomeres efficiently.[45] Topoisomerase III is related to telomere length regulation in both yeast[46] and ataxia telangiectasia fibroblasts[47] by as yet unknown mechanisms. Both type II and type III topoisomerases interact with Rec A[34] and with RecQ helicases like Sgs1 from yeast and its human homologs, the genes for Werner's syndrome and Bloom's syndrome.[48] Together, these data suggest that topoisomerases together with different helicases may play an important role in the maintenance of the loop structure of telomeres and that inhibition of topoisomerases might trigger a DNA damage response again by liberation of the G-rich telomeric overhang.

The model presented so far indicates that there should be two different enzyme systems competing for access to a free overhang, namely, the p53 system (or a detector enzyme upstream of p53) and, in immortal cells, telomerase. Evidently, the p53 system has no access to the overhang during replication of the telomeric end. At least, signal transduction to the cell cycle block or apoptosis effectors is interrupted. One might speculate whether telomerase binds the overhang during S phase. On the other hand, some data suggest that the p53 detector might have prime access to an overhang released "untimely." In many cell types with a functional p53 pathway, apoptosis rather than telomere elongation is the result of overexpression of mutant TRF2.[31] Moreover, treatment of telomerase-negative human fibroblasts with single-stranded G-rich DNA fragments resulted in long-lasting inhibition of proliferation, whereas the same treatment of immortal, wtp53 glioblastoma cells blocked proliferation only transiently. Importantly, glioblastoma cells that overcame the block showed enhanced telomerase activity and elongated telomeres.[43] Chronic or acute oxidative stress triggered the same sequence of events. Not only the transient inhibition of proliferation, but also the induction of telomerase was p53 dependent. Oxidative damage to mutp53 cells did not activate the telomerase above ground level, even if it led to slow telomere shortening. These data suggest that the p53 system not only can remove cells from the proliferating pool by triggering cell cycle arrest or apoptosis, but also might rescue them from a proliferation block by hyperactivating telomerase. In other circumstances, however, the presence of telomerase might also modify a p53-dependent response. Inhibition of telomerase renders tumor cells with functional p53 more susceptible to cell cycle arrest and apoptosis due to treatment with topoisomerase inhibitors, whereas telomerase-expressing fibroblasts are less sensitive to the same inhibitors as their mortal, telomerase-negative counterparts.[8] It appears that interaction of the p53 and the telomerase systems can go both ways.

REFERENCES

1. STEIN, G.H. *et al.* 1999. Differential roles for cyclin-dependent kinase inhibitors p21 and p16 in the mechanisms of senescence and differentiation in human fibroblasts. Mol. Cell. Biol. **19:** 2109–2117.
2. HARLEY, C.B. *et al.* 1990. Telomeres shorten during ageing of human fibroblasts. Nature **345:** 458–460.
3. OLOVNIKOV, A.M. 1971. Principle of marginotomy in template synthesis of polynucleotides. Dokl. Akad. Nauk. **201:** 1496–1499.
4. ALLSOPP, R.C. *et al.* 1995. Models of initiation of replicative senescence by loss of telomeric DNA. Exp Gerontol. **31:** 235–243.
5. BODNAR, A.G. *et al.* 1998. Extension of life-span by introduction of telomerase into normal human cells. Science **279:** 349–352.
6. PRESCOTT, J.C. 1999. Telomerase: Dr. Jekyll or Mr. Hyde? Curr. Opin. Gen. Dev. **9:** 368–373.
7. FENG, J. *et al.* 1995. The RNA component of human telomerase. Science **269:** 1236–1241.
8. LUDWIG, A. *et al.* 1999. Ribozyme cleavage of telomerase mRNA sensitizes tumor cells to doxorubicin and induces delayed crisis. EMBO J. Submitted.
9. RODEMANN, H.P. *et al.* 1989. Selective enrichment and biochemical characterization of seven human skin fibroblast cell types *in vitro.* Exp. Cell Res. **180:** 84–95.
10. CHEN, Q. *et al.* 1995. Senescence-like growth arrest induced by hydrogen peroxide in human diploid fibroblast F65 cells. Proc. Natl. Acad. Sci. USA **91:** 4130–4134.
11. TOUSSAINT, O. *et al.* 1998. Reciprocal relationships between the resistance to stresses and cellular aging. Ann. N. Y. Acad. Sci. **851:** 450–465.
12. VON ZGLINICKI, T. *et al.* 1995. Mild hyperoxia shortens telomeres and inhibits proliferation of human fibroblasts: a model for senescence? Exp. Cell Res. **220:** 186–193.
13. PALMERO, I. *et al.* 1998. p19ARF links the tumour suppressor p53 to Ras. Nature **395:** 125–126.
14. VON ZGLINICKI, T. *et al.* 1995. Mitochondrial water loss and aging of cells. Cell Biochem. Funct. **13:** 181–187.
15. VON ZGLINICKI, T. *et al.* 1995. Lipofuscin accumulation and aging of fibroblasts. Gerontology **41:** 95–108.
16. SARETZKI, G. *et al.* 1998. Similar gene expression pattern in senescent and hyperoxic-treated fibroblasts. J. Gerontol. Biol. Sci. **53A:** B438–B442.
17. VAZIRI, H. *et al.* 1997. ATM-dependent telomere loss in aging human diploid fibroblasts and DNA damage lead to the post-translational activation of p53 protein involving poly(ADP-ribose)polymerase. EMBO J. **16:** 6018–6033.
18. HENLE, E.S. *et al.* 1999. Sequence-specific DNA clevage by Fe2+-mediated fenton reactions has possible biological implications. J. Biol. Chem. **274:** 962–971.

19. OIKAWA, S. *et al.* 1999. Site-specific DNA damage at GGG sequence by oxidative stress may accelerate telomere shortening. FEBS Lett. **453:** 365–368.
20. PETERSEN, S. *et al.* 1998. Preferential accumulation of single-stranded regions in telomeres of human fibroblasts. Exp. Cell Res. **239:** 152–160.
21. SITTE, N. *et al.* 1998. Accelerated telomere shortening in fibroblasts after extended periods of confluency. Free Rad. Biol. Med. **24:** 885–893.
22. VON ZGLINICKI, T. *et al.* 2000. Accumulation of single-strand breaks is the major cause of telomere shortening in human fibroblasts. Free Rad. Biol. Med. **28:** 64–74.
23. GRIFFITH, J.D. *et al.* 1999. Mammalian telomeres end in a large duplex loop. Cell **97:** 503–514.
24. MARTENS, U.M. *et al.* 1998. Short telomeres on human chromosome 17p. Nature Genet. **18:** 76–80.
25. MAKAROV, V.L. *et al.* 1997. Long G tails at both ends of human chromosomes suggest a C strand degradation mechanism for telomere shortening. Cell **88:** 657–666.
26. MAKAROV, V.L. *et al.* 1993. Nucleosomal organization of telomere-specific chromatin in rat. Cell **73:** 775–787.
27. BROCCOLI, D. *et al.* 1997. Human telomeres contain two distinct Myb-related proteins, TRF1 and TRF2. Nature Genet. **17:** 231–235.
28. GRIFFITH, J. *et al.* 1998. TRF1 promotes parallel pairing of telomeric tracts *in vitro*. J. Mol. Biol. **278:** 79–88.
29. VAN STEENSEL, B. *et al.* 1997. Control of telomere length by the human telomeric protein TRF1. Nature **385:** 740–743.
30. VAN STEENSEL, B. *et al.* 1998. TRF2 protects human telomeres from end-to-end fusions. Cell **92:** 401–413.
31. KARLSEDER, J. *et al.* 1999. p53- and ATM-dependent apoptosis induced by telomeres lacking TRF2. Science **283:** 1321–1325.
32. SHIBATA, T. *et al.* 1982. D loop cycle. A circular reaction sequence which comprises formation and dissociation of D-loops and inactivation and reactivation of superhelical closed circular DNA promoted by RecA protein of *Escherichia coli*. J. Biol. Chem. **257:** 13981–13986.
33. REGISTER, J.C. *et al.* 1988. Direct visualization of RecA protein binding to and unwinding duplex DNA following the D-loop cycle. J. Biol. Chem. **263:** 11029–11032.
34. CUNNINGHAM, R.P. *et al.* 1981. Homologous pairing and topological linkage of DNA molecules by combined action of *E. coli* RecA protein and topoisomerase I. Cell **24:** 213–223.
35. BAUMANN, P. *et al.* 1996. Human Rad51 protein promotes ATP-dependent homologous pairing and strand transfer reactions *in vitro*. Cell **87:** 757–766.
36. WRIGHT, W.E. *et al.* 1997. Normal human chromosomes have long G-rich telomeric overhangs at one end. Genes Dev. **11:** 2801–2809.
37. BRIDGES, B.A. 1999. DNA repair: Polymerases for passing lesions. Current Biol. **9:** R475–R477.
38. JOHNSON, R.E. *et al.* 1999. hRad30 mutations in the variant form of *Xeroderma Pigmentosum*. Science **285:** 263–265.
39. MENEGHINI, R. *et al.* 1976. T4 endonuclease V-sensitive sites in DNA from ultraviolet-irradiated human cells. Biochim. Biophys. Acta **425:** 428–437.
40. CORDEIRO-STONE, M. *et al.* 1999. Analysis of DNA replication forks encountering a pyrimidine dimer in the template to the leading strand. J. Mol. Biol. **289:** 1207–1218.
41. CHIN, L. *et al.* 1999. p53 deficiency rescues the adverse effects of telomere loss and cooperates with telomere dysfunction to accelerate carcinogenesis. Cell **97:** 527–538.
42. ALLSOPP, R.C. *et al.* 1995. Evidence for a critical telomere length in senescent human fibroblasts. Exp. Cell Res. **219:** 130–136.
43. SARETZKI, G. *et al.* 1999. Telomere shortening triggers a p53-dependent cell cycle arrest via accumulation of G-rich single stranded DNA fragments. Oncogene **18:** 5148–5158.
44. MICHISHITA, E. *et al.* 1998. DNA topoisomerase inhibitors induce reversible senescence in hormal human fibroblasts. Biochem. Biophys. Res. Commun. **253:** 667–671.
45. YOON, H.J. *et al.* 1998. DNA topoisomerase II cleavage of telomeres in vitro and in vivo. Biochim. Biophys. Acta **1395:** 110–120.

46. Kim, R.A. *et al*. 1995. Effects of yeast topoisomerase III on telomere structure. Proc. Natl. Acad. Sci. USA **92:** 2667–2671.
47. Fritz, E. *et al.* 1997. Overexpression of a truncated human topoisomerase III partially corrects multiple aspects of the ataxia-telangiectasia phenotype. Proc. Natl. Acad. Sci. USA **94:** 4538–4542.
48. Duguet, M. 1997. When helicase and topoisomerase meet. J. Cell Sci. **110:** 1345–1350.

Replicative Senescence and Oxidant-Induced Premature Senescence

Beyond the Control of Cell Cycle Checkpoints

QIN M. CHEN[a]

Department of Pharmacology, University of Arizona, Skaggs Pharmaceutical Science Building, 1703 E. Mabel Street, Tucson, Arizona 85721, USA

ABSTRACT: Normal human diploid fibroblasts (HDFs) undergo replicative senescence inevitably in tissue culture after a certain number of cell divisions. A number of molecular changes observed in replicative senescent cells occur in somatic cells during the process of aging. Genetic studies on replicative senescence indicate the control of tumor suppression mechanisms. Despite the significance of replicative senescence in aging and cancer, little is known about the central cause of the complex changes observed in replicative senescent cells. The interest in the phenomenon has intensified in recent years, since damaging agents, certain oncogenes and tumor suppressor genes have been found to induce features of senescence in early passage young HDFs or in immortalized tumor cells. The reported features of senescence are summarized here in order to clarify the concept of replicative senescence or premature senescence. The experimental results of extending the replicative life span by reducing ambient oxygen tension or by N-tert-butyl-alpha-phenylnitrone (PBN) argue a role of oxidative damage in replicative senescence. By inducing premature senescence with a pulse treatment of H_2O_2, we can study the role of the cell cycle checkpoint proteins p53, p21, p16 and Rb in gaining each feature of senescence. Although p53 and Rb control G1 arrest and Rb appears to control cell enlargement, activation of the senescent associate β-galactosidase, loss of cell replication and multiple molecular changes observed in premature senescent or replicative senescent cells are likely controlled by mechanisms beyond the cell cycle checkpoints.

INTRODUCTION

Aging is an inevitable process for any living organism. Despite the fact that mortality rate increases with age, aging *per se* is rarely a cause of death in humans. Instead, aging results in an increased risk of developing life-threatening diseases. The rate of these diseases often increases exponentially with age in a population over 40 years old. Heart disease, cancer, stroke, and lung disease are the major causes of death in the aging population. In addition, the quality of life often decreases because of the degeneration of muscle, arthritis, osteoporosis, or neurodegenerative diseases in older persons. So why is aging associated with an increased risk of many diseases? In asking this question, we need to know what causes aging.

[a]Phone: 520-626-9126; fax: 520-626-2466.
chen@pharmacy.arizona.edu.

Aging is a complex process involving changes in multiple parameters. Most of these changes are generally determined by chronologic time. Although gray hair, loss of bone density, and degeneration of muscle are certainly measurements of a healthy aging process, one change is often not sufficient to be the cause or judgment of aging. Longevity comparison between animal species supports that aging is a result of a genetic program. Individual differences within a species or a family argue that environmental or nutritional factors can critically influence longevity or the process of aging. Despite voluminous characteristic studies on aging-associated changes and enormous efforts of finding the "fountain of youth," the central question of what causes aging has not been satisfactorily addressed. Many theories have been proposed to explain the cause of aging. A few can actually explain the process of aging as a whole.

The Free Radical Theory of Aging has often been cited by many biologists and gerontologists.[1,2] This theory, proposed by Denham Harman in 1956, holds that normal aging results from random deleterious damage to tissue by free radicals produced during normal aerobic metabolism.[3] The proposal was based on experimental observation that exposing laboratory animals to radiation appeared to age the animals very rapidly and caused an increase in free radical levels.[3] Oxygen-free radicals are thought to be ubiquitous.[4–6] *In vitro* studies using isolated mitochondria demonstrate that about 1–2% of consumed oxygen is converted to superoxide.[2,6] Based on this evidence, it is predicted that mitochondria produce free radicals *in vivo* as a result of incomplete electron transport. Old animals contain defective mitochondria and can produce higher levels of oxidants than can their young counterparts.[7] In addition to mitochondrial respiration, radiation is known to produce oxidants.[4,8] Certain environmental toxins, pharmacologic agents, and metabolism of xenobiotics can also produce oxidants.[4,8] Our immune system produces oxidants to bombard pathogens during inflammatory reactions. Stroke, atherosclerosis, myocardial infarction, and surgery can cause ischemic reperfusion, a process known to produce bursts of oxidants. Several laboratories have demonstrated that tissues from aged individuals or aged experimental animals accumulate oxidative DNA damage, protein oxidation, and lipid peroxidation.[1,6,9,10] Using houseflies as an experimental model, Sohal's group[11] reported that increasing the expression of the antioxidant defense enzymes superoxide dismutase and catalase using transgenic approaches can prolong the life span of Drosophila. These experimental results provide strong support for the Free Radical Theory of Aging. However, the question of how oxidants cause aging or aging-associated changes in humans remains unsolved. An important impediment to addressing this question is the lack of appropriate experimental systems. A number of experimental models have been proven useful in defining aging-associated changes and for finding the "wires and switches" of the changes. Replicative senescence of human diploid fibroblasts (HDFs) in culture has been used to identify aging-associated changes in humans at the molecular level.

WHAT IS REPLICATIVE SENESCENCE?

Normal (i.e., non-tumor) human cells lose replicative potential and inevitably become senescent as a result of serial passages in tissue culture.[12–17] This phenomenon was first reported in HDFs by Hayflick's group in 1961.[18] Upon completion of replicative capacity, the cells remain intact and metabolically active in a stable condition for months and years[19] (unpublished work from Pereira-Smith's and several laboratories). This seemingly infinite nonreplicative state is termed senescence. Senescence is determined by the number of replications that the cells have been through instead of chronologic time. In general, fibroblasts from a human fetus can replicate 50–55 population doublings (PDs). Because senescence is achieved through cell replication, the process is referred to as replicative senescence.

Senescent cells cannot replicate but instead become enlarged and show multiple molecular changes. Differing from quiescent cells, senescent cells cannot replicate when stimulated with any physiologic mitogens.[13,14,16,17] Senescent cells show an enlarged cell volume, an increase in cell surface area, and altered morphology.[13,14] In addition, recent studies found that senescent cells elevate the activity of a unique neutral β-galactosidase (termed senescence associated β-galatosidase or SA β-gal).[20] Because SA β-gal activation is not related to growth arrest (many measurements of senescence in the past were related to growth arrest) and can be determined with cells *in situ*, it serves as an ideal biomarker for qualitative and quantitative determination of senescence. Therefore, senescence can be measured by activation of SA β-gal in addition to cell enlargement and inability to replicate.

At the molecular level, senescence is associated with changes in the expression of a large number of genes.[21–23] Senescent cells show increased levels of cyclin-dependent kinase inhibitors p21 and p16, which are negative regulators of cell proliferation. However, many senescence-associated genes do not appear to be related to growth arrest. In addition, senescent cells downregulate the expression of many genes and are unable to express certain genes upon specific stimulation. For example, senescent cells can not express c-fos,[24] EPC-1 (early PDL cDNA-1),[25] cdc-2 kinase,[26] and proliferating cell nuclear antigen (PCNA)[27] in response to serum stimulation. Heat shock response is attenuated in senescent cells.[28–30] Such a diversity of senescence-associated changes indicates a multiplicity of upstream control mechanisms.

At least five lines of evidence support the idea that replicative senescence *in vitro* is related to organismal aging *in vivo*. The replicative life span of a given cell strain negatively correlates with the age of the donor in general.[13,14,17] Despite a recent study with HDFs challenging this relationship,[31] a negative correlation was observed with several cell types including fibroblasts,[32] lens epithelial cells, keratinocytes, corneal endothelial cells, and T lymphocytes.[33] The growth potential of primary cultures declines at a rate of about 0.2 PDs per year of donor age.[33] Second, cells derived from individuals with inherited premature aging syndromes, such as Werner's or Hutchinson-Gilford, have a reduced replicative life span.[13,14,17] Third, the longevity of a species influences the replicative life span of the cells *in vitro*.[12–14,17] Cells from long-lived species can replicate more PDs than can those from short-lived species. An average life expectancy is 70 years for humans and 3 years for rodents. On average, embryo fibroblasts from humans can replicate 50–55 PDs, whereas those from rodents can only replicate a maximum of 30 PDs on average. The fourth line of evidence comes from a study of the senescent biomarker SA β-gal in tissues from aged donors. SA β-gal is found in skin biopsy samples from elderly individuals.[20,33] Finally, at the molecular level, a number of genes change their expression during cellular senescence following the pattern of aging *in vivo*.[16,17,21] A good example is collagenase-1 gene, which is often overexpressed in senescent cells and tissues of elderly individuals. Based on these correlations, HDFs have been an attractive model for aging studies.

Recent studies from several laboratories found that senescent features can be induced to occur in early passage cells. With H_2O_2, early passage HDFs surviving the treatment appear to be indistinguishable from replicative senescent cells by irreversible growth arrest, morphology, and many other characteristics. Others found that hyperoxia,[34,35] inhibitors of histone deacetylase,[36] hyperactive ras gene,[37] and a phosphatidylinositol 3-kinase inhibitor[38] can also cause early passage HDFs to enter a state resembling senescence. This review focuses on studies from our laboratory and discusses the relationship between cell cycle checkpoints and senescent features.

RELATIONSHIP BETWEEN REPLICATIVE SENESCENCE AND OXIDATIVE DAMAGE

What causes replicative senescence? Normal (i.e., non-tumor) cells from most rodent species can eventually escape from senescence and become immortalized in culture. By contrast, to date, spontaneous immortalization has not been observed with any normal human cells from donors with a normal genetic background. This species difference argues that replicative senescence is a result of a genetic program. Fusion of senescent cells and early passage cells results in heterokaryons that are incapable of synthesizing DNA.[39] mRNA from senescent cells, when introduced into early passage cells by microinjection, can cause early passage cells to develop features of senescence.[40] Several chromosomes have been found to carry senescence genes.[41–48] These findings support that replicative senescence is likely a result of a genetic program or changes in gene expression. However, the finding that replicative senescence is marked with shortened telomeres and the hypothesis that telomeres serve as a timer of replicative senescence[49,50] bring up a contradictory view to the genetic control of senescence.

We hypothesize that accumulation of oxidative damage contributes to replicative senescence. Oxidative damage can be measured by formation of 8-oxo-2'-deoxyguanosine (oxo⁸dG) in DNA or free 8-oxoguanine base (oxo⁸Gua) released by cells. Hydroxyl radicals attack DNA and cause formation of oxo⁸dG and 20 other adducts.[51] The repair enzyme formamidopyrimidine glycosylase (Fpg protein) excises oxo⁸Gua bases from DNA.[52,53] We found that replicative senescent cells contain about 30% more oxo⁸dG in DNA and produce four times more free oxo⁸Gua bases.[54] These results indicate that senescent cells have accumulated oxidative DNA damage. They also indicate an increased rate of free radical generation in senescent cells. A recent study by Allen et al.[55] confirmed that senescent cells produce high levels of oxidant radicals and H_2O_2.

Most cells inside our body are well insulated from environmental oxygen exposure. The oxygen tension in most areas of our body is about 3%. In taking cells outside our body and exposing them to air, the cells experience the much higher oxygen tension of 20%. Oxygen molecules can be converted into free radicals through chemical and biological reactions. By reducing oxygen tension, we can reduce the level of free radicals. When cultured under 3% oxygen tension, HDFs can achieve 20 more PDs than can those cultured under normoxia conditions before senescence.[54,56,57]

A few chemical compounds can extend the replicative life span of HDFs. N-tert-butyl-alpha-phenylnitrone (PBN), a spin trap agent and an antioxidant, has been shown to delay memory loss in aged gerbils.[58] Compared to most antioxidant vitamins, PBN is a biphasic compound that exhibits solubility in lipid and water and is stable under the culture condition.[59] When PBN was included in the culture medium of presenescent cells, treated cells replicated 4–7 more PDs before reaching senescence than did untreated cells.[54] PBN reduced the rate of free radical production in near senescent cells (Paler-Martínez et al., unpublished data). These experimental results support the hypothesis that oxidative damage contributes to replicative senescence. Definitive experiments employing the overexpression of antioxidant enzymes are critical in addressing whether oxidative damage is a cause of replicative senescence.

INDUCTION OF PREMATURE SENESCENCE WITH H_2O_2

Oxidants are known as toxins. Oxidant toxicity has been studied extensively in many animal species, tissues, and cell types. The major biochemical changes found at the cellular level include glutathione depletion, depletion of ATP, elevation of cytosolic free calcium, and activation of several families of kinases. Activation of the transcription factors AP-1, NF-κB, the heat shock factor, and p53 contributes to changes in the expression of a number of genes in cells exposed to oxidants. Cellular outcomes of these biochemical and molecular changes include growth arrest, apoptosis, or necrosis. Because most of the studies are performed with immortalized cell lines or tumor cells, the relationship between these changes and cellular senescence is not known.

We tested the hypothesis that oxidants could cause the cellular and molecular changes observed in senescent cells. Early passage young HDFs were exposed briefly (2 hours) to H_2O_2 at a concentration range that did not appear to kill the majority of cells. These treated cells were then kept under culture conditions to study the long-term consequence of H_2O_2 exposure. We observed that surviving HDFs could not replicate when stimulated with serum or growth factors.[60] These cells cannot synthesize DNA in response to mitogenic stimulation even if they are allowed to recover for 3 weeks.[61] The cell surface area increases to a great extent at 4–7 days after treatment, and the cells appear morphologically indistinguishable from replicative senescent cells after 1 week.[60] We further compared H_2O_2-treated early passage cells with replicative senescent cells by measuring activities of ornithine decarboxylase, thymidine kinase, and SA β-gal. H_2O_2-treated early passage cells showed reduced activity of ornithine decarboxylase or thymidine kinase similar to replicative senescent cells.[60] TABLE 1 summarizes senescent features observed in H_2O_2-treated early passage cells.

CELL CYCLE ARREST IS A CRITICAL FEATURE OF SENESCENCE

Senescent cells are irreversibly growth arrested. The majority of senescent cells show a G1 DNA content.[62] This indicates the possibility that the G1 cell cycle checkpoints are critical controls for senescence. Senescent cells contain an activated p53 transcription factor,[63–65] elevated levels of p21[66] and p16,[67] and cannot hyper-

TABLE 1. Senescent features observed in H_2O_2-treated early-passage HDFs

G1 arrest

Incapable of replicating when G1 arrest is abolished

Cell enlargement

Reduced saturation density

Unable to activate ornithine decarboxylase

Unable to activate thymidine kinase

Activation of SA β-gal

Elevated p21 and p16 proteins

Unable to phosphorylate Rb

Elevated mRNA of

    collagenase-1

    apolipoprotein J

    fibronectin

    α_1(I)-procollagen

    osteonectin

    SM-22

    SS-9

phosphorylate Rb protein in response to mitogenic stimulation.[68] Activation of these cell cycle checkpoints composes a mechanism of growth arrest in senescent cells.

Growth arrest can be reversible or irreversible. In many reported studies with a variety of tumor or immortalized cell types, the cells often undergo reversible growth arrest in response to DNA damaging agents and other insults. Although multiple changes can occur in cells undergoing reversible growth arrest, activation of p53, p21, p16, and Rb ultimately contributes to G1 cell cycle arrest.[69,70] It is thought that these checkpoint proteins halt cell proliferation to allow DNA repair to occur.[71–73] By contrast, with replicative senescent cells or premature senescent cells, growth arrest appears to be irreversible. In the process of replicative senescence, p21 mRNA and protein are elevated in late passage cells. When the cells are senescent, p21 levels return to normal, but levels of p16 mRNA and protein become elevated to a great extent.[67,74] Both p21 and p16 are elevated during induction of premature senescence by H_2O_2 or by hyperactive ras, which can produce reactive oxygen-free radicals.[37,75,76] These two genes may cooperate to inhibit Rb phosphorylation and maintain growth arrest in an irreversible state.

Signals leading to elevation of p21 and p16 need to be elucidated in replicative senescent or premature senescent cells. The p21 gene contains p53 cis-elements in the promoter region. The p53 protein level increases in response to multiple stress signals as a result of reduced proteolytic degradation of p53 protein.[77,78] Elevation of p53 protein or activation of the DNA binding activity of p53 can cause an increase in the transcription of the p21 gene. In the absence of p53, the p21 gene can also increase its transcription in response to DNA damaging agents.[79,80] The mechanism of such p53-independent p21 activation is not entirely clear. In parallel with the fact that p21 can be induced by various chemical or physical stresses, a recent study found that UVB radiation can induce p16.[81] Whether p16 is elevated in response to other forms of chemical or physical insults has not been explored. Regardless, it is not known at present how p21 and p16 are elevated in replicative senescent cells.

Both p21 and p16 exert functions of cyclin-dependent kinase (CDK) inhibitors.[69,70] The p21 protein is a general inhibitor of CDKs, which phosphorylate Rb and allow G1 to S phase progression.[69,70,82,83] The p16 protein forms a complex with cyclin D and negatively regulates the activity of cyclin D-dependent kinases.[69,70,84,85] In addition, p21 protein also inhibits PCNA-mediated DNA replication, a mechanism by which p21 functions as a growth inhibitor independent of its CDK inhibitor activity.[86,87] Inability to hyperphosphorylate Rb protein may result from elevated p21 and p16 in replicative senescent or premature senescent cells.

ROLE OF p53 AND RB IN EXPRESSION OF SENESCENT BIOMARKERS

Although G1 arrest is an important feature of replicative senescence, inactivating the G1 cell cycle checkpoints does not allow senescent cells to replicate. The addition of p53 and/or Rb antisense oligonucleotides throughout serial passages can prolong the replicative life span of HDFs.[88] Human papillomaviral protein E6 or E7, which enhances proteolytic degradation of p53 or Rb protein in HDFs, can delay replicative senescence when introduced into presenescent cells.[89] None of these treatments can abolish senescence in the majority of HDFs. About 1 in 10^6 cells undergoes immortalization when both E6 and E7 genes are introduced into HDFs (Chen et al., unpublished data).[89] However, when cells are senescent, inactivating the G1 cell cycle checkpoints by microinjecting SV-40 large T antigen, which can inactivate p53 and Rb, allows some cells to gain the ability to synthesize DNA, but these cells fail to replicate.[90,91] Taken together, these results suggest that irreversible growth arrest observed in replicative senescent cells is likely controlled by mechanisms beyond p53 and Rb.

H_2O_2 causes prolonged G1 arrest and premature senescence. Treatment with H_2O_2 results in transient elevation of p53 and prolonged inhibition of Rb phosphorylation.[61] HPV E6 gene expression reduces basal levels of p53 and prohibits induction of p53 by H_2O_2.[61] In parallel, HPV E7 protein enhances proteolytic degradation of Rb protein and presumably disturbs the interaction of Rb with its cellular partners including E2F transcription factor.[61,92,93] Most cells expressing E6, E7, or both genes failed to undergo G1 arrest when cell cycle distribution was measured within two-cell cycle time frames after treatment.[61] However, these cells cannot replicate when cell number is monitored over weeks.[61] These data indicate the existence of growth control mechanisms beyond p53 and Rb.

SA β-gal has been used as a biomarker of cellular senescence. Using HPV E6 and E7 genes, we determined whether p53 and Rb controlled SA β-gal activation in H_2O_2-treated early passage cells. Like wild-type cells, cells expressing E6, E7, or both genes showed a dose-dependent increase in the proportion of SA β-gal–positive cells in response to H_2O_2 treatment.[94,95] Therefore, HPV E6 and E7, although able to reduce p53 and Rb in HDFs, do not affect the induction of senescent biomarker SA β-gal by H_2O_2.

TABLE 2. Protein or mRNA species overexpressed in replicative senescent HDFs

Growth regultory genes	Extracellular matrix proteins
p21 (66)	fibronectin (98,99)
p16 (67,74)	α_1(I)-procollagen (100,116)
Statin (113)	α_2(I)-procollagen (101)
Cyclin D1 (114)	Osteonectin (100)
Cyclin E (114)	Lysyl oxidase-related protein (117)
Protein synthesis factor	**Secreted proteases or protease regulators**
Elongation factor I alpha (115)	Collagenase-1 (102,103)
	Stromelysin-1 (103)
	Tissue plasminogen activator (104)
Function unclear genes	Urokinase-type plasminogen activator (104)
Apolipoprotein J (100)	Plasminogen activator inhibitor-1 (104,118)
EGF-like protein (121)	
WS3-10 (SM22) (100,101)	
SS-9 (100)	
IGF-binding protein-3 (101)	**Intracellular proteases**
Hic-5 (122)	Calpain (119)
LPC-1 (123)	Cathepsin B (120)
SAG (124)	

NOTE: Numbers in parentheses are references.

Senescence differs from growth arrest by cell enlargement. In addition to an apparent increase in cell surface area, replicative senescent or H_2O_2-treated early passage cells are 7–9 times bigger in volume.[75] How these cells become enlarged has not been well studied. Because cell enlargement is coupled to irreversible growth arrest, we tested whether p53 or Rb controls cell enlargement following H_2O_2 treatment. IMR-90 cells expressing the E7 but not the E6 gene were reluctant to increase cell size after H_2O_2 treatment.[96] Mutant E7 genes in which the Rb binding site has been destroyed cannot prohibit H_2O_2-induced cell enlargement.[96] This evidence strongly argues that Rb protein but not p53 controls cell enlargement.

MOLECULAR CHANGES UNRELATED TO GROWTH ARREST

Many genes increase their expression when HDFs undergo replicative senescence (TABLE 2). A few genes are upregulated in a pattern consistent with growth arrest (TABLE 2). The majority of genes do not appear to be related to growth arrest *per se* (TABLE 2). These nongrowth arrest-related genes may be most relevant to the process of aging *in vivo*, because most somatic cells do not replicate under normal circumstances *in vivo*. Fibroblasts compose connective tissue *in vivo* and are scaffolds for other cells to function. A major function of fibroblasts is to maintain tissue homeostasis by producing extracellular matrix proteins and secreting factors including cytokines, proteases, and protease inhibitors.[97] Replicative senescent cells develop a much thicker fibronectin network than do early passage cells due to elevated expression of the fibronectin gene at mRNA and protein levels.[98,99] Collagens, a large family of glycoproteins, form insoluble fibers in extracellular matrix. Senescent cells show elevated expression of alpha 1(I),[100] alpha 2(I) procollagens,[101] collagenase-1,[102,103] stromelysin-1,[103] and plasminogen activators.[104] Collagenase-1, stromelysin-1, and plasminogen activators can cleave various extracellular matrix proteins, such as collagen type I, II, III, VII, VIII, or X, fibronectin, and laminin.[105–107] Changes in the expression of extracellular matrix proteins and extracellular matrix proteases indicate that the extracellular matrix undergoes remodeling during the process of senescence. If these changes occur *in vivo*, several consequences of disorganization or functional alteration of connective tissue can be predicted.[108,109]

A few genes on the list in TABLE 2 were tested for their expression in early passage HDFs following H_2O_2 treatment. H_2O_2-treated cells showed elevated mRNA of collagenase-1, apolipoprotein J (clusterin), alpha1 (I) procollagen, osteonectin, fibronectin, calcium binding protein SM22, and a novel gene, SS-9.[110] These studies argue that H_2O_2-treated early passage cells resemble replicative senescent cells by molecular changes that are not related to growth arrest *per se*.

CONCLUSION

Like aging *in vivo*, senescence *in vitro* involves multiple changes. One change is often not a sufficient criterion for judging senescence. Although senescence is mainly studied with cells in culture, increasing evidence supports the idea that similar cellular change occurs during the process of aging *in vivo*. In culture, senescence of many differentiated cells attributes to a decrease in functionality. *In vivo*, functional decline is observed in many organs or tissues during the process of aging. For example, decreased contractile function of heart muscle cells is a common phenomenon of aged hearts.[111] The function of the immune system declines in the elderly.[112] Studying the mechanism of senescence of each cell type may allow us to understand the cause of aging and the origin of aging-associated diseases.

Early passage HDFs respond to sublethal H_2O_2 treatment by developing multiple features of replicative senescent cells. Because the cells develop features of senescence within 4–7 days and the time frame for developing senescent features is well defined, H_2O_2-treated HDFs may provide a useful experimental model to study the underlying mechanism of senescence-associated changes. For example, atrophy of dermal tissue and many other tissues is a common event of aging. Elevated collagenase activity is thought to contribute to dermal atrophy. H_2O_2-treated cells provide a convenient system to study the mechanism underlying the constitutive activation of collagenase expression. Our work supports the belief that induction of premature senescence by H_2O_2 is a symphony of changes in gene expression. Historically, geneticists argue that aging results from changes programmed in the genome. Others argue that aging results from random damage from endogenous, environmental, and dietary sources. Our studies show that exogenous stress or oxidative damage can cause changes in gene expression, which contribute to certain features observed in replicative senescence. The hypothesis that damage changes gene expression to cause senescence or aging opens a new venue for further experimental work.

ACKNOWLEDGMENTS

This work was initiated in Dr. Bruce N. Ames' laboratory in the University of California at Berkeley. I am indebted to Jessica Merrett, Tarrah Dilley, Sally Purdom, and Victoria C. Tu for reading the manuscript. This work was supported by the Burroughs Wellcome Foundation and the start-up fund from the Department of Pharmacology, College of Medicine, University of Arizona.

REFERENCES

1. HALLIWELL, B. & J.M.C. GUTTERIDGE. 1989. In Free Radicals in Biology and Medicine, 2nd Ed. : 416–508. Oxford University Press. Oxford.
2. BECKMAN, K.B. & B.N. AMES. 1998. The free radical theory of aging matures. Physiol. Rev. **78:** 547–581.
3. HARMAN, D. 1956. Aging: a theory based on free radical and radication chemistry. J. Gerontol. **11:** 298–300.
4. CADENAS, E. 1989. Biochemistry of oxygen toxicity. Ann. Rev. Biochem. **58:** 79–110.
5. AMES, B.N. & L.S. GOLD. 1991. Endogenous mutagens and the causes of aging and cancer. Mutat. Res. **250:** 3–16.
6. AMES, B.N., M.K. SHIGENAGA & T.M. HAGEN. 1993. Oxidants, antioxidants, and the degenerative diseases of aging. Proc. Natl. Acad. Sci. USA **90:** 7915–7922.
7. HAGEN, T.M., D.L. YOWE, J.C. BARTHOLOMEW *et al.* 1997. Mitochondrial decay in hepatocytes from old rats: membrane potential declines, heterogeneity and oxidants increase. Proc. Natl. Acad. Sci. USA **94:** 3064–3069.
8. KEHRER, J.P. 1993. Free radicals as mediators of tissue injury and disease. Crit. Rev. Toxicol. **23:** 21–48.
9. STADTMAN, E.R. 1992. Protein oxidation and aging. Science **257:** 1220–1224.
10. BARJA, D.E., G. QUIROGA, T.M. LOPEZ & C.R. PEREZ. 1992. Relationship between antioxidants, lipid peroxidation and aging. Exs **62:** 109–123.
11. ORR, W.C. & R.S. SOHAL. 1994. Extension of life-span by overexpression of superoxide dismutase and catalase in Drosophila melanogaster. Science **263:** 1128–11230.
12. HAYFLICK, L. 1991. Aging under glass. Mutat. Res. **256:** 69–80.
13. SMITH, J.R. & D.W. LINCOLN. I. 1984. Aging of cells in culture. Int. Rev. Cytol. **89:** 151–176.
14. GOLDSTEIN, S. 1990. Replicative senescence: the human fibroblast comes of age. Science **249:** 1129–1133.
15. CRISTOFALO, V.J. & R.J. PIGNOLO. 1993. Replicative senescence of human fibroblast-like cells in culture. Physiol. Rev. **73:** 617–638.
16. CAMPISI, J. 1996. Replicative senescence: an old lives' tale? Cell **84:** 497–500.
17. SMITH, J.R. & S.O. PEREIRA. 1996. Replicative senescence: implications for in vivo aging and tumor suppression. Science **273:** 63–67.
18. HAYFLICK, L. & P.S. MOORHEAD. 1961. The serial cultivation of human diploid cell strains. Exp. Cell Res. **25:** 585–621.
19. MATSUMURA, T., Z. ZERRUDO & L. HAYFLICK. 1979. Senescent human diploid cells in culture: survival, DNA synthesis and morphology. J. Gerontol. **34:** 328–334.
20. DIMRI, G., X. LEE, G. BASILE *et al.* 1995. A biomarker that identifies senescent human cells in culture and in aging skin in vivo. Proc. Natl. Acad. Sci. USA **92:** 9363–9367.
21. CAMPISI, J., G. DIMRI & E. HARA. 1996. *In* Handbook of the Biology of Aging. E. Schneider & J. Rowe, Eds. : 121–149. Academic Press. New York.
22. CRISTOFALO, V.J., R.J. PIGNOLO & M.O. ROTENBERG. 1992. Molecular changes with in vitro cellular senescence. Ann .N. Y. Acad. Sci. **663:** 187–194.
23. CAMPISI, J. 1992. Gene expression in quiescent and senescent fibroblasts. Ann. N.Y. Acad. Sci. **663:** 195–201.
24. SESHADRI, T. & J.CAMPISI. 1990. Repression of c-fos transcription and an altered genetic program in senescent human fibroblasts. Science **247:** 205–209.

25. PIGNOLO, R.J., V.J. CRISTOFALO & M.O. ROTENBERG. 1993. Senescent WI-38 cells fail to express EPC-1, a gene induced in young cells upon entry into the G0 state. J. Biol. Chem. **268:** 8949–8957.

26. STEIN, G.H., L.F. DRULLINGER, R.S. ROBETORYE et al. 1991. Senescent cells fail to express cdc2, cycA, and cycB in response to mitogen stimulation. Proc. Natl. Acad. Sci. USA **88:** 11012–11016.

27. CHANG, C.D., P. PHILLIPS, K.E. LIPSON et al. 1991. Senescent human fibroblasts have a post-transcriptional block in the expression of the proliferating cell nuclear antigen gene. J. Biol. Chem. **266:** 8663–8666.

28. LIU, A.Y., H.S. CHOI, Y.K. LEE & K.Y. CHEN. 1991. Molecular events involved in transcriptional activation of heat shock genes become progressively refractory to heat stimulation during aging of human diploid fibroblasts. J. Cell. Physiol. **149:** 560–566.

29. LUCE, M.C. & V.J. CRISTOFALO. 1992. Reduction in heat shock gene expression correlates with increased thermosensitivity in senescent human fibroblasts. Exp. Cell Res. **202:** 9–16.

30. GUTSMANN-CONRAD, A., A.R. HEYDARI, S. YOU & A. RICHARDSON. 1998. The expression of heat shock protein 70 decreases with cellular senescence in vitro and in cells derived from young and old human subjects. Exp. Cell Res. **241:** 404–413.

31. CRISTOFALO, V.J., R.G. ALLEN, R.J. PIGNOLO et al. 1998. Relationship between donor age and the replicative lifespan of human cells in culture: a reevaluation. Proc. Natl. Acad. Sci. USA **95:** 10614–10619.

32. SCHNEIDER, E.L. & Y. MITSUI. 1976. The relationship between in vitro cellular aging and in vivo human age. Proc. Natl. Acad. Sci. USA **73:** 3584–3588.

33. FARAGHER, R.G.A. & D. KIPLING. 1998. How might replicative senescence contribute to human ageing? BioEssays **20:** 985–991.

34. HONDA, S. & M. MATSUO. 1983. Shortening of the in vitro lifespan of human diploid fibroblasts exposed to hyperbaric oxygen. Exp. Gerontol. **18:** 339–345.

35. VON ZGLINICKI, T., G. SARETZKI, W. DOCKE & C. LOTZE. 1995. Mild hyperoxia shortens telomeres and inhibits proliferation of fibroblasts: a model for senescence? Exp. Cell Res. **220:** 186–193.

36. OGRYZKO, V.V., T.H. HIRAI, V.R. RUSSANOVA et al. 1996. Human fibroblast commitment to a senescence-like state in response to histone deacetylase inhibitors is cell cycle dependent. Mol. Cell. Biol. **16:** 5210–5218.

37. SERRANO, M., A.W. LIN, M.E. MCCURRACH et al. 1997. Oncogenic ras provokes premature cell senescence associated with accumulation of p53 and p16INK4a. Cell **88:** 593–602.

38. TRESINI, M., M. MAWAL-DEWAN, V.J. CRISTOFALO & C. SELL. 1998. A phosphatidylinositol 3-kinase inhibitor induces a senescent-like growth arrest in human diploid fibroblasts [see comments]. Cancer Res. **58:** 1–4.

39. NORWOOD, T.H., W.R. PENDERGRASS, C.A. SPRAGUE & G.M. MARTIN. 1974. Dominance of the senescent phenotype in heterokaryons between replicative and post-replicative human fibroblast-like cells. Proc. Natl. Acad. Sci. USA **71:** 2231–2235.

40. LUMPKIN, C.K.J., J.K. MCCLUNG, O.M. PEREIRA-SMITH & J.R. SMITH. 1986. Existence of high abundance antiproliferative mRNA's in senescent human diploid fibroblasts. Science **232:** 393–395.

41. SANDHU, A.K., K. HUBBARD, G.P. KAUR et al. 1994. Senescence of immortal human fibroblasts by the introduction of normal human chromosome 6. Proc. Natl. Acad. Sci. USA **91:** 5498–5502.

42. HENSLER, P.J., L.A. ANNAB J.C. BARRETT & O.M. PEREIRA-SMITH. 1994. A gene involved in control of human cellular senescence on human chromosome 1q. Mol. Cell. Biol. **14:** 2291–2297.

43. NING, Y., J.L. WEBER, A.M. KILLARY et al. 1991. Genetic analysis of indefinite division in human cells: evidence for a cell senescence-related gene(s) on human chromosome 4. Proc. Natl. Acad. Sci. USA **88:** 5635–5639.

44. VOJTA, P.J., P.A. FUTREAL, L.A. ANNAB et al. 1996. Evidence for two senescence loci on human chromosome 1. Genes, Chromosomes & Cancer **16:** 55–63.

45. UEJIMA, H., K. MITSUYA, H. KUGOH et al. 1995. Normal human chromosome 2 induces cellular senescence in the human cervical carcinoma cell line SiHa. Genes, Chromosomes & Cancer **14:** 120–127.

46. OHMURA, H., H. TAHARA, M. SUZUKI *et al.* 1995. Restoration of the cellular senescence program and repression of telomerase by human chromosome 3. Jpn. J. Cancer Res. **86:** 899–904.

47. FUJITA, T., J.L. MANDEL, T. SHIRASAWA *et al.* 1995. Isolation of cDNA clone encoding human homologue of senescence marker protein-30 (SMP30) and its location on the X chromosome. Biochim. Biophys. Acta **1263:** 249–252.

48. MORELLI, C., T. SHERRATT, C. TRABANELLI *et al.* 1997. Characterization of a 4-Mb region at chromosome 6q21 harboring a replicative senescence gene. Cancer Res. **57:** 4153–4157.

49. HARLEY, C.B., A.B. FUTCHER & C.W. GREIDER. 1990. Telomeres shorten during ageing of human fibroblasts. Nature **345:** 458–460.

50. BODNAR, A.G., M. OUELLETTE, M. FROLKIS *et al.* 1998. Extension of life-span by introduction of telomerase into normal human cells [see comments]. Science **279:** 349–352.

51. AMES, B.N. 1991. Oxygen radicals and 8-hydroxyguanine in DNA. Jpn. J. Cancer Res. **82:** 1460–1461.

52. GROLLMAN, A.P. & M. MORIYA. 1993. Mutagenesis by 8-oxoguanine: an enemy within. Trends Genet. **9:** 246–249.

53. HENLE, E.S. & S. LINN. 1997. Formation, prevention, and repair of DNA damage by iron/hydrogen peroxide. J. Biol. Chem. **272:** 19095–19098.

54. CHEN, Q., A. FISCHER, J.D. REAGAN *et al.* 1995. Oxidative DNA damage and senescence of human diploid fibroblast cells. Proc. Natl. Acad. Sci. USA **92:** 4337–4341.

55. ALLEN, R.G., M. TRESINI, B.P. KEOGH *et al.* 1999. Differences in electron transport potential, antioxidant defenses, and oxidant generation in young and senescent fetal lung fibroblasts (WI-38). J. Cell. Physiol. **180:** 114–122.

56. BALIN, A.K., B.P. GOODMAN, H. RASMUSSEN & V.J. CRISTOFALO. 1976. The effect of oxygen tension on the growth and metabolism of WI-38 cells. J. Cell. Physiol. **89:** 235–249.

57. PACKER, L. & K. FUEHR. 1977. Low oxygen concentration extends the lifespan of cultured human diploid cells. Nature **267:** 423–425.

58. CARNEY, J.M., R.P. STARKE, C.N. OLIVER *et al.* 1991. Reversal of age-related increase in brain protein oxidation, decrease in enzyme activity, and loss in temporal and spatial memory by chronic administration of the spin-trapping compound N-tert-butyl-alpha-phenylnitrone. Proc. Natl. Acad. Sci. USA **88:** 3633–3636.

59. CARNEY, J.M. & R.A. FLOYD. 1991. Protection against oxidative damage to CNS by alpha-phenyl-tert-butyl nitrone (PBN) and other spin-trapping agents: a novel series of nonlipid free radical scavengers. J. Mol. Neurosci. **3:** 47–57.

60. CHEN, Q. & B.N. AMES. 1994. Senescence-like growth arrest induced by hydrogen peroxide in human diploid fibroblast F65 cells. Proc. Natl. Acad. Sci. USA **91:** 4130–4134.

61. CHEN, Q.M., J.C. BARTHOLOMEW, J. CAMPISI *et al.* 1998. Molecular analysis of H2O2-induced senescent-like growth arrest in normal human fibroblasts: p53 and Rb control G1 arrest but not cell replication. Biochem. J. **332:** 43–50.

62. SHERWOOD, S.W., D. RUSH, J.L. ELLSWORTH & R.T. SCHIMKE. 1988. Defining cellular senescence in IMR-90 cells: a flow cytometric analysis. Proc. Natl. Acad. Sci. USA **85:** 9086–9090.

63. KULJU, K.S. & J.M. LEHMAN. 1995. Increased p53 protein associated with aging in human diploid fibroblasts. Exp. Cell Res. **217:** 336–345.

64. ATADJA, P., H. WONG, J. GARKAVTSEV *et al.* 1995. Increased activity of p53 in senescing fibroblasts. Proc. Natl. Acad. Sci. USA **92:** 8348–8352.

65. VAZIRI, H., M.D. WEST, R.C. ALLSOPP *et al.* 1997. ATM-dependent telomere loss in aging human diploid fibroblasts and DNA damage lead to the post-translational activation of p53 protein involving poly(ADP-ribose) polymerase. EMBO J. **16:** 6018–6033.

66. NODA, A., Y. NING, S.F. VENABLE *et al.* 1994. Cloning of senescent cell-derived inhibitors of DNA synthesis using an expression screen. Exp. Cell Res. **211:** 90–98.

67. ALCORTA, D.A., Y. XIONG, D. PHELPS *et al.* 1996. Involvement of the cyclin-dependent kinase inhibitor p16 (INK4a) in replicative senescence of normal human fibroblasts. Proc. Natl. Acad. Sci. USA **93:** 13742–13747.

68. STEIN, G.H., M. BEESON & L. GORDON. 1990. Failure to phosphorylate the retinoblastoma gene product in senescent human fibroblasts. Science **249:** 666–669.
69. ELLEDGE, S.J. & J.W. HARPER. 1994. Cdk inhibitors: on the threshold of checkpoints and development. Curr. Opin. Cell Biol. **6:** 847–852.
70. SHERR, C.J. & J.M. ROBERTS. 1995. Inhibitors of mammalian G1 cyclin-dependent kinases. Genes Dev. **9:** 1149–1163.
71. KASTAN, M.B. & S.J. KUERBITZ. 1993. Control of G1 arrest after DNA damage. Environ. Health Perspect. **5:** 55–58.
72. HARTWELL, L.H. & M.B. KASTAN. 1994. Cell cycle control and cancer. Science **266:** 1821–1828.
73. KASTAN, M.B. 1996. Signalling to p53: where does it all start? Bioessays **18:** 617–619.
74. STEIN, G.H., L.F. DRULLINGER, A. SOULARD & V. DULIC. 1999. Differential roles for cyclin-dependent kinase inhibitors p21 and p16 in the mechanisms of senescence and differentiation in human fibroblasts. Mol. Cell. Biol. **19:** 2109–2117.
75. CHEN, Q.M., K. PROWSE, V.C. TU & M. LINSKENS. 2000. Uncoupling the senescent phenotype from telomere shortening in oxidant-treated fibroblasts. Manuscript submitted.
76. LEE, A.C., B.E. FENSTER, H. ITO et al. 1999. Ras proteins induce senescence by altering the intracellular levels of reactive oxygen species. J. Biol. Chem. **274:** 7936–7940.
77. PRIVES, C. 1998. Signaling to p53: breaking the MDM2-p53 circuit. Cell **95:** 5–8.
78. FUCHS, S.Y., V. ADLER, M.R. PINCUS & Z. RONAI. 1998. MEKK1/JNK signaling stabilizes and activates p53. Proc. Natl. Acad. Sci. USA **95:** 10541–10546.
79. JOHNSON, M., D. DIMITROV, P.J. VOJTA et al. 1994. Evidence for a p53-independent pathway for upregulation of SDI1/CIP1/WAF1/p21 RNA in human cells. Mol. Carcinog. **11:** 59–64.
80. MACLEOD, K.F., N. SHERRY, G. HANNON et al. 1995. p53-dependent and independent expression of p21 during cell growth, differentiation, and DNA damage. Genes Dev. **9:** 935–944.
81. AHMED, N.U., M. UEDA & M. ICHIHASHI. 1999. Induced expression of p16 and p21 proteins in UVB-irradiated human epidermis and cultured keratinocytes. J. Dermatol. Sci. **19:** 175–181.
82. HARPER, J.W., G.R. ADAMI, N. WEI et al. 1993. The p21 Cdk-interacting protein Cip1 is a potent inhibitor of G1 cyclin-dependent kinases. Cell **75:** 805–816.
83. HARPER, J.W., S.J. ELLEDGE, K. KEYOMARSI et al. 1995. Inhibition of cyclin-dependent kinases by p21. Mol. Biol. Cell **6:** 387–400.
84. KOH, J., G.H. ENDERS, B.D. DYNLACHT & E. HARLOW. 1995. Tumour-derived p16 alleles encoding proteins defective in cell-cycle inhibition. Nature **375:** 506–510.
85. SHERR, C.J. 1996. Cancer cell cycles. Science **274:** 1672–1677.
86. WAGA, S., G.J. HANNON, D. BEACH & B. STILLMAN. 1994. The p21 inhibitor of cyclin-dependent kinases controls DNA replication by interaction with PCNA. Nature **369:** 574–578.
87. LI, R., S. WAGA, G.J. HANNON et al. 1994. Differential effects by the p21 CDK inhibitor on PCNA-dependent DNA replication and repair. Nature **371:** 534–537.
88. HARA, E., H. TSURUI, A. SHINOZAKI et al. 1991. Cooperative effect of antisense-Rb and antisense-p53 oligomers on the extension of life span in human diploid fibroblasts, TIG-1. Biochem. Biophys. Res Commun. **179:** 528–534.
89. SHAY, J.W., S.O. PEREIRA & W.E. WRIGHT. 1991. A role for both RB and p53 in the regulation of human cellular senescence. Exp. Cell Res. **196:** 33–39.
90. IDE, T., Y. TSUJI, S. ISHIBASHI & Y. MITSUI. 1983. Reinitiation of host DNA synthesis in senescent human diploid cells by infection with Simian virus 40. Exp. Cell Res. **143:** 343–349.
91. GORMAN, S.D. & V.J. CRISTOFALO. 1985. Reinitiation of cellular DNA synthesis in BrdU-selected nondividing senescent WI-38 cells by simian virus 40 infection. J. Cell. Physiol. **125:** 122–126.
92. BOYER, S.N., D.E. WAZER & V. BAND. 1996. E7 protein of human papilloma virus-16 induces degradation of retinoblastoma protein through the ubiquitin-proteasome pathway. Cancer Res. **56:** 4620–4624.

93. VOUSDEN, K. 1993. Interactions of human papillomavirus transforming proteins with the products of tumor suppressor genes. Faseb J. **7:** 872–879.

94. CHEN, Q.M., J.P. LIU & J.B. MERRETT. 1999. Apoptosis or senescence-like growth arrest: influence of cell cycle position, p53, p21 and bax in H_2O_2 response of normal human fibroblasts. Biochem. J. **347:** 543–551.

95. CHEN, Q.M., V.C. Tu & J.P. Liu. 1999. Induction of senescent associated beta galactosidase by H2O2 in human diploid fibroblasts with downregulated p53 and Rb. Biogerontology. In press.

96. CHEN, Q.M., V.C. Tu, J. CATANIA et al. 1999. Induction of senescent morphology by H2O2 Rb function and redistribution of focal adhesion proteins. Manuscript submitted.

97. SMITH, R.S., T.J. SMITH, T.M. BLIEDEN & R.P. PHIPPS. 1997. Fibroblasts as sentinel cells. Synthesis of chemokines and regulation of inflammation. Am. J. Pathol. **151:** 317–322.

98. SHEVITZ, J., C.S. JENKINS & V.B. HATCHER. 1986. Fibronectin synthesis and degradation in human fibroblasts with aging. Mech. Ageing Dev. **35:** 221–232.

99. KUMAZAKI, T., R.S. ROBETORYE, S.C. ROBETORYE & J.R. SMITH. 1991. Fibronectin expression increases during in vitro cellular senescence: correlation with increased cell area. Exp. Cell Res. **195:** 13–19.

100. GONOS, E.S., A. DERVENTZI, M. KVEIBORG et al. 1998. Cloning and identification of genes that associate with mammalian replicative senescence. Exp. Cell Res. **240:** 66–74.

101. MURANO, S., R. THWEATT, R.J. SHMOOKLER et al. 1991. Diverse gene sequences are overexpressed in werner syndrome fibroblasts undergoing premature replicative senescence. Mol. Cell. Biol. **11:** 3905–3914.

102. WEST, M.D., S.O. PEREIRA & J.R. SMITH. 1989. Replicative senescence of human skin fibroblasts correlates with a loss of regulation and overexpression of collagenase activity. Exp. Cell Res. **184:** 138–147.

103. MILLIS, A.J., M. HOYLE, H.M. McCUE & H. MARTINI. 1992. Differential expression of metalloproteinase and tissue inhibitor of metalloproteinase genes in aged human fibroblasts. Exp. Cell Res. **201:** 373–379.

104. WEST, M.D., J.W. SHAY, W.E. WRIGHT & M.H. LINSKENS. 1996. Altered expression of plasminogen activator and plasminogen activator inhibitor during cellular senescence. Exp. Gerontol. **31:** 175–193.

105. BIRKEDAL-HANSEN, H. 1995. Proteolytic remodeling of extracellular matrix. Curr. Opin. Cell Biol. **7:** 728–735.

106. SHAPIRO, S.D. 1998. Matrix metalloproteinase degradation of extracellular matrix: biological consequences. Curr. Opin. Cell Biol. **10:** 602–608.

107. RIFKIN, D.B. 1992. Plasminogen activator expression and matrix degradation. Matrix Suppl. **1:** 20–22.

108. CAMPISI, J. 1998. The role of cellular senescence in skin aging. J. Invest. Dermatol. Sym. Proc. **3:** 1–5.

109. CAMPISI, J. 1997. Aging and cancer: the double-edged sword of replicative senescence. J. Am. Geriatr. Soc. **45:** 482–488.

110. DUMONT, P., Q.M. CHEN, M. BURTON et al. 2000. Induction of senescent biomarkers by sublethal oxidative stress in normal human fibroblasts. Free Radical Biol. Med. **28:** 361–373.

111. LAKATTA, E.G., G. GERSTENBLITH & M.L. WEISFELDT. 1997. In Heart Disease: A Textbook of Cardiovascular Medicine. Vol. 2, 5th Ed. E. Braunwald, Ed. : 1687–1703. W.B. Saunders Co. Phildelphia, PA.

112. MILLER, R.A. 1996. The aging immune system: primer and prospectus. Science **273:** 70–74.

113. WANG, E. 1985. A 57,000-mol-wt protein uniquely present in nonproliferating cells and senescent human fibroblasts. J. Cell Biol. **100:** 545–551.

114. DULIC, V., L.F. DRULLINGER, E. LEES et al. 1993. Altered regulation of G1 cyclins in senescent human diploid fibroblasts: accumulation of inactive cyclin E-Cdk2 and cyclin D1-Cdk2 complexes. Proc. Natl. Acad. Sci. USA **90:** 11034–1038.

115. GIORDANO, T., D. KLEINSEK & D.N. FOSTER. 1989. Increase in abundance of a transcript hybridizing to elongation factor I alpha during cellular senescence and quiescence. Exp. Gerontol. **24:** 501–513.

116. LINSKENS, M.H., J. FENG, W.H. ANDREWS *et al.* 1995. Cataloging altered gene expression in young and senescent cells using enhanced differential display. Nucleic Acids Res. **23:** 3244–3251.
117. SAITO, H., J. PAPACONSTANTINOU, H. SATO & S. GOLDSTEIN. 1997. Regulation of a novel gene encoding a lysyl oxidase-related protein in cellular adhesion and senescence. J. Biol. Chem. **272:** 8157–81560.
118. MU, X.C., L. STAIANO-COICO & P.J. HIGGINS. 1998. Increased transcription and modified growth state-dependent expression of the plasminogen activator inhibitor type-1 gene characterize the senescent phenotype in human diploid fibroblasts. J. Cell. Physiol. **174:** 90–98.
119. BLUMENTHAL, E.J., A.C. MILLER, G.H. STEIN & A.M. MALKINSON. 1993. Serine/threonine protein kinases and calcium-dependent protease in senescent IMR-90 fibroblasts. Mech. Ageing Dev. **72:** 13–24.
120. DIPAOLO, B.R., R.J. PIGNOLO & V.J. CRISTOFALO. 1992. Overexpression of the two-chain form of cathepsin B in senescent WI-38 cells. Exp. Cell Res. **201:** 500–505.
121. LECKA-CZERNIK, B., C.K. LUMPKIN, JR. & S. GOLDSTEIN. 1995. An overexpressed gene transcript in senescent and quiescent human fibroblasts encoding a novel protein in the epidermal growth factor-like repeat family stimulates DNA synthesis. Mol. Cell. Biol. **15:** 120–128.
122. SHIBANUMA, M., J. MASHIMO, T. KUROKI & K. NOSE. 1994. Characterization of the TGF beta 1-inducible hic-5 gene that encodes a putative novel zinc finger protein and its possible involvement in cellular senescence. J. Biol. Chem. **269:** 26767–26774.
123. PIGNOLO, R.J., M.O. ROTENBERG, J.H. HORTON & V.J. CRISTOFALO. 1998. Senescent WI-38 fibroblasts overexpress LPC-1, a putative transmembrane shock protein. Exp. Cell Res. **240:** 305–311.
124. WISTROM, C. & B. VILLEPONTEAU. 1992. Cloning and expression of SAG: a novel marker of cellular senescence. Exp. Cell Res. **199:** 355–362.

Poly(ADP-Ribosyl)ation, Genomic Instability, and Longevity

ALEXANDER BÜRKLE[a]

Deutsches Krebsforschungszentrum, Abteilung Tumorvirologie, Heidelberg, Germany

ABSTRACT: Poly(ADP-ribosyl)ation is a DNA strandbreak-driven posttranslational modification of nuclear proteins that is catalyzed by poly(ADP-ribose) polymerase-1 (PARP-1), with NAD^+ serving as substrate. Recently, additional PARP isoforms were described that seem to account for a minor fraction of cellular poly(ADP-ribose) synthesis. We have previously described a correlation between poly(ADP-ribosyl)ation capacity of mononuclear leukocytes of various mammalian species and species-specific life span. Likewise, lymphoblastoid cell lines derived from human centenarians display a higher poly(ADP-ribosyl)ation capacity than do controls. At the functional level, recent data show that PARP-1 is a key regulator of alkylation-induced sister-chromatid exchange, imposing a negative control commensurate with the enzyme activity. PARP-1 activity may therefore be responsible for tuning the rate of genomic instability events that are provoked by the constant attack of endogenous and exogenous genotoxins to a level appropriate for the longevity potential of a given organism or species.

INTRODUCTION

A number of aging theories imply that genome damage is a driving force of the ageing process.[1] This postulate is based on the finding that genomes of organisms are continuously exposed to exogenous and endogenous DNA-damaging agents. Important endogenous sources of DNA damage are reactive oxygen intermediates (ROI), which not only are produced in mitochondria as by-products of electron transport in the respiratory chain, but also may arise by several other mechanisms: nitric oxide metabolites, lipid peroxidation products, and reducing sugar compounds. The continuous attack of the genome by DNA-damaging agents can lead to insidious, time-dependent damage accumulation if not fully counteracted by defense and repair systems. The resulting unrepaired, persistent DNA damage and/or the products of improper repair could well interfere with gene expression or DNA replication and ultimately disturb cellular and tissue homeostasis severely. This scenario implies that long-lived species or individuals should be endowed with more proficient defense/repair systems than short-lived ones, to allow for better maintenance of genomic integrity and stability over time.

[a]Present address and address for correspondence: Alexander Bürkle, MD, University of Newcastle upon Tyne, Department of Gerontology, IHE, Wolfson Research Centre, Newcastle General Hospital, Westgate Road, Newcastle upon Tyne, NE4 6BE, UK. Phone: +44-191-256-3324; fax: +44-191-219-5074.

Alexander.Buerkle@ncl.ac.uk

My group has focused on one of the immediate early cellular responses to oxidative and other types of DNA damage, that is, the catalytic activation of the 113-kDa nuclear enzyme poly(ADP-ribose) polymerase-1 (PARP-1)[2,3] (EC 2.4.2.30). This highly conserved enzyme is present in most eukaryotes, with the notable exception of the yeasts *Saccharomyces cerevisiae* and *S. pombe*. Upon binding to DNA strand-breaks, PARP-1 uses NAD^+ as substrate to covalently modify a number of nuclear proteins with poly(ADP-ribose). The DNA-binding domain of PARP-1, located at the aminoterminus, binds to single- or double-strand breaks in DNA. DNA binding is mediated via two zinc fingers and causes immediate and drastic activation of the catalytic center located in the carboxyterminal NAD^+-binding domain of the enzyme. Several "acceptor" proteins for the covalent modification with poly(ADP-ribose) have been identified, among which are histones and PARP-1 itself. In living cells, PARP-1 is actually the major acceptor, and this "automodification" occurs preferentially on a specific domain which is located between those for DNA binding and enzyme catalysis. The presence of poly(ADP-ribose) in the nuclei of living cells is restricted to the presence of DNA strand breaks, as this polymer displays a very short half-life under conditions of DNA breakage because of its rapid degradation by poly(ADP-ribose) glycohydrolase. Studies from many laboratories have demonstrated an important biological role of poly(ADP-ribosyl)ation in the recovery of proliferating cell cultures from DNA damage as induced, for example, by alkylation or ionizing radiation, and this has been linked mechanistically with involvement of PARP-1 in DNA base-excision repair.[4–6] More recently, analogous *in vivo* experiments have become feasible with the establishment of PARP-1–deficient mice.[7,8] Exposure of such mice to alkylating agents or γ-radiation resulted in a phenotype of greatly enhanced acute lethality,[7,8] the critical target tissue apparently being the intestinal mucosa, a highly proliferative tissue. The relevant molecular mechanism by which poly(ADP-ribosyl)ation exerts its effects on DNA repair has not yet been clarified. In addition, much evidence has accumulated for a role of PARP-1 in the maintenance of genomic stability under conditions of genotoxic stress (see below).

In striking contrast to the cytoprotective function of PARP-1 activation in proliferating cells exposed to low doses of genotoxic agents, there is a risk of cellular suicide by PARP-1 (over)activation due to depletion of NAD^+ pools. By using one of the available PARP-1 knockout mouse lines, this cell death mechanism was exemplified in nonroliferative cell types such as pancreatic islet cells[9,10] and neurons.[11,12]

Recently, the presence of several additional polypeptides synthesizing poly(ADP-ribose) was shown. These new members of the "PARP family" are collectively being held responsible for about 15% of the total poly(ADP-ribose) production in mouse fibroblasts.[13–17] Future work should clarify the respective functions of these novel poly(ADP-ribose) polymerases.

POLY(ADP-RIBOSYL)ATION AND LONGEVITY

Several years ago, we first addressed the question of a possible relation between poly(ADP-ribosyl)ation and longevity. We measured maximal poly(ADP-ribosyl)ation capacity in permeabilized mononuclear leukocytes from 13 mammalian species of different life span.[18] Our data yielded a strong positive correlation with the species-specific life span, in agreement with a previous study by Pero and colleagues[19] who had employed a different assay system. Interestingly, the differences in poly(ADP-ribosyl)ation capacity that we detected in the various species tested could not be explained by different levels of PARP protein, but rather by differences in specific enzyme activity. In contrast to the positive correlation with the longevity of species, there was a decline in PARP activity in mononuclear leukocytes with donor age, both in humans and in rats.[18]

More recently, we undertook a cooperative study with Schächter's group to analyze the relation between poly(ADP-ribosyl)ation and longevity in humans.[20] Our strategy was to see if human beings with proven longevity (i.e., centenarians), who often maintain a remarkably good status of general physical health until very old age, would also display differences in PARP activity as compared with individuals from the general population, the vast majority of whom will die well before reaching the age of 100 years. We therefore determined the poly(ADP-ribosyl)ation capacity of Epstein-Barr virus-immortalized lymphoblastoid cell lines that had been established from peripheral blood samples from a French population of centenarians and of adults aged 20–70 years (controls), respectively, and detected significantly higher values in the cell lines derived from centenarians compared with those from controls. *Specific* PARP activity was actually a more powerful parameter to discriminate between centenarian and control samples than was PARP activity per cell.[20] Again, this observation was suggestive of qualitative differences in PARP protein and provided further evidence for the notion that longevity is associated with a high poly(ADP-ribosyl)ation capacity.

Intriguingly, it was reported recently that telomeres in cells from PARP-1 knockout mice are significantly shorter than those from wild-type mice,[21] which may imply a possible role for PARP-1 in delaying the onset of replicative senescence. But clearly, more work is needed to determine whether or not the longevity-dependent variability in cellular poly(ADP-ribosyl)ation capacity that we described is functionally related to a control of telomere length by PARP-1.

ON THE BIOLOGIC FUNCTION OF POLY(ADP-RIBOSYL)ATION

To obtain deeper insight into the biologic function(s) of poly(ADP-ribosyl)ation, we developed over the last decade two complementary molecular genetic strategies: (1) suppression of cellular poly(ADP-ribose) formation by overexpression of a dominant negative PARP-1 mutant,[22] and (2) augmentation of poly(ADP-ribose) formation by overexpression of wild-type, active PARP-1.[23] This was done in stably transfected, SV40-transformed Chinese hamster cells in which transgene expression is either constitutive or conditional, in the latter case mediated by the glucocorticoid-inducible Mouse Mammary Tumor Virus (MMTV) promoter.

In the stably transfected cell clone COM3, conditional expression of the dominant negative PARP-1 mutant led to about 90% reduction of poly(ADP-ribose) levels after γ-irradiation and sensitized the cells to the cytotoxic effect of γ-irradiation or the monofunctional alkylating agent MNNG, without interfering with normal cellular proliferation or replication of a supertransfected polyomavirus replicon.[24] We concluded that the poly(ADP-ribose) synthesis that normally occurs after γ-irradia-

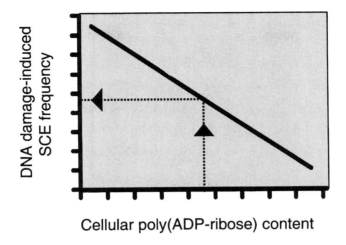

FIGURE 1. PARP-1 is a key regulator of alkylation-induced SCE formation. Previous results by other groups had shown that abrogation of poly(ADP-ribose) formation by pharmacologic inhibitors, nicotinamide (precursor) starvation, overexpression of dominant negative PARP-1, or PARP-1 gene knockout leads to potentiation of DNA damage-induced SCE formation. Our present results reveal that overexpression of active PARP-1, which is associated with poly(ADP-ribose) overaccumulation, leads to suppression of DNA damage-induced SCE formation. Therefore, it appears that PARP-1 is a key regulator of alkylation-induced SCE formation, imposing a control that is strictly negative and commensurate with the enzyme activity level. (NB: In the graph, this relationship has been depicted as a *straight line* for the sake of simplicity. However, the quantitative features of this relationship are unknown as yet.)

tion or MNNG treatment contributes to cellular recovery from the inflicted damage. It is well established that DNA damage not only does cause cytotoxicity but also is an important trigger of various manifestations of genomic instability, such as the formation of chromosomal breaks and aberrations, sister-chromatid exchanges (SCEs), gene rearrangements, deletions, or amplifications. We could show that overexpression of the dominant negative PARP-1 in COM3 cells potentiates carcinogen-induced amplification of chromosomally integrated SV40 DNA sequences[25] as well as alkylation-induced shuttle-vector mutagenesis.[26] These results provided evidence for a role of poly(ADP-ribosyl)ation in antagonizing the induction of genomic instability by DNA damage. Our results are perfectly in line with data from other groups who studied SCE, which is a very sensitive indicator of genetic damage and a key mechanism in the induction of genomic instability in general. These groups showed, by employing a variety of strategies,[7,8,27,28] that spontaneous and alkylation-induced SCEs are more frequent if cellular PARP(-1) activity is abrogated. However, it should be noted that PARP-1 abrogation experiments alone cannot clarify whether this enzyme is just one component that is constitutively necessary for an SCE-limiting or anti-recombinogenic effect to occur or whether it actually exerts a regulatory function (see below).

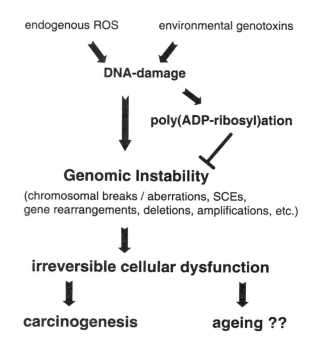

FIGURE 2. Model of the role of PARP-1 in the maintenance of genomic stability. *Arrows*, stimulation; *inverted T*, inhibition. For details see text.

In a complementary approach, we could show that overexpression of full-length, wild-type human PARP-1 in hamster cells leads to above-normal levels of poly(ADP-ribose) in intact cells.[23] To our surprise, this was not associated with improved survival after genotoxic stress, but rather with mild sensitization of the cells to the cytotoxic effects of MNNG, thus suggesting that the normal level of cellular poly(ADP-ribose) is optimal with respect to cell survival under conditions of genotoxic stress.[23] On the other hand, when studying SCE formation as a biologic endpoint, we recently observed that SCE induction by MNNG is strongly suppressed by inducible PARP-1 overexpression (Meyer *et al.*, submitted manuscript). Viewed together with the aforementioned, published data showing that abrogation of PARP(-1) activity leads to upregulation of carcinogen-induced SCEs,[7,8,27,28] our results reveal that PARP-1 is an important regulator of alkylation-induced SCE formation, imposing a control that is strictly negative and commensurate with the enzyme activity level (Fig. 1).

In conclusion, the picture is emerging that PARP-1 is not "merely" acting as a guardian of the genome,[29] but may actually be a key factor responsible for *tuning* the rate of genomic instability events that are provoked by the constant attack by endogenous and exogenous DNA-damaging agents, to a level that is just appropriate for the longevity potential of a given organism or species (Fig. 2).

ACKNOWLEDGMENTS

I thank Professor Harald zur Hausen for his continuous interest and support. I thank the former and present lab members who contributed to the work reviewed in this chapter (Drs. Sascha Beneke, Karlheinz Grube, Jan-Heiner Küpper, Ralph Meyer, Marie-Laure Muiras, Léon van Gool, as well as Mr. Marcus Müller) for their commitment and excellent work and many colleagues from other institutes for their willingness to collaborate. Our own work cited in this chapter was supported by grants from the Deutsche Forschungsgemeinschaft (Bu 698/2-1, -2, -3, and -4) and from the EU Commission (Concerted Action Programme on "Molecular gerontology: the identification of links between ageing and the onset of age-related diseases [MOLGERON]"; BMH1 CT94 1710).

REFERENCES

1. RATTAN, S.I.S. & O. TOUSSAINT, Eds. 1996. Molecular Gerontology – Research Status and Strategies. Plenum. New York.
2. ALTHAUS, F.R. & C. RICHTER. 1987. ADP-ribosylation of proteins: enzymology and biological significance. Mol. Biol. Biochem. Biophys. 37: 1–126.
3. BÜRKLE, A. 1996. Maintaining the stability of the genome. In Molecular Gerontology –Research Status and Strategies. S.I.S. Rattan & O. Toussaint, Eds. :25–36. Plenum. New York.
4. MOLINETE, M. et al. 1993. The poly(ADP-ribose) polymerase DNA-binding domain blocks alkylation-induced DNA repair synthesis in living cells. EMBO J. 12: 2109–2117.
5. TRUCCO, C. et al. 1998. DNA repair defect in poly(ADP-ribose) polymerase deficient cell lines. Nucl. Acids Res. 26: 2644–2649.
6. DANTZER, F. et al. 1999. Role of poly(ADP-ribose) polymerase in base excision repair. Biochimie 81: 69–75.
7. MÉNISSIER-DE MURCIA, J. et al. 1997. Requirement of poly(ADP-ribose) polymerase in recovery from DNA damage in mice and in cells. Proc. Natl. Acad. Sci. USA 94: 7303–7307.
8. WANG, Z.-Q. et al. 1997. PARP is important for genomic stability but dispensable in apoptosis. Genes Dev. 11: 2347–2358.
9. HELLER, B. et al. 1995. Inactivation of the poly(ADP-ribose) polymerase gene affects oxygen radical and nitric oxide toxicity in islet cells. J. Biol. Chem. 270: 11176–11180.
10. BURKART, V. et al. 1999. Mice lacking the poly(ADP-ribose) polymerase gene are resistant to pancreatic beta-cell destruction and diabetes development induced by streptozotocin. Nat. Med. 5: 314–319.
11. ELIASSON, M.J. et al. 1997. Poly(ADP-ribose) polymerase gene disruption renders mice resistant to cerebral ischemia. Nat. Med. 3: 1089–1095.
12. MANDIR, A.S. et al. 1999. Poly(ADP-ribose) polymerase activation mediates 1-methyl-4-phenyl-1,2,3,6,-tetrahydropyridine (MPTP)-induced parkinsonism. Proc. Natl. Acad. Sci. USA 96: 5774–5779.
13. SHIEH, W.M. et al. 1998. Poly(ADP-ribose) polymerase null mouse cells synthesize ADP-ribose polymers. J. Biol. Chem. 273: 30069–30072.
14. SMITH, S. et al. 1998. Tankyrase, a poly(ADP-ribose) polymerase at human telomeres. Science 282: 1484–1487.
15. AMÉ, J.C. et al. 1999. PARP-2, A novel mammalian DNA damage-dependent poly(ADP-ribose) polymerase. J. Biol. Chem. 274: 17860–17868.
16. BERGHAMMER, H. et al. 1999. pADPRT-2: a novel mammalian polymerizing(ADP-ribosyl)transferase gene related to truncated pADPRT homologues in plants and Caenorhabditis elegans. FEBS Lett. 449: 259–263.

17. JOHANSSON, M. 1999. A human poly(ADP-ribose) polymerase gene family (ADPRTL): cDNA cloning of two novel poly(ADP-ribose) polymerase homologues. Genomics **57:** 442–445.
18. GRUBE, K. & A. BÜRKLE. 1992. Poly(ADP-ribose) polymerase activity in mononuclear leukocytes of 13 mammalian species correlates with species-specific life span. Proc. Natl. Acad. Sci. USA **89:** 11759–11763.
19. PERO, R.W. *et al.* 1985. Gamma-radiation induced ADP-ribosyl transferase activity and mammalian longevity. Mutat Res. **142:** 69–73.
20. MUIRAS, M.-L. *et al.* 1998. Increased poly(ADP-ribose) polymerase activity in lymphoblastoid cell lines from centenarians. J. Mol. Med. **76:** 346–354.
21. D'ADDA DI FAGAGNA, F. *et al.* 1999. Functions of poly(ADP-ribose) polymerase in controlling telomere length and chromosomal stability. Nat. Genet. **23:** 76–80.
22. KÜPPER, J.-H. *et al.* 1990. Inhibition of poly(ADP-ribosyl)ation by overexpressing the poly(ADP-ribose) polymerase DNA-binding domain in mammalian cells. J. Biol. Chem. **265:** 18721–18724.
23. VAN GOOL, L. *et al.* 1997. Overexpression of human poly(ADP-ribose) polymerase in transfected hamster cells leads to increased poly(ADP-ribosyl)ation and cellular sensitization to γ irradiation. Eur. J. Biochem. **244:** 15–20.
24. KÜPPER, J.-H. *et al.* 1995. *Trans*-dominant inhibition of poly(ADP-ribosyl)ation sensitizes cells against γ-irradiation and *N*-methyl-*N'*-nitro-*N*-nitrosoguanidine but does not limit DNA replication of a polyomavirus replicon. Mol. Cell. Biol. **15:** 3154–3163.
25. KÜPPER, J.-H. *et al.* 1996. *Trans*-dominant inhibition of poly(ADP-ribosyl)ation potentiates carcinogen-induced gene amplification in SV40-transformed Chinese hamster cells. Cancer Res. **56:** 2715–2717.
26. TATSUMI-MIYAJIMA, J. *et al.* 1999. *Trans*-dominant inhibition of poly(ADP-ribosyl)ation potentiates alkylation-induced shuttle-vector mutagenesis in Chinese hamster cells. Mol. Cell. Biochem. **193:** 31–35.
27. MORGAN, W.F., & J.E. CLEAVER. 1982. 3-Aminobenzamide synergistically increases sister-chromatid exchanges in cells exposed to methyl methanosulfate but not to ultraviolet light. Mutat. Res. **104:** 361–366.
28. SCHREIBER, V. *et al.* 1995. A dominant-negative mutant of human poly(ADP-ribose) polymerase affects cell recovery, apoptosis, and sister chromatid exchange following DNA damage. Proc. Natl. Acad. Sci. USA **92:** 4753–4757.
29. JEGGO, P.A. 1998. DNA repair: PARP – another guardian angel? Curr. Biol. **8:** R49–R51.

Aging and Longevity

A Paradigm of Complementation between Homeostatic Mechanisms and Genetic Control?

CHARIKLIA PETROPOULOU, NIKI CHONDROGIANNI, DAVINA SIMÕES,
GEORGIA AGIOSTRATIDOU, NATALIA DROSOPOULOS, VIOLETTA KOTSOTA,
AND EFSTATHIOS S. GONOS[a]

Laboratory of Molecular and Cellular Aging, Institute of Biological Research and Biotechnology, National Hellenic Research Foundation, Athens 116 35, Greece

ABSTRACT: Aging is a universal and inevitable phenomenon that affects nearly all animal species. It can be considered the product of an interaction between genetic, environmental, and lifestyle factors, which in turn influence longevity that varies between and within species. It has been proposed not only that the aging process is under genetic control, but that it can also be considered a result of the failure of homeostasis due to the accumulation of damage. This review article discusses these issues, focusing on the function of genes that associate with aging and longevity, as well as on the molecular mechanisms that control cell survival and maintenance during aging.

INTRODUCTION: THEORIES OF AGING

Aging is a universal and inevitable phenomenon that affects nearly all animal species. It can be considered to be the product of an interaction between genetic, environmental, and lifestyle factors that in turn influence longevity, which varies between and within species. Evolutionary theories can provide an insight into understanding why aging occurs and the role genetic factors play in this process. Many theories have been proposed in an attempt to explain how aging occurs. However, only three can offer a plausible explanation about the cause of this phenomenon.[1]

The first of these theories proposes that aging is a necessary adaptation, because without aging the turnover and renewal of population would be impaired. Natural selection depends on the existence of novel variation among offspring, and this can only occur if a generation is replaced by a new generation so that a novel combination of genes can arise. Therefore, any species consisting of individuals that could survive indefinitely would become extinct and would be replaced by other species that were able to evolve. However, Darwinian natural selection can only operate if more offspring are produced than can survive to adulthood and reproduce, and selec-

[a]Address for correspondence: Efstathios S. Gonos, Ph.D., Laboratory of Molecular and Cellular Aging, Institute of Biological Research and Biotechnology, National Hellenic Research Foundation, 48 Vas. Constantinou Avenue, Athens 116 35, Greece. Phone: 30 1 7273756; fax: 30 1 7251827.

sgonos@eie.gr

tion depends on the greater chance of survival of specific genetic variants within a whole population.[2]

This misunderstanding of Darwin's natural selection, together with the fact that in a natural environment death occurs through accidents before aging is evident, led to the second theory of aging by Medawar.[3] He proposed that late-acting deleterious mutations, which are otherwise neutral, are responsible for aging. Because in general very few individuals survive to be old, most offspring are born to young parents. Therefore, a deleterious mutation that acted late in life would not be subject to natural selection. In other words, the accumulation of sufficient neutral mutations would give the individual, over long periods of time, a late-acting phenotype. On the same basis, Williams[4] proposed that there is an important class of pleiotropic mutations that have beneficial effects on young animals but harmful effects on older ones. Such mutations would be selected because most offspring are born to young parents, but their accumulation would result in aging and death. This theory of early- and late-acting mutations, however, can only apply to organisms that already have a finite life span and thus what it is actually trying to explain is the evolution of longevity.

The third theory, the disposable soma theory, links these evolutionary observations together.[5,6] It states first that in natural populations (except in humans) most deaths occur accidentally; second that long-term survival depends on somatic maintenance, and maintenance processes are energetically costly; and third that it is disadvantageous to invest a larger fraction of metabolic resources in long-term survival than is necessary for the organism to survive in reasonably good condition. The disposable soma theory suggests that aging is a result of a trade-off between late survival and early reproduction and to achieve maximum Darwinian fitness the optimum investment in maintenance results in finite survival. This results in genetic control of longevity, whereas the effects of environmental factors are also compatible and support the disposable soma theory.

The evolutionary theories described briefly above suggest that the aging process is not only under genetic control, but it can also be considered a consequence of the failure of homeostasis. This review article discusses all these issues, that is, the various genes involved in aging, longevity, and survival as well as the molecular mechanisms that are dysfunctioning during aging, such as protein degradation and failure of proteolysis.

GENES ASSOCIATED WITH SENESCENCE

Most cells in the human body, with the exception of cells in the germ line and some stem cell types, have limited replication potential. This is to some extent reflected by the limited replicative potential of primary human cells in tissue culture, defined as replicative senescence.[7] The *in vitro* culture of normal human diploid fibroblasts (HDF) has long served as a model system for studying this phenomenon. Replicative senescence causes the selective repression of a few positive-acting, growth regulatory genes that have been shown to be important for cell-cycle progression. Among the genes repressed in senescent cells are thymidine kinase, ornithine decarboxilase, thymidylated synthase, dihydrofolate reductase (DHFR), ribonucleotide reductase, proliferating cell nuclear antigen (PCNA) genes, and several histone genes. Other genes have been observed to be overexpressed during cellular aging and inhibition of cell proliferation. These include prohibitin, which has been shown to block DNA synthesis in proliferating normal fibroblasts and HeLa cells; interleukin-α, a known potent inhibitor of cell proliferation that has been found to be present at high levels in senescent human endothelial cells; tumor necrosis factor-α (TNF-α), which is differentially regulated with age; vimentin, which has been found to be overexpressed in senescent HDF; and mortalin gene (*mot-1*), which was found in normal cells and is capable of inducing cellular senescence in NIH3T3 cells.[8]

Different approaches have been applied to further investigate the genes involved in the regulation of cellular aging. Goldstein and colleagues[9,10] have cloned 18 genes that are overexpressed in Werner's syndrome fibroblasts and in normal HDF and are encoding predominantly ECM proteins. Linkens and colleagues,[11] when comparing different cell types and growth conditions by enhanced differential display, cloned 12 known and 11 novel genes. Additionally, we have cloned 8 genes, which encode for fibronectin, osteonectin, α1-procollagen, SM22, apolipoprotein J (apoJ), cytochrome-*c* oxidase, GTP-binding protein-α, and a novel gene, and which are overexpressed in senescent human and rat embryo fibroblasts and osteoblasts.[12] Continuation on this line of research showed that the RNA expression levels of fibronectin are directly associated with donor age, suggesting that fibronectin could be a marker of *in vivo* aging in fibroblasts.[13]

Studies of the retinoblastoma (RB) protein in senescent cells show that RB is present in the unphosphorylated form. One mechanism by which RB functions may be mediated is through its interaction with the transcription factor E2F. E2F has been shown to positively regulate many of the genes required for DNA synthesis, and it forms a complex with the unphosphorylated form of RB, leading to downregulation of E2F activity. Therefore, it is possible that RB may contribute to the arrest of senescent cells by remaining in its active, negative growth regulatory state. E2F repression appears to be caused by overexpression of p21 and possibly p16, which are inhibitors of cyclin-dependent protein kinases (cdks).[14] p21 binds and inactivates several different kinds of cyclin complexes at different points in the cell cycle.[15] Overexpression of p21 has been shown to contribute to growth arrest in senescent cells.[16] The p16 family is believed to inhibit cell growth in cells possessing functional RB, by not allowing cyclins to bind to cdk4 and cdk6.[17,18] Cells lacking functional RB contain high levels of p16 RNA and protein, suggesting a negative feedback loop by which RB might regulate p16 expression in late G1 phase.[19] Senescent cells express higher levels of p21 and p16 compared to young proliferating cells. The physiologic inducer of p16 has not yet been identified; however, p21 expression has been shown to be induced by p53 tumor suppressor gene and transforming growth factor-β (TGF-β), therefore responding to both intracellular and extracellular signals for cell cycle arrest.[20–22]

GENES ASSOCIATED WITH HUMAN LONGEVITY

During the last two decades, a great interest has been developed in identifying genes associated with longevity in humans. Centenarians, the best example of successful aging, have an important role to play in this kind of study.

One strong candidate for such a longevity gene for humans is the ε2 allele of the apoE gene. apoE has three common alleles, ε2, ε3, and ε4. In a study of the French population,[23] it was found that French centenarians had about half the proportion of ε4 and twice the proportion of ε2 of the whole population. This lower frequency of apoE-ε4 in centenarians is consistent with its risk factor status for heart disease, whereas the increase of ε2 provides evidence for a long-term protective effect of this allele when acting late in life. In addition, people carrying the ε4 allele are more susceptible to developing Alzheimer's disease (AD), whereas those with the ε2 allele have a decreased risk for AD. In the same study, it was also found that a variant of angiotensin-converting enzyme (ACE), which confers a predisposition to coronary heart disease, is surprisingly more frequent in centenarians, with a significant increase of the homozygous genotype (DD). Both these associations may provide evidence of pleiotropic age-dependent effects on longevity. Another study, in Italian centenarians,[24] showed that the frequency of apoB with low tandem repeats (apo B-VNTR) is 50% of that in young controls, but this difference was not found in French and Finnish centenarians. Another member of the apolipoprotein family genes, apoJ, was found to be overexpressed during senescence.[12] In this study, it was suggested that apoJ has a protective function, since only the surviving cells express the gene during apoptosis. Because apoJ is similar in terms of overall structure motifs, function, and tissue distribution with apoE, it should be further examined as a candidate gene associated with human longevity.

CELL DEATH AND CELL SURVIVAL SIGNALS

Every organism faces a number of factors during life that damage the cell either at the DNA or the protein level. Depending on the extent and location of damage, each cell activates the appropriate defense mechanism, such as apoptosis during DNA damage or chaperone activity when the proteins are under oxidation stress.

It is now well understood that activation of apoptosis and/or survival pathways is a multifactorial process. The critical cellular decision between life and death depends on the complex that predominates in the cell. A good example of this concept is the *bcl-2* family of genes. Overexpression of *bcl-2* protects against apoptosis caused by growth-factor withdrawal exposure to chemotherapeutics, γ- and ultraviolet irradiation or toxins, viral infection, and oncogene expression. Today, at least 15 Bcl-2 related proteins have been identified in mammalian cells. All family members share at least one of the four conserved motifs (BH1–BH4). However, some of these proteins promote apoptosis, whereas the rest have an anti-apoptotic function.[25] Yang and Korsmeyer[26] suggested that the relative levels of pro- and anti-apoptotic Bcl-2 family proteins function as a "rheostat" regulating the apoptotic threshold of the cell. The antagonistic function of these proteins is based on their ability to form heterodimers. The anti-apoptotic proteins Bcl-$_{xL}$ and Bcl-2 form heterodimers with pro-apoptotic Bax. An excess of Bax promotes death, but overexpression of *bcl-*$_{xL}$ or *bcl-2* neutralizes this effect.[27]

Another important issue is that homeostasis in mammalian cells depends on survival and death signals from the extracellular environment. Raff[28] suggested as an extreme view that cells in higher animals depend on signals from other cells to avoid killing themselves. Newly formed oligodendrocytes were tested both as purified populations and as single cells isolated from the rat optic nerve. They were dying rapidly with the characteristics of apoptosis when they were cultured in the absence of serum and exogenous signaling molecules. The cells could be saved for at least a few days by the addition of neighboring cells or the addition of growth factors present in the optic nerve. Extracellular signals that promote death or survival include peptide growth factors, cytokines, and interleukins. A well-studied receptor family is that of the tumor necrosis factor receptor (TNFR). It has been shown that the fate of the cell appears to be determined by the composition of the intracellular multiprotein aggregates that are formed after multimerization of the receptors. TNFR I induces cell death when it is associated with TRADD (TNFR-associated death domain-containing protein).[29] In contrast, cell survival is promoted when TNFR I recruits with TRAF2 (TNF receptor-associated factor 2) resulting in the activation of NF-κB.[30]

It is obvious that the fate of each cell in an organism depends on the extent of damage, cell type, stage of development, and differentiation. As the cell exists, apoptosis and survival are in a dynamic equilibrium that constantly changes. Alterations in the external environment of a cell are sensed by extracellular and intracellular receptors, and when the signal has been integrated complicated pathways with positive and negative regulators decide the fate of the cell. As an organism ages, the dynamic equilibrium between life and death is distorted.

As cells age they loose their ability to activate the stress response.[31] Accordingly, survival has necessitated the evolution of stress-induced networks to discover, control, and respond to environmental insults. Most of the newly synthesized proteins that are bound to fold in the cell require the interaction of a variety of proteins known as molecular chaperones. In stressed environments proteins can misfold or even unfold; a common feature of aging in the cell is the accumulation of damaged and abnormally folded proteins.[32] During exposure to stresses, molecular chaperones can prevent the irreversible aggregation of proteins—which tend to unfold—and keep them on the effective folding pathway. They also target non-native or aggregated proteins for degradation and removal from the cell.

Molecular chaperones comprise several families of proteins including the heat shock proteins (Hsps). The Hsps are a group of proteins that were originally identified on the basis of their increased expression in cells exposed to elevated temperatures and since then have been shown to be similarly induced by exposure of cells to a variety of stresses.[33] The genes encoding Hsps are highly conserved and are present even in the smallest of genomes. Many of these genes can be grouped into five families on the basis of sequence homology and characteristic molecular weight: Hsp100, Hsp90, Hsp70, Hsp60 (including the GroEL and the T-complex polypeptide 1 [TCP-1] ring complexes), and the small Hsps that range in size from 12 to 42 kDa. Typically, Hsps function as oligomers and can form large multimeric structures with several different chaperones, co-chaperones, and/or nucleotide exchange factors. All Hsps can prevent the aggregation of at least some unfolded or misfolded proteins. Interestingly, the heat-inducible Hsp90 is not required for the folding of most proteins but is required for a specific subset of proteins that have greater difficulty reaching their native conformations.[34] Unlike other families the small Hsps are only produced under conditions of stress. They function as molecular chaperones by preventing protein aggregation in an ATP-independent manner.[35] Although stress-induced expression of Hsps offers many advantages to the cell, experimental data suggest that Hsps can also have negative impacts on the cell's fitness. It seems that natural selection may have acted to balance the positive and negative impacts of Hsps in setting their level of expression.

The induction of Hsps in response to various stresses is dependent on the activation of a specific transcription factor, the heat shock factor (HSF) which binds to the heat shock element (HSE).[36] So far, little is known about the role and action of the HSF family members. It seems, though, that the diversity of HSFs provides specialization of stress signals a mechanism to differentially regulate the rate of transcription of heat shock genes and accommodates novel interactions with other regulatory factors, thus expanding the link between cell stress and other genetic networks.

Several groups have reported that Hsp72 and Hsp27 can prevent apoptotic cell death, whereas in opposite cases Hsp90 and Hsp70 can promote apoptotic pathways.[37,38] Although Hsp72 can suppress apoptosis induced by stress, which does not cause detectable protein damage, there is still a possibility that stress can cause minor, undetectable protein damage that is sufficient to activate apoptotic cascades. Experimental work suggests that overproduction of anti-apoptotic proteins can prevent cell death under many stressful conditions.[39]

A greatly impaired ability to induce Hsps in response to stresses is a hallmark of aging. As mammals age, damaged proteins progressively accumulate, and both the ability to express Hsps and stress tolerance diminishes. Decreased expression of Hsps upon stress could also be observed in cell cultures. Either cells isolated from aged organisms or cells isolated from young donors and aged in culture show reduced ability to respond to stressed conditions as well as higher rates of apoptotic death upon exposure to stress.[40] Furthermore, as mammals age, individual Hsps loose their ability to influence cellular sensitivity/resistance to stressful conditions that potentially lead to mortality. Clearly, the perspective of the Hsps influencing senescence could shed some light on the understanding of aging. An integrative hypothesis is that the diseases an organism acquires during its life, which involve protein damage and the accumulation of abnormally folded proteins, gradually accumulate and can lead to death. The proteins involved in several age-related disorders—including Alzheimer's and Parkinson's disease—have poorly structured native conformations that can be stabilized further by genetic mutations, leading them to adopt deleterious structures that would in turn result in the formation of insoluble, disease-causing protein aggregates.

PROTEIN DEGRADATION DURING AGING

During aging the cellular homeostatic mechanisms become progressively impaired, thus increasing the homeostatic imbalance followed by death. Among the cellular mechanisms that are considered to play a role in the maintenance of the cell are the proteolytic mechanisms.[2]

Although often neglected, protein degradation is a major intracellular function that is not only responsible for housekeeping of the cell but also for the regulation of important cellular functions, such as homeostasis and survival. There are two major categories of proteolysis: limited proteolysis, which is implicated in the maturation of precursors of proteins, and catabolic proteolysis, which is implicated in the elimination of normal or abnormal proteins. As far as catabolic proteolysis is concerned, the protein breakdown is carried out by several systems that may operate under different physiological conditions. The main cellular proteolytic system is the proteasome, which presents two "subtypes," the ATP-dependent and the ATP-independent proteolytic pathways.[41]

The proteasome is a large, multicatalytic protease found mainly in the cytoplasm and the nucleus. Proteasomes exist as particles of 20S (~700 kDa) and 26S (~2000 kDa). 20S is the catalytic core, and it is composed of 14 different but related subunits, encoded by α-type and β-type genes. The α-type subunits are regulatory, in contrast with the β-type subunits, which are catalytic. It has been proposed that the 20S "core" is active by itself, and it presents at least five different proteolytic activities. The 20S particle is responsible for ATP-independent proteolysis. The 20S core and the regulatory complex PA700, also known as 19S, compose the 26S proteasome. The 26S complex is mainly responsible for ATP-dependent (and ubiquitin-dependent) proteolysis. PA28, also known as 11S, is another activator of the 20S. Both regulators seem to play a pivotal role in the activation of the 20S proteasome.

It is well established that aged mammalian cells accumulate abnormal proteins. This accumulation is determined by their rates of formation and also by their rates of hydrolysis. One of the theories interpreting why this accumulation ocurs is that the proteolytic mechanisms are impaired during aging.[42] In that way, the proteolytic systems that have already been presented are examined in relation with the process of aging.

It is common to observe vesicles of lipofuscin in aged cells. Lipofuscin is a brown-yellow, electron-dense, autofluorescent material that accumulates progressively over time in lysosomes of postmitotic cells. One of the main explanations that has been suggested for this increase is that this accumulation is an effect of aging caused by a decline in intralysosomal degradation and/or a decrease in exocytosis as well as a general dysfunction of the lysosomal proteolytic pathway.[43] Studies of the age-dependent change in lysosomal enzyme activities showed a significant increase in the activity of most of these enzymes. The subcellular distribution study of those enzymes in aged cells revealed their increased activity in the cytosolic fraction, thus suggesting leakage and general lysosomal instability in aged cells.[44] Finally, studies in senescent cells have revealed a clear reduction of lysosomal-mediated protein degradation, thus showing the relation of the lysosomes with the process of aging.[45]

Protein oxidation *in vivo* is a natural consequence of aerobic life. This phenomenon is more severe in aged cells, thus presenting aggregates of oxidized proteins. Mammalian cells exhibit only limited direct repair mechanisms and most oxidized proteins undergo selective proteolysis, through the 20S proteasome.[46] However, the accumulation of oxidized proteins in the aged cells leads to the hypothesis that the proteolytic activity against oxidized and generally abnormal proteins decreases during senescence or aging. Studies of the specific, proteolytic activities of the proteasome have revealed their decrease upon aging. More precisely, it was found that the chymotrypsin-like and the peptidylglutamyl peptide-hydrolizing activity exhibits a clear decrease with age.[47,48] As far as the 26S proteasome is concerned, the obtained results propose a decline in the activities of the protease and not a dysregulation of the system of ubiquitination. These results suggest a decrease in the activity of multicatalytic protease (20S and 26S) during aging, thus explaining the accumulation of impaired and altered proteins during this process.

Presently, it is clear that the proteolytic mechanisms (selective or nonselective) are of great importance for physiological cellular function. The understanding of these mechanisms, of their regulation and their actions in relationship with multiple cellular phenomena among them the process of aging, consist important topics deserving systematic study.

CONCLUSIONS

In this review article, the process of aging from a genetic control aspect as well as the failure of the cellular homeostatic mechanisms that accompany the phenomenon were discussed. Is it possible, however, with the knowledge available so far, to make a distinction between these two components of aging in terms of primary cause? The model that is universally used by the majority of labs to study aging is the *in vitro* proliferation of cells. However, the Hayflick phenomenon observed in cell cultures cannot be considered equivalent but only relevant to aging. The discovery that DNA synthesis inhibitors are activated during cell senescence is in favor of the programmed genetic control, although this inhibition can also be seen as a secondary effect of other more crucial events causing *in vitro* aging. Somatic cells progressively lose telomeric DNA at the ends of the chromosome arms, as the organism ages. A recent report by Shiels and colleagues[49] on Dolly, the world-famous cloned sheep, showed that, although Dolly is physically healthy, its telomeres are shorter than those of age-matched controls (3 years old). In fact, this shorter length is consistent with the age of her progenitor mammary tissue (6 years old). It cannot be concluded that physiological age is represented by telomeres' length, since this could reflect only one among hundreds of parameters. However, it is a case of genetic aging that needs to be investigated further. The process of aging is certainly a complicated phenomenon, regulated by many factors.

Although in biology aging and death should not be confused, these are the only events that are bound to occur from the moment of conception. A victory over death can be achieved by supernaturally detaching the soma from its soul. The soul does not decay. The same applies to the genetic message each one carries. It holds the key to eternal life passing from one generation to the next, leaving the decaying body behind.

> ... the mortal nature ever seeks, as best it can, to be immortal. In one way only can it succeed, and that is by generation ... . Every mortal thing is preserved not by keeping it exactly the same for ever, like the divine, but by replacing what goes off or is antiquated with something fresh, in the semblance of the original. Through this device, a mortal thing partakes of immortality, both in its body and in all other as respects. So do not wonder if everything naturally values its own offshoot; since all are beset by this eagerness and this love with a view to immortality. (Symposium, Plato)

ACKNOWLEDGMENTS

The experimental work presented in this review article was supported by the following grants: European Union Biomed-2 (Genage; BMH4-CT98-3149) and Greek-Italian Collaboration Grant (Aging & Longevity; 6787/19-2-1997).

REFERENCES

1. KIRKWOOD, T.B.L. 1991. Genetic basis of limited cell proliferation. Mut. Res. **256:** 323–328.
2. HOLLIDAY, R. 1996. Understanding Ageing. Cambridge University Press. Cambridge, U.K.
3. MEDAWAR, P.B. 1952. An Unsolved Problem of Biology. Lewis. London.
4. WILLIAMS, G.C. 1957. Pleiotropy, natural selection and the evolution of senescence. Evolution **11:** 398–411.
5. KIRKWOOD, T.B.L. & R. HOLLIDAY. 1979. The evolution of aging and longevity. Proc. R. Soc. London B **205:** 531–546.
6. KIRKWOOD, T.B.L. 1977. Evolution of ageing. Nature **270:** 301–304.
7. HAYFLICK, L. 1965. The limited *in vivo* lifetime of human diploid cell strains. Exp. Cell Res. **25:** 585–621.
8. DERVENTZI, A., S.I.S. RATTAN & E.S. GONOS. 1996. Molecular links between cellular mortality and immortality. Anticancer Res. **16:** 2901–2910.
9. MURANO, S. *et al.* 1991. Diverse gene sequences are overexpressed in Werner syndrome fibroblasts undergoing premature replicative senescence. Mol. Cell. Biol. **11:** 3905–3914.
10. LECKA-CZERNIK, B. *et al.* 1996. Identification of gene sequences overexpressed in senescent and Werner syndrome human fibroblasts. Exp. Gerontol. **31:** 159–174.
11. LINKENS, M.H.K. *et al.* 1995. Cataloging altered gene expression in young and senescent cells using enhanced differential display. Nucl. Acids Res. **23:** 3244–3251.
12. GONOS, E.S., A. DERVENTZI, M. KVEIBORG, *et al.* 1998. Cloning and identification of genes that associate with mammalian replicative senescence. Exp. Cell Res. **240:** 66–74.
13. MONDELLO, C., C. PETROPOULOU, D. MONTI, *et al.* 1999. Telomere length in fibroblasts and blood cells from healthy centenarians. Exp. Cell Res. **248:** 234–242.
14. CAMPISI, J. 1997. The biology of replicative senescence. Eur. J. Cancer **33:** 703–709.
15. XIONG, Y. *et al.* 1993. p21 is a universal inhibitor of cyclin kinases. Nature **366:** 701–704.
16. NODA, A. *et al.* 1994. Cloning of senescent cell-derived inhibitors of DNA synthesis using an expression screen. Exp. Cell Res. **221:** 90–98.
17. GUAN, K. *et al.* 1994. Growth suppression by p18, a p16INK4/MTS1- and p14INK4B/MTS2-related CDK6 inhibitor, correlates with wild-type pRb function. Genes Dev. **8:** 2939–2952.
18. SERRANO, M. *et al.* 1993. A new regulatory motif in cell-cycle control causing specific inhibition of cyclin D/CDK4. Nature **366:** 704–707.
19. HARA, E. *et al.* 1996. Regulation of p16CDKN2 expression and its implications for cell immortalization and senescence. Mol. Cell Biol. **16:** 859–867.
20. GRANA, X. & E.P. REDDY. 1995. Cell cycle control in mammalian cells: role of cyclins, cyclin dependent kinases (CDKs), growth suppressor genes and cyclin-dependent kinase inhibitors (CKIs). Oncogene **11:** 211–219.
21. LIEBERMANN, D.A. *et al.* 1995. Molecular controls of growth arrest and apoptosis: p53-dependent and independent pathways. Oncogene **11:** 199–210.
22. DATTO, M.B. *et al.* 1995. Transforming growth factor beta induces the cyclin-dependent kinase inhibitor p21 through a p53-independent mechanism. Proc. Natl. Acad. Sci. USA **92:** 5545–5549.
23. SCHACHTER, F. *et al.* 1994. Genetic associations with human longevity at the APOE and ACE loci. Nature Genet. **6:** 29–32.
24. DE BENEDICTIS, G. *et al.* 1997. DNA multiallelic systems reveal gene/longevity associations not detected by diallelic systems. The APOB locus. Hum. Genet. **99:** 312–318.
25. ADAMS, J.M. & S. CORY. 1998. The Bcl-2 protein family: arbiters of cell survival. Science **281:** 1322–1326.
26. YANG, E. & S.J. KORSEMAYER. 1996. Molecular thanatoptosis: a discourse on the Bcl-2 family and cell death. Blood **88:** 386–401.

27. DRAGOVICH, T. *et al.* 1998. Signal transduction pathways that regulate cell survival and cell death. Oncogene **17:** 3207–3213.
28. RAFF, M. 1992. Social control on cell survival and cell death. Nature **356:** 397–400.
29. HSU, H. *et al.* 1996. TRADD-TRAF2 and TRADD-FADD interactions define two distinct TNF receptor 1 signal transduction pathways. Cell **84:** 299–308.
30. ARCH, R.H. *et al.* 1998. Tumor necrosis factor receptor-associated factors (TRAFs)-a family of adapter proteins that regulates life and death. Genes Dev. **12:** 2821–2830.
31. LIU, A.Y. *et al.* 1996. Attenuated heat shock transcriptional response in aging: molecular mechanism and implication in the biology of aging. EXS **77:** 393–408.
32. ELLIS, R.J. & F.U. HARTL. 1996. Protein folding in the cell: competing models of chaperonin function. FASEB J. **10:** 20–26.
33. YELLON, D.M. & D.S. LATCHMAN. 1992. Stress proteins and myocardial protection. J. Mol. Cell. Cardiol. **24:** 113–124.
34. NATHAN, D.F., M.H. VOS & S. LINDQUIST. 1997. *In vivo* functions of the *Saccharomyces cerevisiae* Hsp90 chaperone. Proc. Natl. Acad. Sci. USA **94:** 12949–12956.
35. EHRNSPRGER, M., M. GAESTREL & J. BUCHNER. 1998. Structure and function of small heat-shock proteins. *In* Molecular Chaperones in the Life Cycle of Proteins. A.L. Fink & Y. Goto, Eds.: 533–575. Dekker. New York.
36. LIS, J. & C. WU. 1993. Protein traffic on the heat shock promoter: parking, stalling and trucking along. Cell **74:** 1–4.
37. GALEA-LAURI, J. *et al.* 1996. The role of the 90-kDa heat shock protein in cell cycle control and differentiation of the monoblastoid cell line U937. Exp. Cell Res. **226:** 243–254.
38. LIOSSIS, S.N. *et al.* 1997. Overexpression of the heat shock protein 70 enhances the TCR/CD3-and Fas/Apo-1/CD45-mediated apoptotic cell death in Jurkat T cells. J. Immunol. **158:** 5668–5675.
39. MIGNOTTE, B. & J.L. VAYSSIERE. 1998. Mitochondria and apoptosis. Eur. J. Biochem. **252:** 1–15.
40. HEYDARI, A.R. *et al.* 1994. Hsp70 and aging. Experimentia **50:** 1092–1098.
41. COUX, O. *et al.* 1996. Structure and functions of the 20S and the 26S proteasomes. Annu. Rev. Biochem. **65:** 801–847.
42. ROSENBERGER, R.F. 1991. Senescence and the accumulation of abnormal proteins. Mut. Res. **256:** 255–262.
43. TERMAN, A. & U.T. BRUNK. 1998. Lipofuscin: mechanisms of formation and increase with age. APMIS **106:** 265–276.
44. NAKAMURA, Y. *et al.* 1998. Lysosome instability in aged rat brain. Neurosci. Lett. **13:** 215–220.
45. CUERVO, A.M. & J.F. DICE. 1998. How do intracellular proteolytic systems change with age? Front. Biochim. **1:** 25–43.
46. GRUNE, T., T. REINHECKEL & K.J.A. DAVIES. 1997. Degradation of oxidized proteins in mammalian cells. FASEB J. **11:** 526–534.
47. CONCONI, M. *et al.* 1996. Age-related decline of rat liver multicatalytic proteinase activity and protection from oxidative inactivation by heat-shock protein 90. Arch. Bioch. Biophys. **331:** 232–240.
48. HAYASHI, T. & S. GOTO. 1998. Age-related changes in the 20S and the 26S proteasome activities in the liver of male F344 rats. Mech. Ageing Dev. **1:** 55–66.
49. SHIELS, P.G. *et al.* 1999. Analysis of telomere lengths in cloned sheep. Nature **399:** 316–317.

Protein Degradation by the Proteasome and Its Implications in Aging

BERTRAND FRIGUET,[a] ANNE-LAURE BULTEAU, NIKI CHONDROGIANNI, MARIANGELA CONCONI, AND ISABELLE PETROPOULOS

Laboratoire de Biologie et Biochimie Cellulaire du Vieillissement, Université Denis Diderot-Paris 7, 75251 Paris Cedex 05, France

ABSTRACT: Free radical damage to cellular components is believed to contribute to the aging process. Studies on proteins have shown both an age-related decline in several enzyme activities and an age-related accumulation of oxidized forms of protein. Oxidized forms of protein are generally degraded more rapidly than their native counterparts. Indeed, the normal functions of the cell involve the regular elimination of these altered molecules. The proteasome, a multienzymatic proteolytic complex, is the major enzymatic system in charge of cellular "cleansing" and plays a key role in the degradation of damaged proteins. Consequently, proteasome function is very important in controlling the level of altered proteins in eukaryotic cells. Because the steady-state level of oxidized protein reflects the balance between the rate of protein oxidation and the rate of protein degradation, age-related accumulation of altered protein can be due to an increase of free radical–mediated damage, a loss of protease activity, or the combination of both mechanisms. One of the hypotheses put forward to explain the accumulation of altered proteins is the decrease of proteasome activity with age. In this paper, the importance of oxidative damage to proteins and that of their elimination by the proteasome are first described. Then, evidence for a decline of proteasome activity upon aging and upon oxidative stress is provided by studies from our and other laboratories.

The free radical theory of aging, as first proposed by Harman, suggests that the continuous oxidative damage that occurs to cellular components over an organism's life span is a critical factor in the aging process.[1,2] According to this theory, cellular aging is dependent in part on oxidant generation, antioxidant defences, and efficiency of the systems responsible for the elimination of oxidatively damaged cellular components. We are particularly interested in the last point because it addresses the cellular maintenance systems that are believed to be critical in aging as emphasized by the disposable soma theory of aging.[3] The focus of this review will be the implication of the proteasome, the main intracellular proteolytic system, in the degradation of altered proteins in cellular aging.

[a]Address for correspondence: Bertrand Friguet, Laboratoire de Biologie et Biochimie Cellulaire du Vieillissement, Université Denis Diderot-Paris 7, CC 7128, 2 place Jussieu, 75251 Paris Cedex 05, France. Phone and fax: 33-1-44 27 82 34.

bfriguet@paris7.jussieu.fr

MACROMOLECULAR DAMAGES AND THE
FREE RADICAL THEORY OF AGING

Oxygen free radicals are formed during aerobic respiration by the mitochondrial electron transport chain, resulting in the production of the unstable superoxide radical ($O_2^{-\cdot}$). Metal traces are then able to catalyze the formation of highly reactive O_2 free radical species such as the hydroxyl radical that can initiate chain reactions and lead to the production of stable damaged molecules.[4]

The already old free radical theory has been strengthened by the recent genetic studies on *Caenorhabditis elegans* and on *Drosophila melanogaster*.[5] Transgenic flies that overexpress catalase and superoxide dismutase, two antioxidant enzymes, exhibit an extension of life-span and a decrease in damaged protein content as compared to wild type.[6] The life-span extension mutants of *C. elegans*, *age-1*[7] and *daf-2*,[8] have shown elevated activities of antioxidant enzymes, Cu, Zn-superoxide dismutase, and catalase, which confer resistance to hydrogen peroxide and paraquat.[9,10]

Lipids, nucleic acids, and proteins are the main targets for free radical attack. Whereas oxidized nucleic acids can be repaired by specific enzymatic systems, very few enzymes are able to reverse the oxidative modifications on proteins. The two known enzymes are the thioredoxin/thioredoxin reductase system that can reduce the oxidized disulfide bonds[11] and the peptide methionine sulfoxide reductase that can reverse methionine oxidation on peptides and proteins.[12,13]

Damage to macromolecules, and in particular protein, has been implicated in the cellular degeneration that occurs during the aging process and is corroborated by the accumulation of oxidative end-products over time. One of these products is lipofuscin, a yellow-brown intracellular fluorescent pigment that accumulates in postmitotic cells as they age.[14–16] This pigment is constituted of cross-linked lipids and proteins.[17] Experiments with cultured cells have established roles for both oxidative damage and lysosomal protein degradation in the accumulation of lipofuscin.[18,19] Antioxidants inhibit lipofuscin formation, whereas lysosomal protease inhibitors increase it.[2]

Proteins subjected to oxidation are more heat labile and less active than their native counterparts.[20] In proteins, almost all amino acid residues are subject to attack by hydroxyl radicals. Met, Cys, Tyr, Phe, Trp, and His residues are the most susceptible, whereas Pro, Arg, and Lys are particularly sensitive to oxidation by metal-catalyzed oxidation and are converted to carbonyl derivatives.[4] Protein carbonyl content can be used to monitor protein damage through highly sensitive procedures that have been developed for detection and quantification of protein carbonyl groups, a widely used method being based on the reactivity of protein carbonyls with DNPH (2, 4-dinitrophenylhydrazine).[21,22] These methods are useful for revealing the presence of carbonyls generated not only by direct oxidation of side chains of amino acid residues but also upon glycation reaction and conjugation with aldehydes that are produced during lipid peroxidation.[4]

There is a significant increase in the amount of oxidized protein as a function of age in rat hepatocytes,[23] in human erythrocytes,[20] in human brain,[24] and in whole body flies.[25] Protein carbonyl content in cultured human dermal fibroblasts increases exponentially with the age of fibroblast donor.[4] More recently, we have shown that in skin epidermis as well as in keratinocytes in culture, there is an age-related increase of protein carbonyl content. Using specific antibodies, we have demonstrated that the carbonyl derivatives were generated not only by direct oxidation of amino acids but also by protein glycation and by reaction with lipid peroxidation products such as 4-hydroxy-2-nonenal.[26]

The substantial reduction in the activities of important enzymes and accumulation of damaged proteins that occur during aging are thought to alter the cellular integrity. Carney *et al.*[27] have shown in gerbils that there is a direct correlation between the age-related spatial and temporal memory loss and the accumulation of oxidized proteins in brain. Moreover, accumulation of oxidized proteins is also observed in numerous age-related diseases including Alzheimer's disease, amyotrophic lateral sclerosis, cataractogenesis, rheumatoid arthritis, progeria, and Werner's syndrome.[28]

One of the outstanding questions is whether all cellular proteins are equally sensitive to oxidation or are these modifications selective, affecting only a few target proteins. The activities of both enzymes, glutamine synthetase and glucose-6-phosphate deshydrogenase, decline with age, and it has been suggested by the authors that this decrease is directly correlated with the oxidative modifications of these two enzymes.[23] Recently, Sohal and co-workers have identified in aging flies two mitochondrial proteins that are specific targets of oxidative damages, aconitase and adenine nucleotide translocase.[29,30]

ELIMINATION OF OXIDATIVELY DAMAGED PROTEINS BY THE PROTEASOME

Carbonyl formation is irreversible, and thus the only way to eliminate altered proteins is the degradation process. Indeed, it has been shown that oxidatively modified proteins are more susceptible to proteolysis than their native counterparts.[31–34] J. Rivett has demonstrated that oxidatively modified glutamine synthetase can be selectively degraded by a nonlysosomal pathway. This and other studies performed on red blood cells,[35] on cultured liver epithelial cells,[36] and on hematopoietic cells[37] have shown that degradation of oxidized proteins is achieved by a proteasome-dependent, but ATP/ubiquitin-independent pathway. However, it appears that degradation of oxidatively modified proteins in lens epithelial cells requires both ATP/ubiquitin-independent (20S proteasome) and ATP/ubiquitin-stimulated (26S proteasome) pathways to avoid cataract formation.[38,39] More recently, the 20S proteasome has been shown to degrade nuclear oxidized protein by an efficient interaction between the 20S proteasome and the poly-ADP ribose polymerase.[40] In summary, the proteasome seems to be the main actor for degradation of abnormal intracellular cytosolic and nuclear proteins.

Protein degradation is a major intracellular complex function that allows protein turnover in all organisms. This mechanism is a highly selective process and plays a pivotal role in the regulation of important cellular functions such as homeostasis and survival. There are two major categories of proteolysis: the lysosomal pathway and the nonlysosomal or cytosolic pathway. Lysosomal protein degradation has been implicated in the maturation and activation of protein precursors as well as in the breakdown of endocytosed proteins and degradation of intracellular proteins in response to starvation.[41–43] In general, lysosomal proteolysis is mainly a nonselective proteolytic mechanism, although it is important for the cellular maintenance. The cytosolic pathway has been implicated in intracellular protein turnover and in elimination of abnormal and damaged proteins and is subdivided mainly into two proteolytic systems. The calpains (calcium-dependent neutral proteases) are Ca^{2+}-dependent and ATP-independent endoproteases that preferentially degrade membrane and cytoskeletal proteins as well as certain transcription factors.[44,45] Moreover, the proteasome carries out the ATP-dependent proteolytic pathway that catalyzes the degradation of transcriptional and cell cycle regulatory proteins. This proteolytic complex also plays a pivotal role in the processing of precursor proteins, such as the NF-κB transcription factor[46] and in antigen presentation.[47] It can also function without ATP and ensures the elimination of abnormal proteins.

The proteasome is a large multicatalytic protease mainly found in the cytoplasm and the nucleus of eukaryotic cells, but also present in archebacteria.[48] Proteasomes exist as 20S (~700 kDa) and 26S (~2000 kDa) particles. The 20S particle is the catalytic core of the proteasome and is composed of 14 different but related subunits, encoded by α-type and β-type genes. The α-type subunits are devoid of proteolytic activity and therefore believed to be regulatory subunits. In contrast, the β-type subunits are carrying the catalytic activities. The 20S proteasome is a cylinder-shaped particle appearing as a stack of four rings, two outer α-rings and two inner β-rings, that are associated in the order αββα.[49,50] Every α-ring and every β-ring is composed by seven different α and β subunits, respectively.[51–53] This structure sequesters the catalytic sites from potential protein substrates, suggesting that the entrance of the substrate into the proteasome is a highly regulated process.

The 20S proteasome acts independently of the presence of ATP or ubiquitin and is characterized by three main proteolytic activities with distinct specificities against short synthetic peptides: the trypsin-like activity (that cleaves after basic residues like arginine or lysine), the chymotrypsin-like activity (that cleaves after large hydrophobic residues like tyrosine or phenylalanine), and the peptidylglutamyl-peptide hydrolase activity (that cleaves after acidic residues like glutamate).[54,55] Apart from these specific activities, two additional endopeptidase activities appear to be present: the BrAAP activity (that cleaves after branched-chain amino acids) and the SNAAP activity (that cleaves after small neutral amino acids).[56] The identity and nature of these catalytic sites is not yet completely understood, but the crystal structure of the yeast proteasome[57] and studies with proteasome mutants[58,59] have shown that one β subunit is responsible for one type of catalytic activity.[60] Moreover, these β subunits have a threonine in their NH_2-terminal position. The hydroxyl group acts as a nucleophile in the active site and defines a new type of protease as compared with serine or cysteine proteases.[57,58] Degradation mediated by proteasome is processive; protein substrates are cleaved at multiples sites and generate short peptides, which are rapidly hydrolyzed by cytoplasmic exopeptidases.[61]

In vitro, the 20S proteasome can be converted from a latent to an active form by various nonphysiological treatments such as incubation with polylysine or fatty acids, heating, freezing, addition of a low concentration of SDS, and dialysis against water.[62–65] The activation presumably results from conformational rearrangements of this multisubunit protein complex, which may modulate its activity by opening the central cavity.[66–68] Such *in vitro* effects are unlikely to have direct physiological significance but probably mimic proteasome activation by regulator proteins.[69] The activity of the proteasome can be modulated by its catalytic subunits composition. Three β subunits (LMP7, LMP2, and MECL-1) are induced in cells upon treatment with γ-interferon[70] and replaced the three catalytic subunits X, Y, and Z, respectively, in newly synthesized proteasome. This modified proteasome exhibits a new catalytic profile and elicits differences in the pattern of peptides products.[71]

The 20S proteolytic core and the regulatory complex PA700, also known as the 19S regulator, associate to form the 26S proteasome. PA700 contains 20 different subunits. Some of these subunits are ATPases (at least six) and others are non-ATPases (at least 14) that play an important role in the function of the 26S proteasome.[72] The 19S regulatory particle is arranged in two structural complexes, the lid, required for ubiquitin-dependent degradation and the base, which binds to the 20S proteasome α ring.[73] In contrast with the 20S proteasome, the 26S complex acts in an ATP- and (but not always) ubiquitin-dependent manner. Ubiquitin is a small polypeptide, consisting of 76 amino acids, that is present in apparently all eucaryotic cells and is highly conserved from yeast to human.

PA28, also known as the 11S regulator, is another activator of the 20S proteasome.[74,75] In contrast with the assembly of the 20S proteasome and the 19S regulator, which occurs in an ATP-dependent manner, the assembly of 20S core with the 11S regulator does not require ATP hydrolysis. PA28 is composed of two related proteins, the PA28α and the PA28β that assemble into a heterohexameric complex, the PA28(αβ)$_3$.[72] Although this complex greatly stimulates multiple peptidase activities of the 20S, this activator does not seem to play a central role in the initial cleavage of protein substrates, but may act by stimulating the degradation of polypeptides in intermediate size that are generated by the 26S proteasome.[76]

PROTEASOME INACTIVATION UPON AGING AND UPON OXIDATIVE STRESS

Aging is characterized by a progressive and irreversible decline of various physiological functions of an organism in the latter part of its life. As an organism ages, there is an increase in oxidatively damaged protein that may be related to different factors: an increase in the rate of damaging radical formation, a decrease of the efficiency of anti-oxygen radical defense systems, and/or a decrease in the rate of repair or removal of damaged molecules. The proteasome is known to be the major proteolytic system involved in the removal of oxidized proteins. Thus, investigations have focused on a possible decline in the proteasome activity with age that may explain the age-related accumulation of oxidized protein. When assayed in crude homogenates, it is difficult to distinguish between the proteasome peptidase activities and other intracellular proteases activities. Even by using specific inhibitors or assay conditions that permit us to selectively measure the proteasome activity, this activity may still be affected by the presence of endogeneous inhibitors, activators, and competing denatured protein substrates. Nevertheless, previous evidence has been provided that neutral protease activity that was assumed to be the proteasome is declining with age.[23,77] We first purified the proteasome from the liver of young (8-month-old) and old (24-month-old) male Fischer 344 rats and assayed the three main peptidase activities (chymotrypsin-like, trypsin-like, and peptidylglutamyl-peptide hydrolase) with fluorogenic peptides. Among these three activities, the peptidylglutamyl-peptide hydrolase activity was found to decline in old animals to 50% of the activity observed with proteasome purified from young animals.[78] The recent report from Shibatani and Ward[79,80] is consistent with this finding. Indeed, they have found that the SDS-activated peptidylglutamyl-peptide hydrolase activity decreased 40%, whereas the chymotrypsin-like activity increased 15% in 26-month-old rat liver homogenates as compared to 7-month-old rat liver homogenates. A 35% decline with age for the peptidylglutamyl-peptide hydrolase activity of the proteasome isolated from bovine lens was also observed by Wagner and Margolis.[81] In addition, bovine lens proteasome trypsin-like activity was also found to decrease with age to a value approximately 40% of that observed for proteasome purified from young animals. More recently, Hayashi and Goto[82] have also reported that in rat liver there is an age-related decline in the peptidylglutamyl-peptide hydrolase activity and in the trypsin-like activity, although to a lesser extent, affecting both 20S and 26S proteasomes. Because in rat liver, the amount of proteasome did not appear to change with age, these authors have made the hypothesis that posttranslational modifications or subunit replacement is possibly responsible for the decrease in the activities.

In a previous report, we also showed that old female Wistar-Lou rats, under self selection of nutriments, decrease their protein intake and keep their liver proteasome peptidylglutamyl-peptide hydrolase activity as high as for the young rats.[83] This observation may be related to other findings that have been associated with dietary restriction of proteins or calories such as the reduction of oxidative damage and the increase in the level of antioxidant enzymes. In the same study, 2D gel electrophoresis has provided evidence of purified proteasome preparations that the age-related decline of proteasome activity can be associated with modifications affecting two different proteasome subunits.

Oxidative damage to protein has been implicated in age- and disease-related impairment of cellular functions and is known to affect protein turnover. For these reasons, the fate of proteasomes subjected to oxidative processes has been investigated. The proteasome is susceptible to inactivation by certain oxidative processes. Conconi et al.[78] have shown that metal-catalyzed oxidation of purified 20S proteasome in vitro results in alterations in certain hydrolase activities of the proteasome (mainly the trypsin-like and peptidylglutamyl peptide hydrolase activities). These alterations in proteasome activities appear dependent on whether the 20S proteasome is in the latent versus the active form before treatment with free radical–generating systems.[84] This finding may be related to the differential susceptibility of the 20S and 26S proteasomes to oxidative inactivation as observed by Reinheckel et al.[85] Indeed, in this study it was found that the 20S proteasome was more resistant to oxidative inactivation than the 26S proteasome, which was readily inhibited. When the 20S proteasome is in a latent form, metal-catalyzed oxidation resulted in activation of the peptidase activities.[84] This observation is supported by the finding of Strack et al.,[86] who have shown that $FeSO_4$/EDTA/ascorbate treatment activates peptidase and casein-hydrolyzing activities of the 20S proteasome. When FAO hepatoma cells are treated with iron and ascorbate to induce metal catalyzed oxidation, the trypsin-like and the peptidylglutamyl peptide hydrolase activities of the proteasome are inhibited. These results indicate that oxidative stress can target the proteasome within the cell and impair its function. Interestingly, binding of heat-shock protein 90 (HSP90) or α-crystallin was found to protect the trypsin-like and the Z-Leu-Leu-Leu-AMC-hydrolyzing activities of the proteasome against oxidative inactivation.[84] Furthermore, depletion or overexpression of HSP90 resulted in either decreased or increased protection of proteasome trypsin-like activity within the cell. The implication of α-crystallin in protection against oxidative stress has already been reported by Mehlen *et al.*[87] They have shown that overexpression of this protein confers resistance to oxidative stress–induced cytotoxicity in murine L929 fibroblasts. As reported by Fukuda *et al.*,[88] HSP90 may have a crucial role in increasing resistance to injury caused by iron-overload induced oxidative stress in rat renal tubular cells. Finally, according to Wagner and Margolis,[81] there is an age-dependent association of bovine lens proteasome with both HSP90 and α-crystallin. The observed decreased association of HSP90 and α-crystallin with bovine lens proteasome might result in a decreased protection of the proteasome against oxidative inactivation and therefore explain, at least in part, the age-related decline in proteasome peptidase activities.

The proteasome is also the target for modification by the lipid peroxidation product 4-hydroxy-2-nonenal (HNE), resulting in the selective inactivation of its trypsin-like activity.[89] Physiological relevance to HNE-mediated impairment of proteasome function has been provided by Okada *et al.*,[90] since they have reported that oxidative stress in kidney cells causes a transient impairment of intracellular proteolysis via covalent binding of HNE to proteasomes. In addition, proteins can be cross-linked by HNE (which is a bifunctional reagent) and become resistant to proteolysis by the proteasome, as observed with the model protein glucose-6-phosphate dehydrogenase.[91] HNE cross-linked proteins were also found to be potent noncompetitive inhibitors of the proteasome.[92] These findings suggest a plausible mechanism for the observed accumulation of damaged protein during the progression of certain pathological processes and aging.

A recent study on proteasome activity in human epidermis and in cultured keratinocytes has shown that the age-related decline of proteasome peptidase activities can be explained by a decreased proteasome content. This finding suggests that the proteasome is downregulated in aging. Indeed, in keratinocyte cultures, as well as in epidermis, there is an inverse relationship between the aging marker β-galactosidase and proteasome content.[26] Additional support for downregulation of proteasome in aging has been provided by the very recent report of Lee *et al.*[93] In this study, analysis of the gene expression profile of the aging process in skeletal muscle of mice have shown a significant decrease of the proteasome gene expression. These alterations are almost completely prevented by caloric restriction, leading the authors to the conclusion that the aging process may be slowed down in caloric restricted animals through increased protein degradation and subsequent decrease of protein damage.

SUMMARY AND CONCLUSION

Accumulation of oxidatively damaged proteins is a hallmark of cellular aging. Protein turnover is essential to preserve cell function, and the main proteolytic system in charge of cytosolic protein degradation is the proteasome, a multicatalytic proteolytic complex that recognizes and selectively degrades damaged and ubiquitinated proteins. One of the hypotheses put forward to explain the accumulation of altered proteins is the decrease of proteasome activity with age. It has been shown that proteasome from rat liver exhibited a decreased proteolytic activity, mainly peptidylglutamyl-peptide hydrolase activity, in old animals as compared to young ones. In contrast, in old rats under self selection of nutriments, which lowers protein intake, liver proteasome peptidylglutamyl-peptide hydrolase activity was kept as high as for young rats. Reduction of oxidative damage associated with dietary restriction of calories or proteins may be related with this observation. Proteasome peptidase activities (trypsin-like and peptidylglutamyl peptide hydrolase) are themselves sensitive to oxidation both *in vitro* and in cultured hepatoma cells. Interestingly, heat shock protein 90 was found to act as a specific protector of certain proteasome peptidase activities. The proteasome is also a target for modification by the lipid peroxidation product 4-hydroxy-2-nonenal (HNE), resulting in inactivation of its trypsin-like activity. In addition, HNE-cross-linked protein is resistant to degradation by the proteasome, and it has been shown that such cross-linked proteins act as potent noncompetitive inhibitors of the proteasome. Moreover, increased levels of oxidized, glycated, and HNE-modified proteins have been observed in human epidermal cells from old donors together with a decline in proteasome chymotrypsin-like and peptidylglutamyl-peptide hydrolase activities and a decreased proteasome content. Therefore, in human epidermis, the age-related decline of the proteasome peptidase activities can be explained, at least in part, by a decrease in the cellular amount of proteasome, suggesting that proteasome may be downregulated in aged cells.

ACKNOWLEDGMENTS

The work from our laboratory is supported by funds from the MENRT (Institut Universitaire de France et Université Denis Diderot-Paris) and the Fondation pour la Recherche Médicale.

REFERENCES

1. HARMAN, D. 1956. Aging: a theory based on free radical and radiation chemistry. J. Gerontol. **11:** 298–300.
2. BECKMAN, K.B. & B.N. AMES. 1998. The free radical theory of aging matures. Physiol. Rev. **78:** 547–581.
3. KIRKWOOD, T.B. 1987. Immortality of the germ-line versus disposability of the soma. Basic. Life. Sci. **42:** 209–218.
4. STADTMAN, E.R. 1992. Protein oxidation and aging. Science **257:** 1220–1224.
5. HEKIMI, S., B. LAKOWSKI, T.M. BARNES, *et al.* 1998. Molecular genetics of life span in *C. elegans*: how much does it teach us? Trends Genet. **14:** 14–20.
6. ORR, W.C. & R.S. SOHAL. 1994. Extension of life-span by overexpression of superoxide dismutase and catalase in *Drosophila melanogaster*. Science **263:** 1128–1130.
7. FRIEDMAN, D.B. & T.E. JOHNSON. 1988. Three mutants that extend both mean and maximum life span of the nematode, *Caenorhabditis elegans*, define the *age-1* gene. J. Gerontol. **43B:** 102–109.
8. KENYON, C., J. CHANG, E. GENSCH, *et al.* 1993. A *C. elegans* mutant that lives twice as long as wild type. Nature **366:** 461–464.
9. LARSEN, P.L. 1993. Aging and resistance to oxidative damage in *Caenorhabditis elegans*. Proc. Natl. Acad. Sci. USA **90:** 8905–8909.

10. VANFLETEREN, J.R. 1993. Oxidative stress and aging in *Caenorhabditis elegans*. Biochem. J. **292:** 605–608.
11. HOLMGREN, A. 1989. Thioredoxin and glutaredoxin systems. J. Biol. Chem. **264:** 13963–13966.
12. BROT, N. & H. WEISSBACH. 1982. The biochemistry of methionine sulfoxide residues in proteins. Trends Biochem. Sci. **7:** 137–139.
13. BROT, N. & H. WEISSBACH. 1983. Biochemistry and physiological role of methionine sulfoxide residues in proteins. Arch. Biochem. Biophys. **223:** 271–281.
14. CHIO, K.S. & A.L. TAPPEL. 1969. Synthesis and characterization of the fluorescent products derived from malonaldehyde and amino acids. Biochemistry **7:** 2821–2826.
15. TAPPEL, A.L. 1980. Free Radicals in Biology. Vol. **4:** 2–47. Academic Press. New York.
16. SOHAL, R.S. 1981. Relationship between metabolic rate, lipofuscin accumulation and lysosomal enzyme activity during aging in the housefly, *Musca domestica*. Exp. Gerontol. **16:** 347–355.
17. TSUCHIDA, M., T. MIURA & K. AIBARA. 1987. Lipofuscin and lipofuscin-like substances. Chem. Phys. Lipids **44:** 297–325.
18. IVY, G.O., S. KANAI, M. OHTA, *et al.* 1991. Leupeptin causes an accumulation of lipofuscin-like substances in liver cells of young rats. Mech. Ageing Dev. **57:** 213–231.
19. BRUNK, U.T., C.B. JONES & R.S. SOHAL. 1992. A novel hypothesis of lipofuscinogenesis and cellular aging based on interactions between oxidative stress and autophagocytosis. Mutat. Res. **275:** 395–403.
20. OLIVER, C.N., B.W. AHN, E.J. MOERMAN, *et al.* 1987. Age-related changes in oxidized proteins. J. Biol. Chem. **262:** 5488–5491.
21. LEVINE, R.L., D. GARLAND, C.N. OLIVER, *et al.* 1990. Determination of carbonyl content in oxidatively modified proteins. Methods Enzymol. **186:** 464–478.
22. LEVINE, R.L., J.A. WILLIAMS, E.R. STADTMAN, *et al.* 1994. Carbonyl assays for determination of oxidatively modified proteins. Methods Enzymol. **233:** 346–357.
23. STARKE-REED, P.E. & C.N. OLIVER. 1989. Protein oxidation and proteolysis during aging and oxidative stress. Arch. Biochem. Biophys. **275:** 559–567.
24. SMITH, C.D., J.M. CARNEY, P.E. STARKE-REED, *et al.* 1991. Excess brain protein oxidation and enzyme dysfunction in normal aging and in Alzheimer disease. Proc. Natl. Acad. Sci. USA **88:** 10540–10543.
25. SOHAL, R.S., S. AGARWAL, A. DUBEY, *et al.* 1993. Protein oxidative damage is associated with life expectancy of houseflies. Proc. Natl. Acad. Sci. USA **90:** 7255–7259.
26. PETROPOULOS, I., M. CONCONI, X. WANG, *et al.* Increase of oxidatively modified proteins is associated with decrease of proteasome activity and content in aging epidermal cells. J. Gerontol. Biol. Sci. In press.
27. CARNEY, J.M., P.E. STARKE-REED, C.N. OLIVER, *et al.* 1991. Reversal of age-related increase in brain protein oxidation, decrease in enzyme activity, and loss in temporal and spatial memory by chronic administration of the spin trapping compound *N*-tetra-butylphenylnitrone. Proc. Natl. Acad. Sci. USA **88:** 3633–3636.
28. BERLETT, B.S. & E.R. STADTMAN. 1997. Protein oxidation in aging, disease, and oxidative stress. J. Biol. Chem. **272:** 20313–20316.
29. YAN, L. J., R.L. LEVINE & R.S. SOHAL. 1997. Oxidative damage during aging targets mitochondrial aconitase. Proc. Natl. Acad. Sci. USA **94:** 11168–11172.
30. YAN, L.J. & R.S. SOHAL. 1998. Mitochondrial adenine nucleotide translocase is modified oxidatively during aging. Proc. Natl. Acad. Sci. USA **95:** 12896–12901.
31. RIVETT, A.J. 1985. Preferential degradation of the oxidatively modified form of glutamine synthetase by intracellular mammalian proteases. J. Biol. Chem. **260:** 300–305.
32. RIVETT, A.J. 1985. Purification of a liver alkaline protease which degrades oxidatively modified glutamine synthetase: characterization as a high molecular weight cysteine proteinase. J. Biol. Chem. **260:** 12600–12606.
33. DAVIES, K.J. & A.L. GOLDBERG. 1987. Oxygen radicals stimulate intracellular proteolysis and lipid peroxidation by independent mechanisms in erythrocytes. J. Biol. Chem. **262:** 8220–8226.
34. DAVIES, K.J. & A.L. GOLDBERG. 1987. Proteins damaged by oxygen radicals are rapidly degraded in extracts of red blood cells. J. Biol. Chem. **262:** 8227–8234.

35. PACIFICI, R.E., D.C. SALO & K.J. DAVIES. 1989. Macroxyproteinase (M.O.P) a 670 kDa proteinase that degrades oxidatively denatured proteins in red blood cells. Free Rad. Biol. Med. **7:** 521–536.
36. GRUNE, T., T. REINHECKEL, M. JOSHI, *et al.* 1995. Proteolysis in cultured liver epithelial cells during oxidative stress. Role of the multicatalytic proteinase complex, proteasome. J. Biol. Chem. **270:** 2344–2351.
37. GRUNE, T., T. REINHECKEL & J.A. DAVIES. 1996. Degradation of oxidized proteins in K562 human hematopoietic cells by proteasome. J. Biol. Chem. **271:** 15504–15509.
38. TAYLOR, A. & K. J. DAVIES. 1987. Protein oxidation and loss of protease activity may lead to cataract formation in aged lens. Free Rad. Biol. Med. **3:** 371–377.
39. SHANG, F. & A. TAYLOR. 1995. Oxidative stress and recovery from oxidative stress are associated with altered ubiquitin conjugating and proteolytic activities in bovine lens epithelial cells. Biochem. J. **307:** 297–303.
40. ULLRICH, O., T. REINHECKEL, N. SITTE, *et al.* 1999. Poly-ADP ribose polymerase activates nuclear proteasome to degrade oxidatively damaged histones. Proc. Natl. Acad. Sci. USA **96:** 6223–6228.
41. DICE, J.F. 1987. Molecular determinants of protein half-lives in eukaryotic cells. FASEB J. **5:** 349–357.
42. MORTIMORE, G.E. 1987. Lysosomes: their role in protein breakdown. H. Glaumann & F.J. Ballard, Eds.: 415–444. Academic Press. New York.
43. DICE, J.F. 1990. Peptide sequences that target cytosolic proteins for lysosomal proteolysis. Trends Biochem. Sci. **15:** 305–309.
44. MELLONI, E. & S. PONTREMOLI. 1989. The calpains. Trends Neurosci. **12:** 438–444.
45. CROALL, D.E. & G.N. DEMARTINO. 1991. Calcium-activated neutral protease (calpain) system: structure, function, and regulation. Physiol. Rev. **71:** 813–847.
46. PALOMBELLA, V.J., O.J. RANDO, A.L. GOLDBERG, *et al.* 1994. The ubiquitin-proteasome pathway is required for processing the NF-κ B₁ precursor protein and the activation of NF-κ B. Cell **78:** 773–785.
47. ROCK, K.L., C. GRAMM & L. ROTHSTEIN, *et al.* 1994. Inhibitors of the proteasome block the degradation of most cell proteins and the generation of peptides presented on MHC class I molecules. Cell **78:** 761–771.
48. COUX, O., K. TANAKA & A.L. GOLDBERG. 1996. Structure and functions of the 20S and 26S proteasomes. Annu. Rev. Biochem. **65:** 801–847.
49. KLEINSCHMIDT, J.A., B. HUGLE, C. GRUND, *et al.* 1983. The 22S cylinder particles of Xenopus laevis. Biochemical and electron microscopic characterization. Eur. J. Cell. Biol. **32:** 143–156.
50. BAUMEISTER, W., B. DAHLMANN, R. HEGERL, *et al.* 1988. Electron microscopy and image analysis of the multicatalytic proteinase. FEBS Lett. **241:** 239–245.
51. GRZIWA, A., W. BAUMEISTER, B. DAHLMANN, *et al.* 1991. Localization of subunits in proteasomes from *Thermoplasma acidophilum* by immunoelectron microscopy. FEBS Lett. **290:** 186–190.
52. PUHLER, G., S. WEINKAUF, L. BACHMANN, *et al.* 1992. Subunit stochiometry and three-dimensional arrangement in proteasomes from *Thermoplasma acidophilum.* EMBO J. **11:** 1607–1616.
53. LOWE, J., D. STOCK, B. JAP, *et al.* 1995. Crystal structure of the 20S proteasome from the archaeon *T. acidophilum* at 3.4 Å resolution. Science **268:** 533–539.
54. RIVETT, A.J. 1989. The multicatalytic proteinase. Multiple proteolytic activities. J. Biol. Chem. **264:** 12215–12219.
55. ORLOWSKI, M. 1990. The multicatalytic proteinase complex, a major extralysosomal proteolytic system. Biochemistry **29:** 10289–10297.
56. ORLOWSKI, M., C. CARDOZO & C. MICHAUD. 1993. Evidence for the presence of five distinct proteolytic components in the pituitary multicatalytic proteinase complex. Properties of two components cleaving bonds on the carboxyl side of branched chain and small neutral amino acids. Biochemistry **32:** 1563–1572.
57. GROLL, M., L. DITZEL, J. LOWE, *et al.* 1997. Structure of 20S proteasome from yeast at 2.4 Å resolution. Nature **386:** 463–471.
58. SEEMULLER, E., A. LUPAS, D. STOCK, *et al.* 1995. Proteasome from *Thermoplasma acidophilum*: a threonine protease. Science **268:** 579–582.

59. ARENDT, C.S. & M. HOCHSTRASSER. 1997. Identification of the yeast 20S proteasome catalytic centers and subunit interactions required for active-site formation. Proc. Natl. Acad. Sci. USA **94:** 7156–7161.
60. BAUMEISTER, W., J. WALZ, F. ZÜHL, *et al.* 1998. The proteasome: paradigm of a self-compartmentalizing protease. Cell **92:** 367–380.
61. GOLDBERG, A.L., T.N. AKOPIAN, A.F. KISSELEV, *et al.* 1997. New insights into the mechanisms and importance of the proteasome in intracellular protein degradation. Biol. Chem. **378:** 131–140.
62. TOKUMOTO, T. & K. ISHIKAWA. 1993. A novel "active" form of proteasomes from *Xenopus laevis* ovary cytosol. Biochem. Biophys. Res. Commun. **192:** 1106–1114.
63. MCGUIRE, M.J., M.L. MCCULLOUGH, D.E. CROALL, *et al.* 1989. The high molecular weight multicatalytic proteinase, macropain, exists in a latent form in human erythrocytes. Biochem. Biophys. Acta. **995:** 181–186.
64. FALKENBURG, P.E. & P.M. KLOETZEL. 1989. Identification and characterization of three different subpopulations of the Drosophila multicatalytic proteinase (proteasome). J. Biol. Chem. **264:** 6660–6666.
65. SAITOH, Y., H. YOKOSAWA & S. ISHII. 1989. Sodium dodecyl sulfate-induced conformational and enzymatic changes of multicatalytic proteinase. Biochem. Biophys. Res. Commun. **162:** 334–339.
66. CONCONI, C., L. DJAVADI-OHANIANCE, W. UERKWITZ, *et al.* 1999. Conformational changes in the 20S proteasome upon macromolecular ligand binding analysed with monoclonal antibodies. Arch. Biochem. Biophys. **362:** 325–328.
67. DJABALLAH, H., A.J. ROWE, S.E. HARDING, *et al.* 1993. The multicatalytic proteinase complex (proteasome): structure and conformational changes associated with changes in proteolytic activity. Biochem. J. **292:** 857–862.
68. FIGUEIREDO-PEREIRA, M.E., W.E. CHEN, H.M. YUAN, *et al.* 1995. A novel chymotrypsine-like component of the multicatalytic proteinase complex optimally active at acidic pH. Arch. Biochem. Biophys. **317:** 69–78.
69. DEMARTINO, G.N. & C.A. SLAUGHTER. 1999. The proteasome, a novel protease regulated by multiple mechanisms. J. Biol. Chem. **274:** 22123–22126.
70. FRUH, K., M. GOSSEN, K. WANG, *et al.* 1994. Displacement of housekeeping proteasome subunits by MHC-encoded LMPs: a newly discovered mechanism for modulating the multicatalytic proteinase complex. EMBO J. **13:** 3236–3244.
71. GACZYNSKA, M., A.L. GOLDBERG, K. TANAKA, *et al.* 1996. Proteasome subunits X and Y alter peptidase activities in opposite ways to the interferon-gamma-induced subunits LMP2 and LMP7. J. Biol. Chem. **271:** 17275–17280.
72. TANAKA, K. 1998. Molecular biology of the proteasome. Biochem. Biophys. Res. Commun. **247:** 537–541.
73. GLICKMAN, M.H., D.M. RUBIN, H. FU, *et al.* 1999, Functional analysis of the proteasome regulatory particle. Mol. Biol. Rep. **26:** 21–28.
74. MA, C.P., C.A. SLAUGHTER & G.N. DEMARTINO. 1992. Identification, purification, and characterization of a protein activator (PA28) of the 20S proteasome (macropain). **267:** 10515–10523.
75. DUBIEL, W., G. PRATT, K. FERRELL, *et al.* 1992. Purification of an 11 S regulator of the multicatalytic protease. J. Biol. Chem. **267:** 22369–22377.
76. KUEHN, L. & B. DAHLMANN. 1996. Proteasome activator PA28 and its interaction with 20 S proteasomes. Arch. Biochem. Biophys. **329:** 87–96.
77. AGARWAL, S. & R.S. SOHAL. 1994. Aging and proteolysis of oxidized proteins. Arch. Biochem. Biophys. **309:** 24–28.
78. CONCONI, M., L.I. SZWEDA, R.L. LEVINE, *et al.* 1996. Age-related decline of rat liver multicatalytic proteinase activity and protection from oxidative inactivation by heat-shock protein 90. Arch. Biochem. Biophys. **331:** 232–240.
79. SHIBATANI, T. & W.F. WARD. 1996. Effect of age and food restriction on alkaline protease activity in rat liver. J. Gerontol. Biol. Sci. **51A:** B175–B178.
80. SHIBATANI, T., M. NAZIR & W.F. WARD. 1996. Alterations of rat liver 20S proteasome activities by age and food restriction. J. Gerontol. Biol. Sci. **51A:** B316–B322.
81. WAGNER, B.J. & J.W. MARGOLIS. 1995. Age-dependent association of isolated bovine lens multicatalytic proteinase complex (proteasome) with heat-shock protein 90, an endogenous inhibitor. Arch. Biochem. Biophys. **323:** 455–462.

82. HAYASHI, T. & S. GOTO. 1998. Age-related changes in the 20S and 26S proteasome activities in the liver of male F344 rats. Mech. Aging. Dev. **102:** 55–66.
83. ANSELMI, B., M. CONCONI, C. VEYRAT-DUREBEX, *et al.* 1998. Dietary self-selection can compensate an age-related decrease of rat liver 20S proteasome activity observed with standard diet. J. Gerontol. Biol. Sci. **53A:** B173–B179.
84. CONCONI, M., I. PETROPOULOS, I. EMOD, *et al.* 1998. Protection from oxidative inactivation of the 20S proteasome by heat-shock protein 90. Biochem. J. **333:** 407–415.
85. REINHECKEL, T., N. SITTE, O. ULLRICH, *et al.* 1998. Comparative resistance of the 20S and 26S proteasome to oxidative stress. Biochem. J. **335:** 637–642.
86. STRACK, P.R., L. WAXMAN & J.M. FAGAN. 1996. Activation of the multicatalytic endopeptidase by oxidants. Effects on enzyme structure. Biochemistry **35:** 7142–7149.
87. MEHLEN, P., X. PREVILLE, P. CHAREYRON, *et al.* 1995. Constitutive expression of human hsp27, Drosophila hsp27, or human alpha B-crystallin confers resistance to TNF- and oxidative stress-induced cytotoxicity in stably murine transfected murine L929 fibroblasts. J. Immunol. **154:** 363–374.
88. FUKUDA, A., T. OSAWA, H. ODA, *et al.* 1996. Oxidative stress response in iron-induced acute nephrotoxicity: enhanced expression of heat shock protein 90. Biochem. Biophys. Res. Commun. **219:** 76–81.
89. CONCONI, M. & B. FRIGUET. 1997. Proteasome inactivation upon aging and on oxidation-effect of HSP 90. Mol. Biol. Rep. **24:** 45–50.
90. OKADA, K., C. WANGPOENGTRAKUL, T. OSAWA, *et al.* 1999. 4-Hydroxy-2-nonenal-mediated impairment of intracellular proteolysis during oxidative stress. Identification of proteasomes as target molecules. J. Biol. Chem. **274:** 23787–23793.
91. FRIGUET, B., E.R. STADTMAN & L.I. SZWEDA. 1994. Modification of glucose-6-phosphate dehydrogenase by 4-hydroxy-2-nonenal. Formation of cross-linked protein that inhibits the multicatalytic protease. J. Biol. Chem. **269:** 21639–21643.
92. FRIGUET, B. & L.I. SZWEDA. 1997. Inhibition of the multicatalytic proteinase (proteasome) by 4-hydroxy-2-nonenal cross-linked protein. FEBS Lett. **405:** 21–25.
93. LEE, C.-K., R.G. KLOPP, R. WEINDRUCH, *et al.* 1999. Gene expression profile of aging and its retardation by caloric restriction. Science **285:** 1390–1393.

Fibroblast Responses to Exogenous and Autocrine Growth Factors Relevant to Tissue Repair

The Effect of Aging

DIMITRIS KLETSAS,[a,b] HARRIS PRATSINIS,[b] IRENE ZERVOLEA,[b] PANAGIOTIS HANDRIS,[b] ELENI SEVASLIDOU,[b] ENZO OTTAVIANI,[c] AND DIMITRI STATHAKOS[b]

[b]Institute of Biology, National Centre for Scientific Research "Demokritos," 153 10, Athens, Greece

[c]Department of Animal Biology, via Berengario 14, University of Modena, 41100 Modena, Italy

ABSTRACT: The aging process is often associated with impaired wound healing, but the cellular and molecular mechanisms implicated are not completely understood. Accordingly, we have investigated the response of human fibroblasts from donors of various ages to platelet-derived and autocrine growth factors, in terms of mitogenicity as well as extracellular matrix synthesis and degradation. Our data indicate that fibroblast responses persist during aging, suggesting the involvement of systemic factors in the regulation of the healing process. In this context, we have found that neutral endopeptidase-24.11, a metalloproteinase controlling the action of neuroendocrine peptides and also of immunocyte chemotaxis, is overexpressed during aging. Finally, the connection between these data and those from *in vitro* aging studies is discussed.

INTRODUCTION

A prominent demonstration of homeostasis in the adult tissue is the regulation of the tissue repair process. Wound repair is a complex procedure, involving an intricate interplay among a variety of cells, growth factors, as well as extracellular matrix proteins and the proteases that degrade them, and is characterized by an orderly sequence of events that can be temporally categorized into three major phases: inflammation, tissue formation, and remodeling. The three stages, however, are not mutually exclusive but rather overlap in time.[1]

Immediately after wound formation, blood platelets aggregate and subsequently secrete a plethora of growth factors that (a) chemotactically attract several cell types into the wound area, (b) stimulate their proliferation, and (c) induce the secretion of

[a]Address for correspondence: Dr. D. Kletsas, Institute of Biology, National Centre for Scientific Research "Demokritos," 153 10, Athens, Greece. Phone: +30 1 6503565; fax: +30 1 6511767.

dkletsas@mail.demokritos.gr

extracellular matrix components. On the other hand, in the last phases of the repair process, as well as in the intact tissue, platelet-derived mitogens are obviously absent. However, the cells still secrete growth factors that regulate the above processes and, ultimately, tissue homeostasis in an autocrine and paracrine manner.

Of the various cell types implicated in wound repair in the connective tissue, fibroblasts are among the most important ones, because they migrate and proliferate in the wound area, they secrete autocrine growth factors, and they synthesize and deposit extracellular matrix (ECM) components—critical for wound closure and contraction—and, finally, also secrete proteases that regulate tissue remodeling.

Aging is one of the major events that affects the wound repair process, often leading to impaired and/or delayed healing. Diminished and delayed connective tissue deposition, reduced fibroblastic activity and growth, as well as less efficient wound contraction characterize the response to tissue injury in the elderly, in comparison to the response in younger individuals. However, the cellular and molecular mechanisms underlying these changes are not completely delineated and are under intense investigation.[2]

The aim of this study was to address the question of whether the differences in the wound repair process between young and aged individuals is also expressed at the cellular level or if they are the consequence of a generalized, systemic failure of homeostasis. In this context, previous studies have shown that *in vitro* senescent fibroblasts are unable to respond to mitogenic stimuli[3] and, furthermore, that they produce increased amounts of collagenases,[4] leading to the hypothesis that the accumulation of senescent cells in the tissues of old individuals could affect local tissue homeostasis.

Here, an alternative approach has been followed: the study of the features of fibroblasts aged *in vivo*, that is, cells derived from donors of different ages. In particular, several important parameters concerning tissue repair have been investigated, namely the ability of these cells to proliferate after mitogenic stimulation, the secretion of autologous growth factors, and the production and secretion of ECM constituents, as well as of collagenases. Finally, since the role of the neuroendocrine system on the repair of the connective tissue has recently received special focus, the involvement of neutral endopeptidase-24.11 (NEP), which regulates the action of the components of this system, is also discussed in this paper.

METHODS

Cell Strains

Ten strains of human fibroblasts were developed in our laboratory from skin explants, one from a neonatal donor and nine from healthy, consenting volunteers of various ages, that is, 7 (two donors), 8, 23, 60, 65, 74, 75, and 100 years old. These cells have been used at early passages.[5] Furthermore, the following commercially available human fibroblast cell strains were used as controls: (a) human lung fibroblasts, one fetal strain (Flow 2002) and one from a 20-year-old adult donor (CCD 19Lu, ECACC, U.K.); (b) human skin fibroblasts, one fetal strain (Detroit 551, ATCC, USA), and one from a 23-year-old adult donor (1.BR3, ECACC).

Conditioned Medium Collection

Confluent cultures of human fibroblasts were incubated with serum-free MEM for 48-hour periods and the conditioned medium (CM) was clarified from cell debris by centrifugation.[6]

DNA Synthesis Assay

Cells were grown until confluency, arrested in MEM/0.1% fetal calf serum (FCS), and stimulated with growth factors (human recombinant (h.r.) PDGF-BB and h.r.TGF-β1) or conditioned media (1:1 with fresh medium) in the presence of [methyl-^{3}H]thymidine. After 48 hours of incubation, the radioactivity incorporated into newly synthesized DNA was estimated as previously described.[7]

Inhibition of CM Mitogenicity with Neutralizing Antibodies

Neutralizing antibodies were preincubated with CM for 3 hours at 37°C, at a concentration that totally inhibits the respective growth factors. Inhibition was estimated by DNA synthesis assay, as described above.

Collagen Synthesis Assay

Human fibroblast cultures were rendered quiescent by serum deprivation for 48 hours, as described above (see DNA synthesis assay). Then they were incubated for 48 hours with medium supplemented with L-[^{3}H]proline, β-aminopropionitrile (βAPN), and ascorbic acid, in the presence or absence of TGF-β1. Collagen synthesis was measured by a modification of the protease-free collagenase method.[8] Triplicate samples were measured, and the results were normalized to the number of cells recovered.

Zymography

Gelatin zymography was performed by using a modification of the procedure of Herron *et al.*[9] Human fibroblast–conditioned media were separated on nonreducing 10% SDS-polyacrylamide gels that contained 0.1% gelatin. After removal of SDS, the gels were incubated in substrate buffer and stained with Coomassie brilliant blue R250. Zones containing proteases appeared as clear bands on a blue background.

Analysis of the NEP Effect on Immunocyte Migration

Invertebate immunocytes were incubated with growth factors in the presence or absence of NEP, and their migration was estimated after computer-assisted microscopic image analysis by using a shape factor (SF) formula.[10]

Analysis of NEP Activity

Confluent cultures of human fibroblasts were scraped, sonicated, and subjected to spectrofluorimetric analysis of NEP activity, as described.[5]

RESULTS AND DISCUSSION

Fibroblast Response to Platelet-Derived Mitogens Persists during Aging

One of the initial events after tissue injury is the aggregation of blood platelets, which leads to the secretion of several growth factors from their α-granules. Most prominent among these factors are platelet-derived growth factor (PDGF), a potent mitogen for cells of mesenchymal origin,[11] and transforming growth factor-β (TGF-β), a true multifunctional factor.[12] Both PDGF and TGF-β control several key functions in the wound site, such as cell migration and proliferation and ECM formation and remodeling.

Concerning cell proliferation, we have previously shown that the response of human fibroblasts to PDGF and TGF-β—as well as to other growth factors such as epidermal growth factor (EGF) and fibroblast growth factor (FGF)—declines sharply as a result of *in vitro* aging after serial passaging.[13] Here, we present evidence on the effect of growth factors on human fibroblasts aged *in vivo*, that is, from donors of various ages. As can be seen in FIGURE 1, the response of adult fibroblasts to PDGF does not decline as an effect of aging. Interestingly, similar responses have also been found regardless of the developmental state of the donor (fetal, neonatal, or adult)

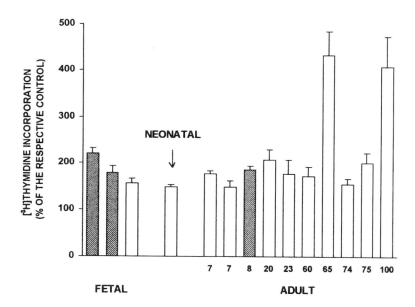

FIGURE 1. Effect of PDGF on DNA synthesis of human fibroblasts. Quiescent cultures of human fetal, neonatal, or adult fibroblasts (numbers indicate the age of the donor) of skin (*open bars*) and lung (*hatched bars*) origin were stimulated with PDGF-BB (3 ng/ml), and DNA synthesis was estimated as described under MATERIALS AND METHODS. Values represent stimulation in comparison to the respective unstimulated culture (mean of three individual experiments, with each experiment performed in triplicate dishes).

and of the tissue of origin (skin or lung). As far as neonatal and adult fibroblasts are concerned, the stimulatory effect of TGF-β does not seem to decline with the age of the donor (not shown here). The only change observed was on the action of TGF-β on fibroblasts from fetal donors, which were strongly inhibited, reflecting the different repair strategies between fetuses and adults (Kletsas *et al.*, submitted for publication).

The above data are in accordance with those of Freedland *et al.*,[14] who also found no evidence for an age-dependent decline of the response of human adult fibroblasts to several growth factors, such as EGF, TNF-α, PDGF, and fetal bovine serum. In the same context, Tesco *et al.*[15] have reported that the growth characteristics of skin fibroblasts from centenarian donors are remarkably similar to those of young controls.

The Effect of Aging on Extracellular Matrix Formation

Growth factors released by platelets, beyond their chemotactic and mitogenic activity, also induce the synthesis of ECM components. Collagen is the most prominent ECM constituent, at least for the connective tissue, and it is mainly secreted by fibroblasts in the wound area.[1] Accordingly, we have studied collagen secretion by human fibroblasts, before (basal levels) and after stimulation with the most potent inducer of ECM synthesis, that is, TGF-β.[12]

As can be seen in FIGURE 2A, beyond some extreme cases due to an expected interdonor variability, collagen secretion by human adult fibroblasts does not seem to change considerably with the age of the donor. Furthermore, the fetal and neonatal cells strains that have been used in this study secrete only slightly higher collagen concentrations from the average secretion by adult-donor fibroblasts. In FIGURE 2B, the stimulation of collagen synthesis and secretion after treatment with TGF-β in comparison to the respective control—that is, unstimulated cultures—is depicted. From these data also there is no evidence for an age-related decline in the secretion of collagen from human fibroblasts.

Another important determinant of tissue repair and remodeling is the local levels of matrix metalloproteinases, and in particular collagenases, that regulate the transition from the final stages of the wound—the granulation tissue—to the mature scar.[1] Accordingly, we have investigated the levels of collagenases secreted by human fibroblasts derived from donors of various ages. By the use of gelatin zymography, we have observed that fibroblasts secrete a major metalloproteinase that in SDS-PAGE migrates at 53 kDa, corresponding to the MMP-1, that is, the interstitial collagenase (FIG. 3). As can be seen, beyond interdonor variabilities, no age-related change was observed in the secretion of MMP-1. Similar levels of MMP-1 were found to be secreted by the fetal and neonatal cell strains studied. These data, at variance with the strong upregulation of collagenase production by human fibroblasts aged *in vitro* after serial passaging,[4] indicate a difference between *in vivo* and *in vitro* aging at this level.

Autocrine Growth Regulation during Aging

At the early, acute phase of the wound, the healing process is mainly dependent on the bulk of growth factors released by blood platelets and other immunocytes, such as activated macrophages. However, at the late stages of the wound and during

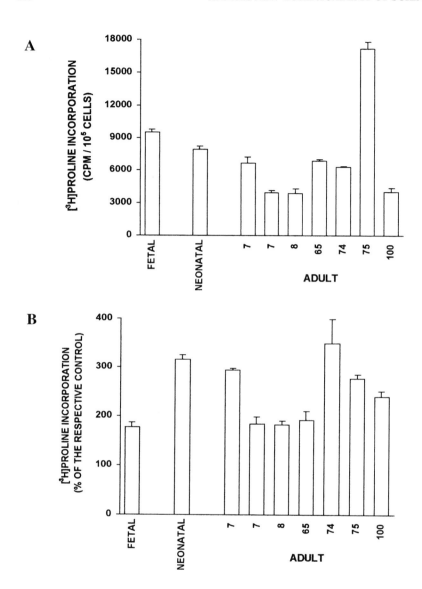

FIGURE 2. Basal and TGF-β–stimulated collagen synthesis of human fibroblasts.
In **(A)** the secreted collagenous protein by quiescent cultures of human fetal, neonatal, or
adult skin fibroblasts (numbers indicate the age of the donor) labeled with [³H]proline was
estimated as described under MATERIALS AND METHODS. Values represent the mean of two
individual experiments, with each experiment performed in triplicate dishes. In **(B)** values
represent collagen synthesis after stimulation of the above cultures with TGF-β (5 ng/l) in
comparison to the respective unstimulated culture (mean of two individual experiments,
with each experiment performed in triplicate dishes).

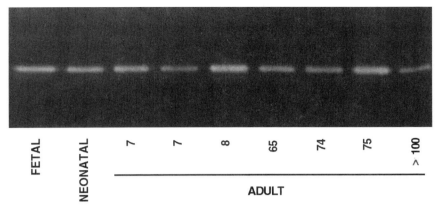

FIGURE 3. Secretion of collagenase-1 by human skin fibroblasts. The collagenase-1 (MMP-1) activity secreted by quiescent cultures of human fibroblasts (strains as in FIG. 2) was estimated by gelatin zymography (see MATERIALS AND METHODS). One of two similar experiments is depicted.

normal tissue turnover, this "potential" of exogenous growth factors is not available. Consequently, at this stage, it is the sum of the factors released by the cells, which act locally in an autocrine or paracrine manner, that is more significant for the homeostatic regulation.

As previously reported, we have shown that human fibroblasts secrete autocrine mitogens that, in the absence of exogenous growth factors, provide them with a certain degree of autonomy in the maintenance of tissue repair and regeneration.[6] Here, we have investigated the possibility that alterations of this autocrine mechanism during the aging process could be involved in the decline of homeostasis in the elderly. Accordingly, we have studied the mitogenic potential of media conditioned by fibroblasts, which contain, by definition, all the secreted growth factors. As can be seen in FIGURE 4A, the mitogenic potential secreted by fibroblasts originating from donors of various ages does not decline as a consequence of aging, indicating that this autocrine mechanism persists during aging *in vivo*.

PDGF and TGF-β, beyond being released by platelets and other immunocytes, have also been observed to be secreted by fibroblasts.[6] In order to test whether the mitogenic action of the conditioned media is due to the presence of one of these two factors (or both), we have treated the conditioned media with pan-specific, neutralizing antibodies against PDGF, TGF-β, or both. As can be seen in FIGURE 4B, both antibodies only partly inhibit the effect of the conditioned medium, indicating (a) that both factors are included in this medium and contribute to its mitogenicity and (b) that additional autocrine mitogens are also secreted by human fibroblasts.

Furthermore, in order to estimate whether the levels of TGF-β change with the age of the donor, we have used the most sensitive bioassay for TGF-β, based on the specific inhibition of mink lung epithelial cells by this factor. Interestingly, our preliminary observations indicate that the levels of TGF-β secreted by human fibroblasts remain relatively constant, regardless of the aging level of the donors (data not shown).

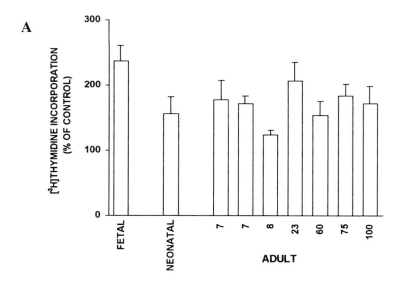

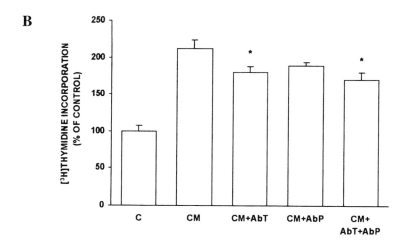

FIGURE 4. Mitogenic activity of human fibroblast-conditioned media. (A) Quiescent cultures of human skin fibroblasts from a 7-year-old donor were stimulated with media conditioned by fetal, neonatal, and adult fibroblast strains (numbers indicate the age of the donor), and DNA synthesis was estimated as described under MATERIALS AND METHODS. Values represent stimulation in comparison to the control, that is, MEM/0.1% FCS (mean of two individual experiments, with each experiment performed in triplicate dishes). **(B)** Quiescent cultures (C) of human fibroblasts were stimulated with autologous conditioned medium (CM) preincubated with anti-TGF-β (AbT) or anti-PDGF (AbP) or both pan-specific antibodies (10 μg/ml), and DNA synthesis was estimated as above. The * corresponds to statistically significant inhibition with reference to CM ($p < 0.05$ after Student's t test analysis).

Neutral Endopeptidase-24.11 Function and Activity during Aging

All the data presented above indicate that there are no major differences, as an effect of the aging of the donor, in the response of human fibroblasts to growth factors that could possibly account for impaired healing in the elderly. It has to be mentioned that only a few, although important, parameters have been studied here. This is becoming increasingly problematic with the growing understanding of the multiplicity of the systems involved in tissue repair. As an example, the interaction between the neuroendocrine and the skin immune systems has been recently receiving increasing attention.[16] One of the important components of the neuroendocrine system is neutral endopeptidase-24.11 (NEP), which regulates the action of small bioactive peptides.[5] As can be seen in FIGURE 5A, we have found that NEP is able to completely inhibit the chemotactic activity of PDGF and TGF-β on invertebrate immunocytes.[17] Because the chemotactic attraction of immunocytes is a key event in the early phases of wound repair, we have studied the activity of NEP in human fibroblasts during development and aging. The results from these studies, depicted in FIGURE 5B, show an increase in NEP activity during the fetal-to-adult transition. Furthermore, an increase in the activity has been found in cells from aged donors, whereas fibroblasts from a centenarian donor display an activity at the level of young cells.[5] Although these data originate from a few donors, Solmi *et al.*[18] provided similar data from a large number of young and aged individuals. The physiological significance of the increased expression of NEP with aging on the regulation of homeostasis obviously requires further investigation.

CONCLUSIONS, EMERGING QUESTIONS, AND FUTURE PROSPECTS

In this report we have studied several important parameters of the wound repair process focusing on the ability of human fibroblasts to respond to platelet-derived and autocrine mitogens as well as the secretion of ECM components and the metalloproteinases that degrade them. Interestingly, when we have used cells from donors of various ages, no age-related decline in any of these aspects has been found that could possibly account for impairments of the wound-healing process in the elderly. Although several parameters have to be further investigated in future studies, these data point to the importance of systemic factors, such as hormone levels, that regulate body homeostasis in general.

As mentioned above, studies on cells aged *in vitro* after serial passaging have shown that these cells are unable to respond to exogenous stimuli and, moreover, secrete large amounts of collagenases. However, the results presented here and by other laboratories (see above in RESULTS AND DISCUSSION) indicate a major difference at this level between cells aged *in vivo* and those aged *in vitro*. Several hypotheses, not mutually exclusive, could be raised to explain this discrepancy:

- *(a)* *in vivo* and *in vitro* aging processes are, in part or in general, two different mechanisms;
- *(b)* the percentage of senescent cells in the tissues of old individuals is negligible; and
- *(c)* the development of the primary culture from tissue explants represents a selection process in favor of the "young" cells.

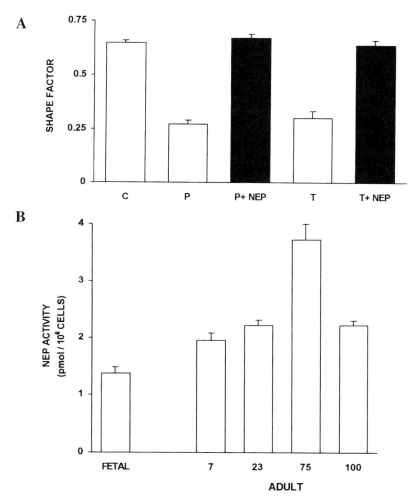

FIGURE 5. NEP function and activity. (A) Effect of NEP (0.25 ng/ml) on the chemo-tactic activity of PDGF (P) and TGF-β (T) on molluscan immunocytes, expressed as shape factor, compared to unstimulated cells (C). **(B)** NEP activity in human skin fibroblasts from fetal and adult donors (numbers indicate the age of the donor) expressed as pmol/10^6 cells. In both **A** and **B**, the mean and standard deviation of three experiments is shown.

The increase in NEP activity in fibroblasts from old donors requires further studies in order to be properly evaluated. These, along with experiments on human immunocytes, could possibly provide information concerning the interplay of the neuroendocrine and the immune system in the processes of tissue repair and of aging in general.

Finally, the *in vitro* study of a complex process such as this of wound repair requires cell assay systems that simulate the conditions of the tissue *in vivo* as closely as possible. Accordingly, we are currently studying the response of fibroblasts from young and aged donors when cultured in three-dimensional gels of polymerized collagen (Zervolea *et al.*, in preparation), as these gels represent the closest *in vitro* approximation of the granulation tissue, that is, one of the last phases of tissue repair. Our preliminary observations indicate qualitative differences in the response of fibroblasts under these conditions, in comparison to the remote from the *in vivo* state cultures on plastic surfaces.

ACKNOWLEDGMENT

This work has been partially supported by the Greek General Secretary of Research and Technology (PENED'95 Programme).

REFERENCES

1. CLARK, R.A.F. 1996. Wound repair: overview and general considerations. *In* The Molecular and Cellular Biology of Wound Repair. R.A.F. Clark, Ed.: 3–50. Plenum. New York.
2. ASHCROFT, G.S. *et al.* 1997. Estrogen accelerates cutaneous wound healing associated with an increase in TGF-β1 levels. Nature Med. **3:** 1209–1215.
3. CRISTOFALLO, V.J. & R.J. PIGNOLO. 1993. Replicative senescence of human fibroblast-like cells in culture. Physiol. Rev. **73:** 617–638.
4. MILLIS, A.J.T. *et al.* 1992. Differential expression of metalloproteinase and tissue inhibitor of metalloproteinase genes in aged human fibroblasts. Exp. Cell Res. **201:** 373–379.
5. KLETSAS, D. *et al.* 1998. Neutral endopeptidase-24.11 (NEP) activity in human fibroblasts during development and ageing. Mech. Ageing Dev. **102:** 15–23.
6. PRATSINIS, H. *et al.* 1997. Autocrine growth regulation in fetal and adult human fibroblasts. Biochem. Biophys. Res. Commun. **237:** 348–353.
7. KLETSAS, D. *et al.* 1995. The growth inhibitory block of TGF-β is located close to the G1/S border in the cell cycle. Exp. Cell Res. **217:** 477–483.
8. PETERKOVFSKY, B. & R. DIEGELMANN. 1971. Use of mixture of proteinase free collagenases for the specific assay of radioactive collagen in the presence of other proteins. Biochemistry **10:** 988–994.
9. HERRON, G.S. *et al.* 1986. Secretion of metalloproteinases by stimulated capillary endothelial cells. II. Expression of collagenase and stromelysin activities is regulated by endogenous inhibitors. J. Biol. Chem. **261:** 2814–2818.
10. KLETSAS, D. *et al.* 1998. PDGF and TGF-β induce cell shape changes in invertebrate immunocytes via specific cell surface receptors. Eur. J. Cell Biol. **75:** 362–366.
11. HELDIN, C.-H. & B. WESTERMARK. 1990. Platelet-derived growth factor: mechanism of action and possible in vivo function. Cell Regul. **1:** 555–566.
12. ROBERTS, A.B. & M.B. SPORN. 1996. Transforming growth factor-β. *In* The Molecular and Cellular Biology of Wound Repair. R.A.F. Clark, Ed.: 275–308. Plenum. New York.
13. KLETSAS, D. & D. STATHAKOS. 1992. Quiescence and proliferative response of normal human embryonic fibroblasts in homologous environment. Effect of ageing. Cell Biol. Int. Rep. **16:** 103–113.
14. FREEDLAND, M. *et al.* 1995. Fibroblast responses to cytokines are maintained during aging. Ann. Plast. Surg. **35:** 290–296.
15. TESCO, G. *et al.* 1998. Growth properties and growth factor responsiveness in skin fibroblasts from centenarians. Biochem. Biophys. Res. Commun. **244:** 912–916.

16. SCHOLZEN, T. *et al.* 1998. Neuropeptides in the skin: interactions between the neuroendocrine and the skin immune systems. Exp. Dermatol. **7:** 81–96.
17. CASELGRANDI, E. *et al.* 2000. Neutral endopeptidase-24.11 (NEP) deactivates PDGF- and TGF-β-induced cell shape changes in invertebrate immunocytes. Cell Biol. Int. In press.
18. SOLMI, R. *et al.* 1996. In vitro study of gingival fibroblasts from normal and inflamed tissue: age-related responsiveness. Mech. Ageing Dev. **92:** 31–41.

The Werner Syndrome

A Model for the Study of Human Aging

JAN O. NEHLIN,[a,b] GUNHILD LANGE SKOVGAARD,[b] AND VILHELM A. BOHR[c]

[b]*Laboratory of Molecular Gerontology and Dermatology, Copenhagen University Hospital, Copenhagen, DK-2100, Denmark*

[c]*Laboratory of Molecular Genetics, Gerontology Research Center, National Institute on Aging, 5600 Nathan Shock Drive, Baltimore, Maryland 21224, USA*

ABSTRACT: Human aging is a complex process that leads to the gradual deterioration of body functions with time. Various models to approach the study of aging have been launched over the years such as the genetic analysis of life span in the yeast *S. cerevisiae*, the worm *C. elegans*, the fruitfly, and mouse, among others. In human models, there have been extensive efforts using replicative senescence, the study of centenerians, comparisons of young versus old at the organismal, cellular, and molecular levels, and the study of premature aging syndromes to understand the mechanisms leading to aging. One good model for studying human aging is a rare autosomal recessive disorder known as the Werner syndrome (WS), which is characterized by accelerated aging *in vivo* and *in vitro*. A genetic defect implicated in WS was mapped to the *WRN* locus. Mutations in this gene are believed to be associated, early in adulthood, with clinical symptoms normally found in old individuals. WRN functions as a DNA helicase, and recent evidence, summarized in this review, suggests specific biochemical roles for this multifaceted protein. The interaction of WRN protein with RPA (replication protein A) and p53 will undoubtedly direct efforts to further dissect the genetic pathway(s) in which WRN protein functions in DNA metabolism and will help to unravel its contribution to the human aging process.

INTRODUCTION

Human aging is defined as the gradual deterioration of tissue and body function that affects all individuals worldwide.[1] Because aging is likely caused by numerous genetic and environmental factors acting simultaneously, it is very difficult to dissect the roles of individual genes in this process. In order to simplify the quest for genes involved in human aging, we and others have focused on the study of very rare genetic disorders known as premature aging syndromes. It is believed that the genes that contribute to rapid aging also participate in the process of normal human aging, albeit at a slower rate. Several premature aging syndromes have been characterized including the Werner syndrome, Hutchinson-Guilford disease (progeria), Bloom's syndrome, Cockayne's syndrome, ataxia telangiectasia, and Down's syndrome. The progeroid syndrome that best mimics the characteristics of normal human aging is

[a]Address for correspondence: Dr. Jan O. Nehlin, Laboratory of Molecular Gerontology and Dermatology, Copenhagen University Hospital, Blegdamsvej 9, Section 6311, Copenhagen, DK-2100, Denmark. Phone: +45 35 456392; fax: +45 35 456315.

nehlinj@rh.dk

the Werner syndrome (WS), and some of the knowledge focused on molecular aspects of this affliction is reviewed below.

THE WERNER SYNDROME

Many, but not all, of the features that distinguish this syndrome are very much reminiscent of the major features observed in normal human aging. Hence, WS has been termed a segmental premature aging syndrome. WS is also considered a chromosome instability syndrome displaying multiple stable chromosomal rearrangements and, to a lesser extent, chromosome breakage. The average life span of WS patients is 47 years of age; and cancer, myocardial, or cerebrovascular ailments are the usual causes of death.[2]

The majority of clinical symptoms of WS are very similar to those observed in elderly subjects and include skin wrinkling, graying of the hair, baldness, cataracts, diabetes type II, atherosclerosis, heart disease, osteoporosis, hypermelanosis, cerebral cortical atrophy, lymphoid depletion, and thymic atrophy. Certain other features are very common in WS patients but unusual in normal aging. These include laryngeal atrophy, soft tissue calcification, trophic ulcers, atrophy of extremities, sarcomas, and tumors of connective tissue.[2–6] In addition, a recent study reported accelerated aging of the central nervous system in WS patients.[7]

MOLECULAR AND CELLULAR FEATURES OF
WERNER SYNDROME CELLS

At the cellular level, WS cells show a number of distinctive characteristics (TABLE 1). A single gene, *WRN*, located on human chromosome 8p12, was found to be mutated or contained deletions in patients diagnosed with WS.[31] Therefore, its deficiency was implicated with causing the age-related diseases observed in afflicted individuals.[2,3] The causes of such rapid body deterioration are germinal mutations in the *WRN* locus.[31–37] It is thought that similar loss of function in the *WRN* locus may take place in the general population as a function of age. Loss of WRN function may be responsible in part for the geneses of diseases of old age. A polymorphism 1367 Cys/Arg in the *WRN* locus is associated with an increase in the risk of myocardial infarction among a Japanese population,[38] but is not associated with longevity.[39]

WRN PROTEIN FEATURES

Initial analysis of the WRN protein (WRNp) sequence predicted that it belongs to the RecQ subfamily of DexH helicases. The central domain contains seven helicase motifs and an ATP-binding motif.[31] The NH_2-terminal domain contains a region with similarity to the $3'–5'$ proofreading exonuclease domain of DNA polymerase I and to RNase D.[40,41] At the COOH terminus, an HRDC domain with possible nucleic acid domain properties[40] and a nuclear localization signal (NLS) that targets WRNp to the nucleus (see below) are present (FIG. 1).

TABLE 1. Molecular and cellular defects in Werner syndrome

Abnormal DNA replication

Slow-growing, extended S-phase of the cell cycle phase[8,9]

Delayed replication[10]

Accumulation of noncycling cells in G_1

Reduced replicative life span, premature replicative senescence[2,11]

Elevated genomic instability

Chromosomal rearrangements (i.e., deletions, insertions) and variegated translocated mosaicism[2,11-17]

Accelerated reduction in telomere length[18,19]

Compromised ligation fidelity (error-prone DNA ligation)[20]

Defective DNA repair

Subtle transcription-coupled DNA repair defects[21]

Elevated number of double-strand breaks after X-ray irradiation[22]

Elevated somatic mutation rate, particularly deletions[17,23,24]

Subtle telomeric repair defect[25]

Hypersensitivity to 4-NQO[14,26]

Sensitivity to S-phase-specific agents such as the topoisomerase I inhibitor camptothecin[27-29]

Defective transcription

Deficient RNA pol II-dependent transcription[30]

Global

Gene specific?

WRN belongs to a subfamily of the RecQ family of helicases.[29,40,42] These include RecQ (*E. coli*), *SGS1* (*S. cerevisiae*), *rqh1+/rad12* (*S. pombe*),[43] K02F3.1, F18C5.2, T04A11.6 (*C. elegans*), FFA-1 (*Xenopus*)(Yan), *Dmblm*, *RECQ5* (*Drosophila*),[42,45] and in humans, *RECQL/Q1*, *RECQ4*, *RECQ5*,[46] and *BLM*, implicated in Bloom's syndrome.[29] Interestingly, RECQ4 has been implicated in some cases of Rothmund-Thomson syndrome, a genetic disorder with some symptoms of premature aging.[47]

EXPRESSION STUDIES AND WRN PROTEIN SUBCELLULAR LOCALIZATION

WRN mRNA levels are significantly reduced in WS cells. In heterozygous WS mutants, defective *WRN* transcripts are downregulated, and relatively low levels of *WRN* mRNA are necessary to prevent the onset of Werner syndrome.[48] The expression of a number of genes in WS fibroblasts shows V patterns similar to that observed in senescent fibroblasts[49] but not identical.[50]

Wild-type *WRN* expression is downregulated in normal quiescent and senescent cultured fibroblasts.[51] A decrease in activity of the *WRN* promoter in cells from WS patients has been found, and basal regulatory elements identified.[52,53]

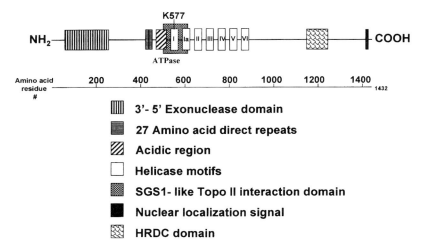

FIGURE 1. Conserved structural and functional domains in the WRN protein. WRN belongs to the RecQ subfamily of helicases. Analysis of WRN showed that WRN is a helicase with an ATPase motif,[31,62–65] an exonuclease motif,[40,41,68] an HRDC domain,[40] and a nuclear localization signal.[59]

WRN gene expression is stimulated by the tumor suppressor protein Rb and repressed by the tumor suppressor protein p53.[52] WRNp levels are upregulated in transformed and immortal cancer cell lines in comparison to normal cells.[54,55] WRNp seems to have an essential function in growth and cellular division. WRN could represent an important marker of replicative capacity and hence, of malignant growth.

Very little is known about the normal levels of *WRN* in any given tissue as human beings age, but presumably *WRN* is highly expressed in organs or tissues with strong proliferative capacity, and then its presence diminishes during the normal aging process.

Knock-out studies of the mouse counterpart of *WRN* have shown that live-born mice are normal during their first year of life, but that WS –/– ES cells show a higher mutation rate and are more sensitive to topoisomerase I inhibitors (i.e. camptothecin) than normal cells.[56] Normal and WS lymphoblastoid cells arrest in S phase after treatment with camptothecin. WS cells undergo induced apoptosis, suggesting the presence of topoisomerase I intermediates.[28,29] However, p53-mediated apoptosis is attenuated in WS cells,[55] suggesting the presence of other apoptotic pathways. Also, defective mouse *WRN* gene expression leads to accumulation of cells that are highly susceptible to Fas-induced apoptosis.[57]

A recombinant WRNp was first described to be nuclear[58] and a nuclear localization signal was mapped to its COOH-terminal end (FIG. 1; see also Ref. 59). The WRNp was subsequently found to be predominantly nucleolar during growing conditions and nuclear during quiescence or when exposed to 4-NQO.[60,61] However, such studies have relied on polyclonal antisera. Recently, it was shown that WRNp is distributed in the nucleoplasm during the interphase but not in the nucleolus.[54]

Further characterization of WRNp subcellular localization during oxidative DNA damage and during the cell cycle, would help us monitor WRNp movement to areas of damage and/or to specific nucleoplasmic compartments.

BIOCHEMICAL CHARACTERIZATION OF THE WERNER PROTEIN

WRNp has been shown to be a DNA-dependent ATPase and a 3′-5′ helicase capable of unwinding short DNA duplexes.[62–66] RPA (Replication protein A), a functional homologue of *E. coli* SSB, stimulated WRN-catalyzed DNA-unwinding.[64,65] SSB completely failed to stimulate WRN helicase to unwind long duplexes, indicating a specific functional interaction between WRNp and hRPA. WRNp and RPA interact both *in vitro* and *in vivo*.[65] WRN helicase catalyzed the unwinding of long duplex DNA substrates of up to 849 bp in a hRPA-dependent fashion.[65]

WRNp was able to bind with higher affinity to single-stranded DNA than to double-stranded DNA and was unable to distinguish substrates damaged by 4-NQO or UV.[66] Recently, it was found that WRNp can unwind G′2 bimolecular tetraplex structures of the fragile X syndrome repeat sequence $d(CGG)_7$ more efficiently than double-stranded DNA.[67]

WRNp has also been shown to have an exonuclease function with 3′-5′ proofreading activity.[40,41,68,69] or in a 5′-3′ direction dependent on duplex unwinding.[70] Recent work from our laboratory[66] confirms the 3′-5′ activity and suggests that the 5′-3′ activity is a contaminating nuclease.

MECHANISMS OF ACTION OF WRN PROTEIN: INTERACTION WITH RPA AND p53

As previously mentioned, RPA efficiently helped WRNp unwind DNA through direct protein–protein association.[65] Replication protein A (RPA, RFA, or hSSB) is a single-stranded DNA binding protein that is necessary for various processes of DNA metabolism, such as DNA replication, DNA repair, recombination, and transcription. RPA homologues have been identified in many organisms. RPA is an abundant heterotrimeric protein complex consisting of subunits of ~70 kDa (RPA1), ~30 kDa (RPA2), and ~14 kDa (RPA3). These proteins bind single-stranded DNA as well as many different proteins.[71–73]

WRN is involved in some aspect of DNA metabolism (FIG. 2). Its functional role, in view of the latest findings, involves the following aspects:

Role in DNA Replication

In the S phase of the cell cycle, chromosomal replication takes place at various sites simultaneously in the nucleus. These sites, called replication foci, contain 300 to 1000 replicating DNA molecules and many proteins necessary for DNA replication and cell cycle progression.[71] Because WRN helicase binds RPA1 (p70), it is possible that WRN unwinds DNA for replication. Immunolocalization of human RPA showed that RPA1 (p70), cyclin A, and cdk2 associate with replication foci during S phase, whereas RPA2 (p34) has another distribution, indicating that p70 and

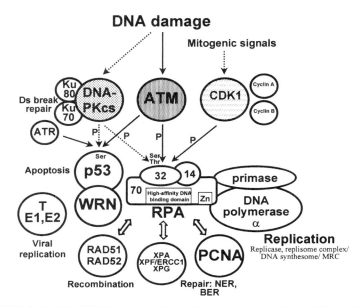

FIGURE 2. Model of WRN-associated proteins suggesting a role in DNA replication, recombination, transcription, and repair. WRN binds RPA[65] and p53.[55] RPA binds to numerous proteins (for details see text and Refs. 71–73, 77–79, 86–88, 97–100, and 102). The timing of these interactions during cellular metabolism remains to be studied in detail.

p34 interact transiently during the cell cycle.[74–76] Assembly of the RPA complex is cell cycle–regulated (see below). Mutational analysis of RPA1 (p70) has helped to define its domains of interaction with DNA and proteins and its essential role in replication.[72,73] In *Xenopus*, a protein known as FFA-1 was identified as being essential for RPA association to replication foci. FFA-1, a helicase and DNA-dependent ATPase, has been shown to be a homologue of Werner protein.[44]

RPA1 (p70) can stimulate the activity of DNA polymerases alfa (pol alfa/primase complex), sigma, and epsilon and increase their processivity during replication. RPA1 and pol alfa interact directly.[77] Simian virus 40 (SV40) and papillomavirus can extend DNA synthesis through influence of RPA function. SV40 T antigen binds to RPA1 (p70) and to DNA polymerase alfa/ primase and in this way initiates replication of the viral genome.[78,79] RPA is required during the elongation step of DNA synthesis.[79,80] WRN shares a number of features with SV40 large T antigen, which is capable of extending life *in vitro*.[78,79]

Role in Transcription

The transcription efficiency in WS cell lines is reduced to 40–60% of the transcription measured in normal cells, both *in vivo* and *in vitro*.[30] This defect was complemented by addition of WRN protein. WRN has transcriptional activator potential and has a role in RNA pol II–driven transcription. A global or regional transcription deficiency in WS may be one of the causes of the clinical symptoms observed in this syndrome.[30] The critical region for transcription activation most likely mapped to the region of 315 to 403 amino acids.[30,81]

WRN could have a role in rDNA transcription since [^{3}H]uridine incorporation into nucleoli is coincident with WRN enrichment and decreases dramatically upon 4-NQO treatment.[60] RPA can also modulate gene expression, that is, repression of the human metallothionein IIA gene.[71]

Role in Recombination

A defect in recombination is suggested by the hyper-rec phenotype observed in some WS cell lines.[2] WRN and SGS1 helicases can suppress the increased homologous and illegitimate recombination in the *S. cerevisiae sgs1* mutant.[82] SGS1 is able to interact with topoisomerase II and III,[83,84] is localized to the nucleolus, and prevents nucleolar fragmentation with age.[85] Also, *rqh1+* from fission yeast is required to suppress inappropiate recombination which is essential for reversible S phase arrest.[43] These reports suggest that WRN, like SGS1, is required for chromosome segregation and controls genomic stability in human cells. *Sgs1⁻* cells show an accumulation of extrachromosomal rDNA circles in old cells, leading to premature aging and shorter life span.[85] The helicase activity of WRN may be required for chromosome segregation, like SGS1.[83,84] SGS1 coordinately functions with topoisomerase III during the decatenation of intertwined chromosomes. Defects in this process may give rise to mutations and chromosomal instability as observed in WS.

RPA also can bind to RAD52 and RAD51, proteins that are essential for double-strand DNA break repair, and this interaction is essential for homologous recombination. RPA stimulates homologous pairing and strand transfer by hRAD51.[86–88]

Role in DNA Damage Repair

DNA repair and/or survival of WS cells appeared to be normal following ultraviolet, gamma-, and X-irradiation.[89] Later work reported subtle transcription-coupled repair deficiency in WS lymphoblasts.[21] In addition, WS cells showed a clear accumulation of chromosomal breaks after X-ray irradiation, and radiation sensitivity increase with age,[22] suggesting that some aspect of double-strand break repair is impaired, although not as severely as in ataxia telangiectasia (AT).

Hypersensitivity to the genotoxic agent 4-nitro-quinoline oxide (4-NQO), which causes oxidative damage, has been reported.[14,26] Hence, it is possible that WRN is involved in some aspect of oxidative DNA damage repair. However, 4-NQO is pleiotropic in causing a wide spectrum of DNA adducts, suggesting that damages to DNA other than oxidative damage should be considered. 4-NQO induces an apoptotic response in WS cells.[26] There is yet no study indicating that WS DNA has increased oxidative damage with age as compared to normal cells, but early studies showed that addition of radical-scavenging enzymes to WS cultures decreased the frequency of chromosome breaks.[13,90] Reintroduction of chromosome 8 did not complement or restore the defects accumulated in WS.[91]

Recently, WS cells were also shown to be sensitive to S phase–specific drugs, that is, camptothecin (inhibitor of topoisomerase 1),[27,28] suggesting that WRN functions primarily during DNA replication, correcting DNA lesions with high fidelity during the progression of the replication fork.[29]

Early reports showed that WS cell lines exhibit normal repair of damage caused by the alkylating agent diepoxybutane and bleomycin, a generator of double-strand breaks.[14] Also, there is evidence for variegated translocation mosaicism with passage number,[2,12] implying that DNA breaks, perhaps single-strand breaks that are not repaired efficiently, are generated with time and represent a good source for recombination events leading eventually to translocations and genomic rearrangements.[2,12–17]

The WRN-interacting protein RPA can bind to double-stranded DNA and with cisplatin and UV-damaged DNA[92,93] and is associated with chromatin.[94] RPA can interact with XPA, ERRC1/XPF, and XPG, indicating that RPA is involved in nucleotide excision repair (NER).[71,93] RPA1 deletions mutants are defective in NER, and RPA has also been found to be required for mismatch repair *in vitro*.[95] Also, there is evidence suggesting an interaction of RPA1 (p70) with uracil-DNA glycosylase involved in base-excision repair, which detects modified bases and initiates DNA repair. Interestingly, 5-hydroxymethyluracil-DNA glycosylase activity is reduced in WS cells, whereas other DNA glycosylase activities remain normal, suggesting that base excision repair of some pyrimidine derivatives is compromised in WS. Senescent fibroblasts also show similar defect.[96]

Both WRN and RPA1 (p70) have been shown to bind to the tumor suppressor protein p53.[55,71] The RPA1–p53 interaction blocks RPA1 binding to ssDNA. p53 binds to damaged DNA such that the RPA1-p53 interaction can help localize RPA near the damage and initiate repair.[71,77] Cellular DNA damage triggers phosphorylation of RPA2 (p32) leading to the disassembly of the replication complex and DNA repair can start.[97] A DNA-dependent protein kinase activity associated with ATM or ATM itself[98,99] (protein mutated in ataxia telangiectasia) and various cyclin-dependent kinases (CDKs) can phosphorylate RPA2 (p32).[97,98,100] Finally, DNA-dependent protein kinase (DNA-PK) inactivates SV40 T antigen function in DNA replication by phosphorylation, and the expression of RPA is reduced after exposure to DNA-damaging agents.[101] DNA-PK consists of a DNA-PK catalytic subunit (DNA-PKcs) and Ku autoantigen (heterodimer of Ku70 and Ku80) and functions in ds break recognition.[102] DNA-PKcs binds RPA1 (p70) but is replaced by Ku autoantigen upon camptothecin treatment, leading to RPA2 (p32) phosphorylation.[100]

Thus, RPA function represents an S-phase checkpoint that is crucial for maintaining genome integrity after damage during cell cycle progression.

FUTURE PERSPECTIVES

A number of studies will be necessary in the years to come to elucidate more specifically the functional role of WRN in human cells, including identification of novel protein-protein interactions during cellular division and cellular stress, as well as its role during the normal aging process. The interaction kinetics between WRN and RPA1 (70 kDa subunit) during processing of DNA damage and the specific type of damage will prove important for our understanding of WRN function.[66,72]

Accumulation of senescent cells has been observed both *in vitro*, as a result of exhaustion of replicative capacity and, *in vivo*, in various tissues as a function of age.[103] Mutations accumulate with age,[22,24,104] and it is possible that *WRN* (or an associated gene in the same pathway) is also affected by mutations as a function of age in the normal population. WRN malfunction induces premature cell senescence, and this leads to changes in tissue metabolism, which eventually leads to deterioration of certain body processes.

Hallmarks of replicative senescence that reflect the loss of proliferative capacity have yielded some insight into the mechanism triggering senescence of WS.[105] It is becoming increasingly evident that more than one signaling pathway or mechanism can lead to senescence or aging. WRN is highly expressed in tumor cell lines that proliferate rapidly[54,55] but is absent in senescent cells.[51] Thus, WRN could become a potential target for novel anticancer drugs.

WRN helicase function may prove necessary in many activities for which RPA1 (p70) activity is required: a DNA repair function tightly associated with replication arrest following DNA damage is gaining interest. Further studies would show whether WRN is the long-sought functional homologue of SV40 T antigen in human cells.

ACKNOWLEDGMENTS

We thank Drs. George Martin, Junko Oshima, and Matthew D. Gray for fruitful discussions and collaboration. We thank the Danish Centre for Molecular Gerontology for support.

REFERENCES

1. MERCK RESEARCH LABORATORIES. 1995. Merck Manual of Geriatrics. 2nd edit. Merck & Co., Inc. Rahway, NJ.
2. SALK, D. 1982. Werner's syndrome: a review of recent research with an analysis of connective tissue metabolism, growth control of cultured cells, and chromosomal aberrations. Hum. Genet. 62: 1–15.
3. EPSTEIN, C.J. *et al.* 1966. Werner's syndrome: a review of its symptomatology, natural history, pathologic features, genetics and relationship to the natural aging process. Medicine 45: 177–221.
4. MURATA, K. & H. NAKASHIMA. 1982. Werner's syndrome: twenty-four cases with a review of the Japanese medical literature. J. Am. Geriat. Soc. 30: 303–308.
5. THWEATT, R. & S. GOLDSTEIN. 1993. Werner syndrome and biological ageing: a molecular genetic hypothesis. BioEssays 15: 421–426.
6. MARTIN, G.M. 1997. Genetics and the pathobiology of ageing. Phil. Trans. R. Soc. London B. 352: 1773–1780.
7. LEVERENZ, J.B. *et al.* 1998. Aging-associated neuropathology in Werner syndrome. Acta Neuropathol. (Berlin) 96: 421–424.
8. TAKEUCHI, F. *et al.* 1982. Prolongation of S phase and whole cell cycle in Werner's syndrome fibroblasts. Exp. Gerontol. 17: 473–480.
9. POOT, M. *et al.* 1992. Impaired S-phase transit of Werner syndrome cells expressed in lymphoblastoid cell lines. Exp. Cell. Res. 202: 267–273.
10. FUJIWARA, Y. *et al.* 1977. A retarded rate of DNA replication and normal level of DNA repair in Werner's syndrome fibroblasts in culture. J. Cell. Physiol. 92: 365–374.
11. SALK, D. *et al.* 1981. Evidence of clonal attenuation, clonal succession, and clonal expansion in mass cultures of aging Werner's syndrome skin fibroblasts. Cytogenet. Cell Genet. 30: 108–117.
12. HOEHN, H. *et al.* 1975. Variegated translocation mosaicism in human skin fibroblast cultures. Cytogenet. Cell Genet. 15: 282–298.
13. NORDENSON, I. 1977. Chromosome breaks in Werner's syndrome and their prevention in vitro by radical-scavenging enzymes. Hereditas 87: 151–154.
14. GEBHART, E. *et al.* 1988. Spontaneous and induced chromosomal instability in Werner syndrome. Hum. Genet. 80: 135–139.

15. SCAPPATICCI, S. *et al.* 1982. Clonal structural chromosomal rearrangements in primary fibroblast cultures and in lymphocytes of patients with Werner's syndrome. Hum. Genet. **62:** 16–24.
16. SCHONBERG, S. *et al.* 1984. Werner's syndrome: proliferation in vitro of clones of cells bearing chromosome translocations. Am. J. Hum. Genet. **36:** 387–397.
17. FUKUCHI, K. *et al.* 1989. Mutator phenotype of Werner syndrome is characterized by extensive deletions. Proc. Natl. Acad. Sci. USA **86:** 5893–5897.
18. SCHULZ, V.P. *et al.* 1996. Accelerated loss of telomeric repeats may not explain accelerated replicative decline of Werner syndrome cells. Hum. Genet. **97:** 750–754.
19. TAHARA, H. *et al.* 1997. Abnormal telomere dynamics of B-lymphoblastoid cell strains from Werner's syndrome patients transformed by Epstein-Barr virus. Oncogene **15:** 1911–1920.
20. RÜNGER, T.M. *et al.* 1994. Hypermutable ligation of plasmid DNA ends in cells from patients with Werner syndrome. J. Invest. Dermatol. **102:** 45–48.
21. WEBB, D.K. *et al.* 1996. DNA repair fine structure in Werner's syndrome cell lines. Exp. Cell Res. **224:** 272–278.
22. WEIRICH-SCHWAIGER, H. *et al.* 1994. Correlation between senescence and DNA repair in cells from young and old individuals and in premature aging syndromes. Mut. Res. **316:** 37–48.
23. FUKUCHI, K. *et al.* 1990. Increased frequency of 6-thioguanine-resistant peripheral blood lymphocytes in Werner syndrome patients. Hum. Genet. **84:** 249–252.
24. KYOIZUMI, S. *et al.* 1998. In vivo somatic mutations in Werner's syndrome. Hum. Genet. **103:** 405–410.
25. KRUK, P.A. *et al.* 1995. DNA damage and repair in telomeres: relation to aging. Proc. Natl. Acad. Sci. USA **92:** 258–262.
26. OGBURN, C.E. *et al.* 1997. An apoptosis-inducing genotoxin differentiates heterozygotic carriers for Werner helicase mutations from wild-type and homozygous mutants. Hum. Genet. **101:** 121–125.
27. OKADA, M. *et al.* 1998. Differential effects of cytotoxic drugs on normal and immortalised B-lymphoblastoid cell lines from normal and Werner's syndrome patients. Biol. Pharm. Bull. **21:** 235–239.
28. POOT, M. *et al.* 1999. Werner syndrome lymphoblastoid cells are sensitive to camptothecin-induced apoptosis in S-phase. Hum. Genet. **104:** 10–14.
29. CHAKRAVERTY, R.K. & I.D. HICKSON. 1999. Defending genome integrity during DNA replication: a proposed role for RecQ family helicases. Bioessays **21:** 286–294.
30. BALAJEE, A.S. *et al.* 1999. The Werner syndrome protein is involved in RNA polymerase II transcription. Mol. Biol. Cell **10:** 2655–2668.
31. YU, C.-E. *et al.* 1996. Positional cloning of the Werner's syndrome gene. Science **272:** 258–262.
32. OSHIMA, J. *et al.* 1996. Homozygous and compound heterozygous mutations at the Werner syndrome locus. Hum. Mol. Gen. **5:** 1909–1913.
33. GOTO, M. *et al.* 1997. Analysis of helicase gene mutations in Japanese Werner's syndrome patients. Hum. Genet. **99:** 191–193.
34. YU, C.-E. *et al.* 1997. Mutations in the consensus helicase domains of the Werner syndrome gene. Am. J. Hum. Genet. **60:** 330–341.
35. MATSUMOTO, T. *et al.* 1997. Mutation and haplotype analyses of the Werner's syndrome gene based on its genomic structure: genetic epidemiology in the Japanese population. Hum. Genet. **100:** 123–130.
36. MEISSLITZER, C. *et al.* 1997. Werner syndrome: characterization of mutations in the *WRN* gene in an affected family. Eur. J. Hum. Genet. **5:** 364–370.
37. MOSER, M.J. *et al.* 1999. WRN mutations in Werner syndrome. Hum. Mutat. **13:** 271–279.
38. YE, L. *et al.* 1997. Association of a polymorphic variant of the Werner helicase gene with myocardial infarction in a Japanese population. Am. J. Med. Genet. **68:** 494–498.
39. CASTRO, E. *et al.* 1999. Polymorphisms at the Werner locus: I. Newly identified polymorphisms, ethnic variability of 1367Cy/Arg, and its stability in a population of Finnish centenarians. Am. J. Med. Genet. **82:** 399–403.

40. MOROZOV, V. *et al.* 1997. A putative nucleic acid-binding domain in Bloom's and Werner's syndrome helicases. Trends Biochem. Sci. **22:** 417–418.

41. HUANG, S. *et al.* 1998. The premature ageing syndrome protein, WRN, is a $3' \rightarrow 5'$ exonuclease. Nat. Genet. **20:** 114–116.

42. KUSANO, K. *et al.* 1999. Evolution of the RECQ family of helicases: a drosophila homolog, Dmblm, is similar to the human bloom syndrome gene. Genetics **151:** 1027–1039.

43. STEWART, E. *et al.* 1997. *rqh1+*, a fission yeast gene related to the Bloom's and Werner's syndrome genes, is required for reversible S phase arrest. EMBO J. **16:** 2682–2692.

44. YAN, H. *et al.* 1998. Replication focus-forming activity 1 and the Werner syndrome gene product. Nat. Genet. **19:** 375–378.

45. SEKELSKY, J.J. *et al.* 1999. *Drosophila* and human RecQ5 exist in different isoforms generated by alternative splicing. Nucleic Acids Res. **27:** 3762–3769

46. KITAO, S. *et al.* 1998. Cloning of two new human helicase genes of the RecQ family: biological significance of multiple species in higher eukaryotes. Genomics **54:** 443–452.

47. KITAO, S. *et al.* 1999. Mutations in RECQ4L cause a subset of cases of Rothmund-Thomson syndrome. Nat. Genet. **22:** 82–84.

48. YAMABE, Y. *et al.* 1997. Down-regulation of the defective transcripts of the Werner's syndrome gene in the cells of patients. Biochem. Biophys. Res. Comm. **236:** 151–154.

49. MURANO, S. *et al.* 1984. Diverse gene sequences are overexpressed in Werner syndrome fibrblasts undergoing premature replicative senescence. Mol. Cell. Biol. **11:** 3905–3914.

50. TODA, T. *et al.* 1998. A comparative analysis of the proteins between the fibroblasts from Werner's syndrome patients and age-matched normal individuals using two-dimensional gel electrophoresis. Mech. Ageing Dev. **100:** 133–143.

51. MATUOKA, K. & T. TAKENAWA. 1998. Downregulated expression of the signaling molecules Nck, c-Crk, Grb2/Ash, PI 3-kinase p110 alpha and WRN during fibroblast aging in vitro. Biochim. Biophys. Acta **1401:** 211–215.

52. YAMABE, Y. *et al.* 1998. Sp1-mediated transcription of the Werner helicase gene is modulated by Rb and p53. Mol. Cell Biol. **18:** 6191–6200.

53. WANG, L. *et al.* 1998. Structure and function of the human Werner syndrome gene promoter: evidence for transcriptional modulation. Nucl. Acids Res. **26:** 3480–3485.

54. SHIRATORI, M. *et al.* 1999. Detection by epitope-defined monoclonal antibodies of Werner DNA helicases in the nucleoplasm and their upregulation by cell transformation and immortalization. J. Cell. Biol. **144:** 1–9.

55. SPILLARE, E.A. *et al.* 1999. p53-mediated apoptosis is attenuated in Werner syndrome cells. Genes Dev. **13:** 1355–1360.

56. LEBEL, M. & P. LEDER. 1998. A deletion within the murine Werner syndrome helicase induces sensitivity to inhibitors of topoisomerase and loss of cellular proliferative capacity. Proc. Natl. Acad. Sci. USA **95:** 13097–13102.

57. WU, J. *et al.* 1998. Effect of age and apoptosis on the mouse homologue of the huWRN gene. Mech. Ageing Dev. **103:** 27–44.

58. MATSUMOTO, T. *et al.* 1997. Impaired nuclear localization of defective DNA helicases in Werner's syndrome. Nat. Genet. **16:** 335–336.

59. MATSUMOTO, T. *et al.* 1998. Characterization of the nuclear localization signal in the DNA helicase involved in Werner's syndrome. Int. J. Mol. Med. **1:** 71–76.

60. GRAY, M.D. *et al.* 1998. Werner helicase is localized to transcriptionally active nucleoli of cycling cells. Exp. Cell Res. **242:** 487–494.

61. MARCINIAK, R.A. *et al.* 1998. Nucleolar localization of the Werner syndrome protein in human cells. Proc. Natl. Acad. Sci. USA **95:** 6887–6892.

62. GRAY, M.D. *et al.* 1997. The Werner syndrome protein is a DNA helicase. Nat. Genet. **17:** 100–103.

63. SUZUKI, N. *et al.* 1997. DNA helicase activity in Werner's syndrome gene product synthesized in a baculovirus system. Nucl. Acids Res. **25:** 2973–2978.

64. SHEN, J.-C. *et al.* 1998. Characterization of Werner syndrome protein DNA helicase activity: directionality, substrate dependence and stimulation by replication protein A. Nucleic Acids Res. **26:** 2879–2885.

65. BROSH, R.M., JR. *et al.* 1999. Functional and physical interaction between WRN helicase and human replication protein A. J. Biol. Chem. **274:** 18341–18350.

66. ORREN, D.K. *et al.* 1999. Enzymatic and DNA binding properties of purified WRN protein: high affinity binding to single-stranded DNA but not to DNA damage induced by 4NQO. Nucleic Acids Res. **27:** 3557–3566.

67. FRY, M. & L.A. LOEB. 1999. Human Werner syndrome DNA helicase unwinds tetrahelical structures of the fragile X syndrome repeat sequence d(CGG)n. J. Biol. Chem. **274:** 12797–12802.

68. KAMATH-LOEB, A.S. *et al.* 1998. Werner syndrome protein.II. Characterization of the integral 3′-5′ DNA exonuclease. J. Biol. Chem. **273:** 34145–34150.

69. SHEN, J.C. *et al.* 1998. Werner syndrome protein.I. Dna helicase and dna exonuclease reside on the same polypeptide. J. Biol. Chem. **51:** 34139–34144.

70. SUZUKI, N. *et al.* 1999. Werner syndrome helicase contains a 5′ → 3′ exonuclease activity that digests DNA and RNA strands in DNA/DNA and RNA/DNA duplexes dependent on unwinding. Nucleic Acids Res. **27:** 2361–2368.

71. WOLD, M.S. 1997. Replication protein A: a heterotrimeric, single-stranded DNA-binding protein required for eukaryotic DNA metabolism. Ann. Rev. Biochem. **66:** 61–92.

72. WALTHER, A.P. *et al.* 1999. Replication protein A interactions with DNA. 1. Functions of the DNA-binding and zinc-finger domains of the 70-kDa subunit. Biochemistry **38:** 3963–3973.

73. BOCHKAREV, A. *et al.* 1999. The crystal structure of the complex of replication protein A subunits RPA32 and RPA14 reveals a mechanism for single-stranded DNA binding. EMBO J. **18:** 4498–4504.

74. ADACHI, Y. & U.K. LAEMMLI. 1992. Identification of nuclear pre-replication centers poised for DNA synthesis in *Xenopus* egg extracts: immunolocalization study of replication protein A. J. Cell. Biol. **119:** 1–15.

75. CARDOSO, M.C. *et al.* 1993. Reversal of terminal differentiation and control of DNA replication: cyclin A and Cdk2 specifically localize at subnuclear sites of DNA replication. Cell **74:** 979–992.

76. MURTI, K.G. *et al.* 1996. Dynamics of human replication protein A subunit distribution and partitioning in the cell cycle. Exp. Cell. Res. **223:** 279–289.

77. BRAUN, K.A. *et al.* 1997. Role of protein–protein interactions in the function of replication protein A (RPA): RPA modulates the activity of DNA polymerase alfa by multiple mechanisms. Biochemistry **36:** 8443–8454.

78. WEISSHART, K. *et al.* 1998. The replication protein A binding site in simian virus 40 (SV40) T antigen and its role in the initial steps of SV40 DNA replication. J. Virol. **72:** 9771–9781.

79. HAN, Y. *et al.* 1999. Interactions of the papovavirus DNA replication initiator proteins, bovine papillomavirus type 1 E1 and simian virus 40 large T antigen, with human replication protein A. J. Virol. **73:** 4899–4907.

80. WALTHER, A.P. *et al.* 1999. A novel assay for examining the molecular reactions at the eukaryotic replication fork: activities of replication protein A required during elongation. Nucleic Acids Res. **27:** 656–664.

81. YE, L. *et al.* 1998. Transcriptional activation by the Werner syndrome gene product in yeast. Exp. Gerontol. **33:** 805–812.

82. YAMAGATA, K. *et al.* 1998. Bloom's and Werner's syndrome genes suppress hyperrecombination in yeast *sgs1* mutant: implication for genomic instability in human diseases. Proc. Natl. Acad. Sci. USA **95:** 8733–8738.

83. GANGLOFF, S. *et al.* 1994. The yeast type I topoisomerase Top3 interacts with Sgs1, a DNA helicase homolog: a potential eukaryotic reverse gyrase. Mol. Cell. Biol. **14:** 8391–8398.

84. WATT, P.M. *et al.* 1995. Sgs1: a eukaryotic homolog of *E. coli* RecQ that interacts with topoisomerase II *in vivo* and is required for faithful chromosome segregation. Cell **81:** 253–260.

85. JOHNSON, F.B. *et al.* 1999. Molecular biology of aging. Cell **96:** 291–302.

86. PARK, M.S. *et al.* 1996. Physical interaction between human RAD52 and RPA is required for homologous recombination in mammalian cells. J. Biol. Chem. **271:** 18996–19000.

87. BAUMANN, P. & S.C. WEST. 1997. The human Rad51 protein: polarity of strand transfer and stimulation by hRP-A. EMBO J. **16:** 5198–5206.
88. GOLUB, E.I. *et al.* 1998. Interaction of human Rad51 recombination protein with single-stranded DNA binding protein, RPA. Nucl. Acids Res. **26:** 5388–5393.
89. HIGASHIKAWA, T. & Y. FUJIWARA. 1978. Normal level of unscheduled DNA synthesis in Werner's syndrome fibroblasts in culture. Exp. Cell Res. **113:** 438–442.
90. SALK, D. *et al.* 1981. Effects of radical-scavenging enzymes and reduced oxygen exposure on growth and chromosome abnormalities of Werner syndrome cultured skin fibroblasts. Hum. Genet. **57:** 269–275.
91. KODAMA, S. *et al.* 1998. Failure to complement abnormal phenotypes of simian virus 40-transformed Werner syndrome cells by introduction of a normal human chromosome 8. Cancer Res. **58:** 5188–5195.
92. PATRICK, S.M. & J.M. TURCHI. 1999. Replication protein A (RPA) binding to duplex cisplatin-damaged DNA is mediated through the generation of single-stranded DNA. J. Biol. Chem. **274:** 14972–14978.
93. STIGGER, E. *et al.* 1998. Functional analysis of human replication protein A in nucleotide excision repair. J. Biol. Chem. **273:** 9337–9343.
94. TREUNER, K. *et al.* 1998. Chromatin association of replication protein A. J. Biol. Chem. **273:** 31744–31750.
95. LIN, Y.-L. *et al.* 1998. The evolutionarily conserved zinc finger motif in the largest subunit of human replication protein A is required for DNA replication and mismatch repair but not for nucleotide excision repair. J. Biol. Chem. **273:** 1453–1461.
96. GANGULY, T. & N.J. DUKER. 1992. Reduced 5-hydroxymethyluracil-DNA glycosylase activity in Werner's syndrome cells. Mutat. Res. **275:** 87–96.
97. TREUNER, K. *et al.* 1999. Phosphorylation of replication protein A middle subunit (RPA32) leads to a disassembly of the RPA heterotrimer. J. Biol. Chem. **274:** 15556–15561.
98. GATELY, D.P. *et al.* 1998. Characterization of ATM expression, localization, and associated DNA-dependent protein kinase activity. Mol. Biol. Cell **9:** 2361–2374.
99. CHENG, X. *et al.* 1996. Ionizing radiation-induced phosphorylation of RPA p34 is deficient in ataxia telangiectasia and reduced in aged normal fibroblasts. Radiother. Oncol. **39:** 43–52.
100. SHAO, R.G. *et al.* 1999. Replication-mediated DNA damage by camptothecin induces phosphorylation of RPA by DNA-dependent protein kinase and dissociates RPA:DNA-PK complexes. EMBO J. **18:** 1397–1406.
101. WANG, Y. *et al.* 1999. Roles of replication protein A and DNA-dependent protein kinase in the regulation of DNA replication following DNA damage. J. Biol. Chem. **274:** 22060–22064.
102. SMITH, G.C.M. & S.P. JACKSON. 1999. The DNA-dependent protein kinase. Genes Dev. **13:** 916–934.
103. DIMRI, G.P. *et al.* 1995. A biomarker that identifies senescent human cells in culture and in aging skin *in vivo*. Proc. Natl. Acad. Sci. USA **92:** 9363–9367.
104. WARNER, H.R. & T.E. JOHNSON. 1997. Parsing age, mutations and time. Nat. Genet. **17:** 368–370.
105. OSHIMA, J. *et al.* 1995. Regulation of *c-fos* expression in senescing Werner syndrome fibroblasts differs from that observed in senescing fibroblasts from normal donors. J. Cell. Physiol. **162:** 277–283.

Calorie Restriction and Age-Related Oxidative Stress

B.J. MERRY[a]

School of Biological Sciences, University of Liverpool, Liverpool, L69 3BX, United Kingdom

ABSTRACT: Calorie restriction (CR) in mammals has been recognized as the best characterized and most reproducible strategy for extending maximum survival, retarding physiological aging, and delaying the onset of age-related pathologic conditions in mammals. The overwhelming majority of studies using CR have used short-lived rodent species, although current work using rhesus and squirrel monkeys will determine whether this paradigm is also relevant to manipulating the rate of primate aging. The mechanism by which restricted calorie intake modifies the rate of aging and pathology has been the subject of much controversy, although an attenuation in the lifetime accumulation of oxidative damage appears to be a central feature. Although the majority of studies have focused on the ability of cells from calorie-restricted animals to scavenge free radicals to explain the slower accrual of oxidative damage with age, it is not established that CR has a consistent effect to upregulate the activity of these enzymes in all tissues. A major effect of calorie-restricted feeding now appears to be on the rate of production or leak of free radicals from the mitochondria. The details of the adaptation and the signaling pathway that induces this effect are currently unknown.

MODEL

Subsequent to the original reports of McCay[1,2] that underfeeding rats extended lifetime survival, this finding has been confirmed using a variety of feeding strategies, dietary compositions, severity of underfeeding, animal species, and strains. A consistent observation is that diets that provide essential nutrients and vitamins, avoid malnutrition, but limit the overall energy intake of the animal (30–60% from *ad libitum* intake) over a protracted period of the postweaning life span, result in an extension in life span.[3–5] This is a highly reproducible and robust manipulation for extending the mean, maximum, or average survivorship of the last decile by 20–50% (for a comprehensive review of the CR paradigm see Yu[6]) but the development of a detailed understanding of how diet retards the rate of aging has been slow.

Degree and Duration of Restriction

The basic parameters of the model give some indication about the properties of the underlying mechanism. Analysis of 24 published studies using rats and mice maintained under varying degrees of severity of energy restriction revealed a signif-

[a]Address for correspondence: Phone: +44 0151 794 5076; fax: +44 0151 794 5017. bm01@liv.ac.uk

icant negative correlation between survival and energy intake (Merry, unpublished observations). It is clear that a dose–response relationship exists over a wide range of energy intakes between the intensity of energy restriction and survival achieved. The shape of the survival curve is unaltered but moved to the right.

An analysis of 36 published studies (Merry, unpublished observations) that used a degree of food restriction in the range of 40–50% *ad libitum* intake revealed a significant positive correlation between survival parameters (mean, maximum, and average last decile survival) and the duration of the restricted feeding regime. The longer an animal is maintained on restricted feeding during the postweaning period of the life span, the greater the survival, confirming the earlier observation of Yu.[6] Thus the underlying mechanism of CR that enhances survival can be varied in intensity, and its effects are accumulative.

The degree to which the effects of CR on survival are retained or reversed by a return to full feeding is controversial. Restricting Fischer344 rats to 60% *ad libitum* intake between 1.5 and 6 months of age extended maximum survival by approximately 15%.[7] This small but significant effect of short-term CR early in life on subsequent survival, insufficient to modify the morbidity profile, is intriguing and suggests a "memory imprint."[7] The adult phase of the survival curve after a return to full feeding, however, was steeper than for *ad libitum*–fed control animals, indicative of a slightly higher rate of aging. Data from the author's laboratory cannot confirm a "memory imprint" for early, short-term CR and suggests that refeeding food-restricted rats partly or fully reverses the beneficial effects on survival.[8]

Therefore, the survival data characterises the underlying mechanism retarding the rate of aging as dynamic, requiring the continuous constraint of low calorie intake to maintain the effect, and is partly reversible by a return to full feeding.

Amelioration of the rate of accrual of oxidative damage with age is considered to offer the best mechanistic explanation to explain this dynamic and wide-ranging effect of CR on survival, age-related physiology and pathology.[9] However, caution is required in an uncritical acceptance of this explanation for knowledge of the *in vivo* generation of reactive oxygen species (ROS) in response to both aging[10] and CR feeding is still very limited.

Oxidative Damage with Aging: Modification by CR

Lipids

Peroxidation. The earliest work reporting age-related oxidative modification of cellular components and the effect of CR centered around the formation of lipofuschin and malondialdehyde.[11,12] These authors reported that brain lipofuscin and malondialdehyde was significantly reduced in mice maintained on 50% *ad libitum* food intake. Lipid peroxidation and the secondary decomposition products include a variety of compounds such as hydrocarbons, epoxides, aldehydes, ketones, and carboxylic acids, many of which are highly toxic, and some, such as pentane and ethane, provide for noninvasive markers of lipid peroxidation.[13]

Koizumi *et al.*[14] were the first to show that extended survival in CR mice (approximately 40% restriction) was associated with a 30% decrease in liver lipid peroxidation at 12 months of age and 13% at 24 months of age.

Yu's group has published a series of studies on oxidative damage in liver microsomes and mitochondrial membranes isolated from the Fischer344 rat as a function of age and feeding regime and confirmed this earlier report.[15–17] These studies have been extended to include cells isolated from the immune system. The functional decline in T-lymphocytes observed during aging in *ad libitum*–fed rats was inversely correlated with the levels of both lipid peroxidation in the plasma and splenic lymphocytes. Caloric restriction was found to partially reverse the age-dependent decrease in T-lymphocyte proliferation and significantly reduce lipid peroxidation in plasma and splenocytes.[18]

Membrane lipid composition. The seminal studies of Yu and co-workers were the first to demonstrate that a CR feeding regime, recognized to extend longevity, modifies the lipid composition of the mitochondria and microsomal membrane, leaving them less susceptible to peroxidation.[15] Calorie-restricted feeding induces a lower unsaturation/saturation index, specifically increasing fatty acids 18:2 and 18:3 and decreasing the content of 20:4, 22:5, and 22:6. These changes are the reverse of the membrane composition changes seen with age in fully fed animals, namely, a decrease in 18:2 and 18:3 and an increase in the highly peroxidizable, unsaturated 20:4, 22:5, and 22:6 fatty acids. This finding has been confirmed by Bailey *et al.*[19] and by Lee[20] for ventricular mitochondrial membranes and in the author's own laboratory (TABLE 1).

Membrane order (fluidity). A major functional consequence of an age-related peroxidation of the plasma and subcellular membranes in fully fed animals is considered to be a loss in membrane fluidity.[21,22] Mitochondrial membranes may be particularly susceptible to such peroxidative damage because of the reactive oxygen species generated in this organelle. Increased membrane rigidity with age may compromise many membrane-associated functions, particularly those of the electron transport chain in the mitochondria and receptor binding, thus impairing cell signalling. Calorie restriction feeding was found to preserve liver mitochondrial and microsomal membrane order with only a small decline in membrane fluidity occurring over 24 months of age.[23] Yu's laboratory has now confirmed this effect of CR for liver,[24,25] brain,[26] and cardiac ventricular mitochondrial membranes.[20]

The membrane composition changes in the polyunsaturated fatty acid component may explain the retention of membrane fluidity with age under CR feeding regimes. The reported higher 18:2 and lower 22:4 fatty acid content of mitochondrial membranes of CR rats should make these membranes more resistant to peroxidation. Coupled with the decreased ROS production under CR feeding, it might be expected that the rate of lipid peroxidation and hence the increase in membrane rigidity with aging would be retarded.[20]

The effect of CR on age-related change in membrane rigidity and reactive oxygen species generation has been studied in a mixed population of brain synaptosomes and mitochondria.[27] Mitochondria and synaptosomes prepared from Brown Norway rats of varying age and dietary status were labeled with 5-nitroxyl stearate, a membrane lipid–specific spin label, which was detected by electron paramagnetic resonance (ESR). The anisotropic motion of the spin probe provided a measure of membrane fluidity. On stimulation of the electron transport chain with 10 mM or 20 mM succinate, the ROS flux generated, and the presumed peroxidation-induced changes in membrane fluidity, were recorded. A significant effect on membrane flu-

TABLE 1. Changes in heart mitochondrial membrane phosphatidylcholine (PC), phosphatidylethanolamine (PE), and cardiolipin (CL) lipid fractions induced by calorie restriction[a] by 6 months of age

Heart	PC		PE		CL[3]	
Fatty acid[b]	Fully fed	DR	Fully fed	DR	Fully fed	DR
14:0	0.9	1.0	0.6	1.0	0.8	1.6
16:0	36.8	37.1	23.7	27.1	28.2	31.1
16:1	1.2	1.2	1.5	1.4	1.4	1.7
16:3	3.5	4.5	3.0	3.5	3.9	4.7
18:0	17.7	17.5	22.5	25.4	9.8	10.5
18:1	8.9	7.2	6.0	5.6	5.6	4.9
18:2	5.9	7.2	4.9	5.3	29.1	27.5
20:1	2.6	2.7	2.0	1.9	3.0	2.7
20:4	7.6	7.6	15.0	11.8	2.9	1.7
22:1	11.5	9.3	7.7	6.9	12.5	11.2
22:5	1.1	1.7	2.3	1.1		
22:6	2.0	1.9	15.0	12.3	3.7	2.3
%sat	55.8	55.6	44.8	50.8	38.7	43.1
%unsat	44.2	44.4	55.2	49.2	61.3	56.9
%MUFA	24.2	20.3	17.0	15.2	16.5	22.7
%PUFA	20.0	23.1	38.2	34.0	44.8	34.1
UI	94.1	99.5	188.7	163.0	139.6	108.5

[a]Calorie-restricted, male Brown Norway rats were fed 50% of the age-matched *ad libitum* food intake from 6 weeks of age.

[b]Total lipid extracted from isolated mitochondria was separated into PE, PC, and CL classes of phospholipids by thin-layer chromatography before saponification and methylation. Fatty acid methyl esters were analyzed by gas–liquid chromatography and identified using known standards whose identity had been confirmed by mass spectrometry. Retention times and peak areas were calculated using the ATI Unicam Chromatography Data Handling System.

[c]Major changes in the PC and CL lipid classes are indicated by the double solid lines and show a transition to a more saturated fatty acid composition with a loss of double bonds in the polyunsaturated fatty acids under CR feeding conditions.

idity, either with aging in *ad libitum* fed animals or between CR and control animals, was not seen until after 16 months of age. Stimulation of state 4 mitochondrial respiration with 20 mM succinate resulted in greater oxy-radical production in 25-month-old animals as compared to younger animals, suggesting increased mitochondrial leakage with age, but CR feeding however, was not found to attenuate the ROS flux on succinate stimulation of mitochondria *in vitro*.

The maintenance of membrane fluidity by a reduction in the rate of oxidative damage should be associated with an attenuation in age-related changes in membrane receptor number and binding affinity.[28] The age-related decline in isoproterenol-stimulated adenylyl cyclase of ventricular membranes in *ad libitum*–fed rats is retarded in CR animals. Similarly, inhibition of isoproterenol-stimulated adenylyl cyclase by the AdoA(1) R agonist, N-6-p-sulfophenyladenosine (SPA) decreased with age and is associated with a decrease in the percentage of high-affinity binding sites for SPA, declining from 55% at 6 months to 23% at 24 months. Calorie-restricted feeding again tempered this age-related loss of high-affinity receptors, the percentage of high-affinity binding sites being 42% at 24 months.[29] Chen *et al.,*[30] studying the effects of CR on $\alpha(1)$-adrenergic receptor function in the parotid gland, concluded that CR may have a modulatory effect on age-related impairment resulting from altered G protein–binding activity. Peroxidation of membrane lipids to change membrane fluidity could underlie such age-related loss of receptor and G-protein coupling.

Proteins

Data from several laboratories supports the contention that accumulation of oxidized protein occurs in a number of tissues during normal aging and that this damage is associated with identifiable behavioral and functional impairments. This age-related accumulation of oxidized protein can be partly reversed in fully-fed animals treated with spin-trapping compounds.[31–37] Metal-catalyzed oxidation of proteins introduces carbonyl groups at lysine, arginine, proline, or threonine in a site-specific manner while oxidative modification converts the side-chain of methionine, histidine, and tyrosine and forms cysteine disulfide bonds. Earlier studies indicated that the age-related, steady-state increase in carbonyl content of whole-tissue homogenates was significantly retarded by CR (60% of *ad libitum*) in C57BL/6Nnia mice.[38]

Data illustrating the physiological relevance of increased tissue concentrations of oxidized protein is limited, although it may be envisaged that such a process could modify transcription factor activity and cell signaling. Within the mouse brain, a marked heterogeneity is seen with aging in both the severity and type of protein oxidative damage.[33] Mouse cortex, hippocampus, striatum, and midbrain all showed a significant increase in protein carbonyl between 8 and 27 months of age while the striatum and hippocampus had the greatest absolute increase. Although mice of 15 months that had been maintained on 40% restricted feeding from 4 months showed a 35% lower carbonyl content for the whole brain, the greatest proportional decrease was recorded for the striatum (48%), cerebellum (34%), midbrain (23%), and cortex (17%). The striatum, which showed the greatest carbonyl increase with age, exhibited the greatest response to restricted feeding, whereas the carbonyl content of the hippocampus was relatively unaffected by calorie restriction.[33] These regions of the brain are considered to be significantly damaged during normal aging resulting in cognitive, memory and motor impairment.[39] The age-related loss of these behavioral functions is retarded by calorie restriction.[33,40] The selective effect of CR on retarding the accretion of protein carbonyl with age in the brain with the retention of behavioral functions is indicative that protein oxidation with age has a physiological relevance.

An important observation is that in mice of 15 months, the steady-state level of brain protein oxidative damage (protein carbonyl content) is reversible over a six-week period (FIG. 1).[33] This dynamic response of protein oxidation content to refeeding is comparable to the effect seen for survival discussed above. Crossover studies indicate that these changes in brain protein carbonyl concentration closely follow body weight changes. Whether changes in behavior can be shown to correlate with such acute changes in brain carbonyl concentrations is uncertain.

The ability to reverse or induce oxidative damage with crossover feeding studies remains controversial. Mice fed *ad libitum,* or a restricted diet (60% *ad libitum* intake) from 4 weeks, were switched feeding regimes for 6 weeks at 18–22 months of age.[41] This study measured oxidative damage in mitochondria isolated from hindlimb skeletal muscle. They found no convincing evidence for a reduction or increase in mitochondrial carbonyl concentration induced by the changed feeding regime. This contrary finding from the same laboratory to their earlier work for the brain[33] may reflect specific tissue differences, a differing cellular response between mitochondrial proteins and whole tissue homogenates, or a loss with age in metabolic flexibility in mice 3–7 months older than used in the initial crossover study.

The nature of the mechanisms that promote oxidation of proteins with aging may not only be tissue specific, but appear to differ in their sensitivity to CR feeding. This observation may be exploited to provide insight into the mechanism of action of CR. Calorie restriction (40% from *ad libitum* in C57BL/6Nnia male mice) prevented the age-related increase in o,o'-dityrosine with age in cardiac and skeletal muscle.[42] No age-related increase in o-tyrosine was observed in liver, brain, skeletal muscle, or heart; and CR did not influence o-tyrosine levels. Tyrosyl radicals, generated by peroxidases and other heme proteins, can react to form o,o'-dityrosine, a stable end product, while o-tyrosine is considered to result from hydroxyl radical conversion of phenylalanine. The authors suggest that one possible explanation for their observations is that the increased generation with age of mitochondrial-derived H_2O_2, interacting with heme proteins such as myoglobin, would generate the tyrosyl radical. The recognized effect of calorie restriction to decrease mitochondrial H_2O_2 production would be to reduce the peroxidation of muscle tyrosine and its age-related increase in muscle.

Mitochondrial Proteins

Proteins vary in their susceptibility to oxidative modification, which results in a loss of catalytic and signaling competence or structural integrity. Consequently, measures of total tissue protein oxidative concentrations may be too imprecise. Mitochondria generate the majority of the cellular superoxide anion radical and hydrogen peroxide and would therefore be expected to experience significant oxidative damage with age.[43] Carbonyl content of hindlimb skeletal muscle mitochondria from fully fed C57Bl/6Nnia measured between 3 and 31 months showed a steady increase in the second year of life, increasing ~150% between 12 and 29 months compared to 32% for whole-muscle homogenates.[41] Mice maintained on 60% of *ad libitum* food intake exhibited no age-related increase in mitochondrial carbonyl increase. Data from a number of laboratories indicates that mitochondria undergo significant oxidative damage during aging, which may impede respiratory function. There is initial data from work with the housefly and *E. coli* that specific enzymes possessing active-site iron–sulfur clusters, which are central to mitochondrial ATP synthesis, may be particularly sensitive to oxidative damage with age.[44] It would be of particular interest to know whether the same is true for mammals and whether calorie-restricted feeding specifically protects such enzymes.

Oxidative Damage to DNA: Modification by CR

Genomic DNA

Free radical–induced reactions with nucleic acid produce a range of modifications including adducts of base and sugar groups, single- and double-strand breaks, and cross-links to proteins.[45] Over 20 different products have been recognized from the interaction of the hydroxyl radical with DNA. The electrochemical properties of several of these products, for example, the adduct 8-oxo-guanine (oxo8gua), and the deoxynucleoside 8-oxo-2,7,-dihydro-2′-deoxyguanosine (oxo8dG) and hydroxy-2′-deoxyguanosine (8-OHdG), have provided sensitive markers for estimating oxidative damage to DNA. Using this marker, Fraga et al.[46] demonstrated significant amounts of oxidative damage with aging in rat tissues. Subsequently, Yu's group studied the effect of CR (40% restriction from ad libitum) on the age-related concentration of this marker in the liver of Fischer344 rats.[47] Even at 3 months of age, a substantial level of 8-OHdG was detectable in nuclear DNA of ad libitum–fed animals, which was significantly reduced by 1.5 months of CR feeding.[48] Interestingly, no age effect on nuclear oxidative damage was seen when the comparison was repeated at 24 months of age, but again CR significantly reduced the oxidative damage of nuclear DNA.

A more detailed study of 8-OHdG accumulation in nuclear DNA of the Fischer344 rat with age and restricted feeding for four organs did not confirm this early effect of restricted feeding.[49] The 8-OHdG concentrations in nuclear DNA were similar in ad libitum–fed and CR animals until after 24 months of age. Only late in the life of ad libitum–fed animals was an increase in this oxidative marker observed. This occurred at 24 months for kidney, 27 months for heart and liver, and not until 30 months of age for the brain. The effect of CR feeding was to delay these late-onset increases until 30 months of age or later. Therefore, CR retarded the onset of the late age-related increase in 8-OHdG concentration but had no effect on this oxidative marker in young and middle-aged animals.

Mitochondrial DNA

In view of the critical role mitochondria play in cellular energy production and their resultant exposure to reactive oxygen, nitrogen, and aldehydic radicals generated as metabolic by-products, the functioning of these organelles with aging has attracted considerable attention. Free radicals and other reactive species damage the highly unsaturated mitochondrial membrane lipids,[6,50] mitochondrial proteins,[51,52] and mtDNA,[53,54] of which the latter is the most intensely studied. Deletion frequency of mtDNA has been shown to increase in brain, liver, heart, and skeletal muscle in rodents with age[55,56]; and in skeletal muscle these deletions appear to be clustered in particular muscle fibers.[57]

Calorie-restricted feeding of Fischer344 male rats significantly retarded in liver the age-related increase in mtDNA deletion frequency observed under ad libitum feeding.[58] At 24 months of age, considerable variability is observed between individual ad libitum–fed animals. No difference in deletion frequency could be demonstrated at 6 months of age between control and CR animals but by 18 months, the age-related increase was clearly inhibited by CR feeding (FIG. 2). This observation extends an earlier report from the same laboratory that mitochondria from CR 24-

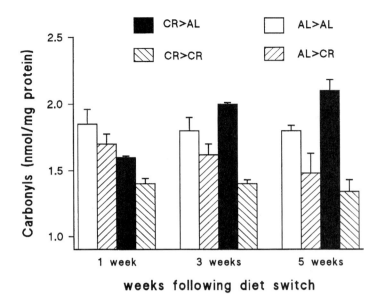

FIGURE 1. Whole-brain protein carbonyl concentrations for 15-month-old C57BL/6Nnia male mice measured over a 5-week period. Study shows the effect of short-term changes in diet on brain protein carbonyl concentrations. Group sizes were two to three mice (3–7 replicate measurements). Animals were either fed *ad libitum* (AL > AL); maintained on 60% of *ad libitum* food intake (AL > CR); or switched to the alternative feeding regime, that is, *ad libitum*–fed mice switched to 60% of *ad libitum* food intake (AL > CR) or *ad libitum*–fed animals switched to calorie-restricted feeding (CR > AL). A significant interaction of chronic and short-term diet ($p < 0.001$) and a two-way interaction of these factors with time ($p < 0.001$) was shown. (Data redrawn from Dubey *et al.*[33])

month-old Fischer344 rats had significantly lower concentrations of oxidative damage when measured by 8-OHdG.[48]

However, this effect of CR on age-related deletion frequency could not be replicated for brain mitochondrial DNA. Although a rise in deletion frequency with age was seen, CR feeding did not modify this increase.[58] The authors speculate that this observation may reflect the greater proportion of postmitotic cells in the brain compared to liver, or mitochondrial turnover which may not be affected by CR. Data on the effects of CR on cell mitochondrial number and turnover has, to the author's knowledge, not been published and represents an important gap in our knowledge of mitochondrial dynamics for this model system.

Two general points are worthy of emphasis with respect to putative mtDNA oxidative damage as a function of aging. First, estimates of age-related mtDNA deletions are highly variable between studies, with estimates varying between a fraction of a percent to 10%. Interpretation of the physiological relevance of such estimates is difficult and highly speculative. Second, in many studies the profile of frequency deletion shows a late-life acceleration. This may indicate acute homeostatic failure or the onset of pathology, rather than a chronic process of accrual throughout the adult life span.

DEFENSE ENZYME ACTIVITY TO ROS: EFFECT OF AGE AND CR

A critical mass of data now exists from which it is possible to conclude that oxidative damage accrues with aging, particularly in short-lived species such as rodents. The consensus from work with CR-feeding regimes is that the rate of accrual of such damage is slower in these longer lived animals. Steady-state measures of oxidative damage represent an equilibrium between rates of oxidant generation, rates of oxidant scavenging, and repair processes.[45] (See Fig. 1 in Beckman and Ames.[10]) Antioxidant defense, including the scavenging enzymes, is the component in this interaction which, because of technical considerations, has been the most intensely studied in CR animals.

The early initial work with mice indicated that in liver homogenates, catalase activity was increased by CR feeding,[14] an observation confirmed for rat hepatic cytosol catalase where CR prevented the age-associated decline observed in control animals. It also prevented the age-related decrease in cellular glutathione and glutathione S-transferase activity observed with aging in fully fed animals.[16] In contrast, however, Luhtala et al.,[59] studying rat skeletal muscle from BN × F344 F1 crosses fed *ad libitum* or 30% food restricted, found that catalase activity increased with aging and this was partially attenuated by restricted feeding. The marked age-related increase reported for glutathione peroxidase activity was completely prevented by CR. Superoxide dismutase activity measured in homogenates of brain, heart, and kidney from C57BL/6Nnia mice showed no clear trend either with aging or CR feeding (60% *ad libitum*).[60] In contrast, catalase activity was higher in CR mice compared to age-matched controls. Glutathione peroxidase showed no consistent profile across organs or with diet. These observations demonstrate that the response of antioxidant-scavenging enzyme activities to CR feeding are species-, strain-, enzyme-, and tissue-specific and generalizations are difficult to make.

Calorie restriction (40% of *ad libitum*) in F344 rats attenuated the age-related decline for the enzyme activities and mRNA concentrations for Cu-Zn superoxide dismutase, catalase, and glutathione peroxidase. Between 21 and 28 months of age, significantly higher activities were recorded in CR rats for these scavenging enzymes.[61] The activity levels correlated with an increase in mRNA concentrations indicative of either transcriptional control or changes in mRNA stability and turnover.

The correlation between mRNA concentrations and enzyme activity levels is not always evident with age and feeding status.[62] Of particular interest is the effect of age and CR on MnSOD activity. At 12 and 24 months significantly higher activity levels were observed in livers of CR Fischer344 rats but this effect of diet did not represent a general trend for nonsignificant differences at 6 months and 30 months were seen. A significant age effect was observed for both *ad libitum* fed and CR for MnSOD RNA in the liver but the significant difference in activity at 12 and 24 months was not reflected in different mRNA concentrations.[62]

MITOCHONDRIAL FREE RADICAL GENERATION

It can be concluded from a number of studies reporting both oxidative damage and defense enzyme activity with age that CR induces significantly lower tissue oxidative damage at an age before a demonstrable difference in defense enzyme activity can be seen. The primary adaptive response induced by CR may be a reduction in the generation of free radicals, the upregulation of scavenging and repair enzymes systems being a secondary response to the feeding regime. It is established that the rate of production of reactive oxygen species from isolated mitochondria in CR animals is significantly reduced for a number of tissues and that this effect can be demonstrated early in life.[38]

Initial explanations for this difference in ROS production of CR animals was based on the assumption that CR animals had a lower metabolic rate. This rationale was developed from the interspecies correlation that as life span increased, so metabolic rate, mitochondrial oxygen consumption, and electron flow decreased. Metabolic rate data, both BMR and 24 hours rates, from McCarter's laboratory when normalized to body mass show no significant differences between age-matched control and CR animals.[63] This is difficult to reconcile with significant reductions in plasma concentrations of insulin (30–40%)[64] and thyroxine (30%) and triiodothyronine (60%),[65] hormones recognized to elevate the metabolic rate and increase superoxide release in mitochondria from *ad libitum*–fed animals.[66,67]

Yu[6] and Barja *et al.*[68] have drawn attention to the lack of a stoichiometric coupling of ROS production to oxygen consumption. Pigeons have a similar metabolic rate and mass to the laboratory rat but a maximum life span eightfold greater. This is associated with a lower ROS production rate from individual mitochondria.[68,69] The rates of superoxide and H_2O_2 production by mitochondria have been widely reported to be highest when the electron chain complexes are most reduced during state 4 respiration, where there is excess substrate and no exogenous ADP.[70,71] Sohal *et al.*[60] have shown for brain, kidney, and heart mitochondria, a lower age-related increase in state 4 respiration when C57BL/6Nnia mice are maintained on 60% of *ad libitum* food intake.

The important observation is that CR feeding reduces ROS production in isolated mitochondria (FIG. 3).[60] More controversial is whether increased rates of ROS occur with aging. Hansford *et al.*[72] were unable to show an increase in H_2O_2 production by rat heart mitochondria with aging and the explanation for this lack of consistency is unknown and may represent differences in mitochondrial preparation and substrate concentrations.

FUNCTIONAL EFFECTS *IN VIVO* OF OXIDATIVE DAMAGE WITH AGING AND CR

Studies on isolated mitochondria have shown that an acute oxidative stress causes an inhibition of mitochondrial respiration and mitochondrial swelling mediated by the Ca^{2+}-dependent inner permeability transition.[73,74] Sastre *et al.*[75] interpret their observations in intact cells as supporting the conclusion from some *in vitro* studies that chronic oxidative stress increases with aging *in vivo*.

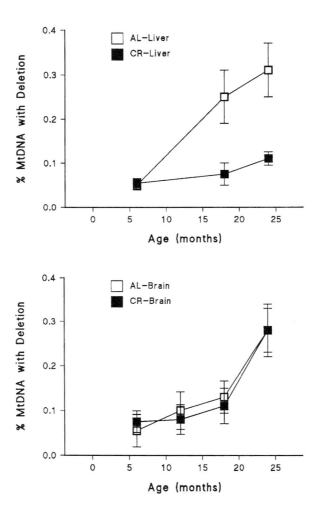

FIGURE 2. Effect of calorie restriction in Fischer344 rats on the frequency of deleted mtDNA in liver and brain mitochondria. The data are shown as the mean ± SEM. Group sizes 13 (6mth AL), 6 (6mth CR), 8 (18mth AL), 6 (18mth CR), 18 (24mth AL), and 13 (24mth CR). A significant age-related increase in frequency deletion was seen for liver between 6- and 28-month-old and 6- and 24-month-old animals ($p < 0.01$). Calorie-restricted feeding significantly depressed the frequency deletions at 18 and 24 months of age ($p < 0.05$ and $p < 0.01$, respectively) in comparison with fully fed animals. Calorie-restricted feeding did not depress the age-related increase in frequency deletions in mtDNA for brain tissue. A significant age effect was seen for both *ad libitum*–fed and calorie-restricted animals ($p < 0.01$). (Redrawn from Kang *et al.*[58])

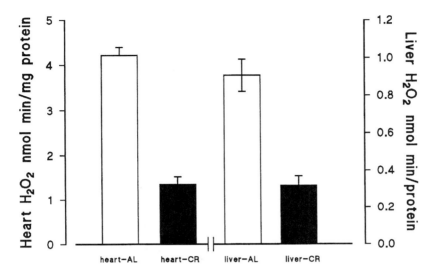

FIGURE 3. Hydrogen peroxide generation from heart and liver mitochondria isolated from 6-month-old male Brown Norway rats. Animals had been fed either *ad libitum* or 50% of the calorie intake of fully fed animals from 6 weeks of age. Calorie-restricted feeding significantly reduced the release of hydrogen peroxide in state 4 respiration with succinate as the substrate when the data is normalized either to milligrams mitochondrial protein or to nanomoles of O used (data not shown). Statistics: Student's t test, heart $p < 0.0001$; liver $p < 0.001$).

Kristal and Yu[76] have demonstrated a protective effect of CR in isolated liver mitochondria from Fischer344 rats for the induction by calcium or calcium in conjunction with the strong oxidant t-butyl hydroperoxide, of the mitochondrial permeability transition. The increased resistance to permeability transition induction was maintained throughout 24 months of life and was age-independent. The authors comment that the observed effects represent one of the largest recognized CR-mediated increases in a parameter related to antioxidant defenses.

If, as a working hypothesis, it is accepted that mitochondrial ROS generation is sufficient to induce oxidative damage leading to functional impairment over time, and that this is ameliorated by CR feeding, the question arises as to the mechanisms by which diet induces this effect. Feuers[77] has published the first study that systematically attempts to identify those components of the electron transport chain (ETC) that lose efficiency with age and are protected by CR feeding. The complexes of the ETC were screened spectrophotometrically in the gastrocnemius of 10-month and 20- to 26-month-old B6C3F1 female mice fed either *ad libitum* or restricted to 60% of the *ad libitum* food intake from 14 weeks. At 10 months of age, CR mice showed a significant decline (71%) in the specific activity of complex I when compared to fully fed animals. Between 10 and 20 months, a significant age-related decline (65%) in specific activity of complex I was recorded for fully fed mice, but no further decline in specific activity was seen in CR mice for complex I. In contrast, the activity of complex II was not affected by diet in 10-month-old mice and neither was an age effect observed in *ad libitum* fed animals. At 20 month, CR animals showed a significant (69%) lower enzyme activity when compared with *ad libitum*–fed animals.

The age- and diet-induced changes in complex III were similar to those observed for complex I. A decline in activity (40%) induced by CR was seen at 10 months of age while in *ad libitum* fed animals, an age-related fall in activity (55%) between 10 and 20 months was observed. A kinetic analysis of complex III revealed a similar pattern for the Michaelis-Menten constant (K_m). At 10 months CR induced a significant decrease in the K_m with no further change observed by 20 months of age. Conversely, for fully fed mice, the K_m increased significantly between 10 and 20 months.

Feuers[77] reports that of the four complexes in *ad libitum*–fed mice, complex IV exhibited the most severe loss of activity with age. Specific activity fell 75% between 10 and 20 months and this trend continued to 26 months of age. As observed for complexes I and III, CR reduced the activity of complex IV (36%) in middle-aged mice of 10 months in comparison to fully fed animals but protected this complex from further, age-dependent decline in activity. The age-related loss of high affinity binding sites for complex IV (68%–48%) in *ad libitum*–fed animals between 10 and 20 months of age was not observed in CR animals, conversely the proportion of high-affinity binding sites increased to 80% at both ages studied. Thus a typical crossover effect is seen with age in complex IV activity. At 20 months of age in CR animals, complex IV specific activity was 121% that of age-matched, *ad libitum*–fed animals. This twofold effect of CR first to depress an enzymatic or physiological activity and second to prevent any further age-related impairment, has been recorded previously for the effect of CR on core temperature, protein synthesis, and peptide elongation rates during protein synthesis.[5,78]

With the exception of complex II, CR decreased enzyme activity of the ETC complexes when measured at 10 months of age. This reduction in total enzyme mass was opposed by a higher catalytic efficiency as indicated by the K_m. In control mice aging resulted in both a loss in enzyme content with decreased catalytic activity and a preferential loss of the high-affinity binding sites for complex IV[77]; these changes were prevented by CR feeding. Feuers interprets these observations as representing partial obstruction of electron flow in the ETC, thereby increasing the potential for free radical generation. It is recognized from the work of Sohal that partial inhibition of complex IV in *Drosophila* increased hydrogen peroxide generation.[79]

The details of the mechanism whereby CR protects the ETC complexes against a loss in total enzyme mass with age and increases the catalytic activity of complex III are unknown. Detailed kinetics of the mitochondrial membrane potential and proton-motive force in CR animals with age are lacking. Of particular interest is the observation of Hansford[80] that for rat heart mitochondria, H_2O_2 production is dependent on both high $\Delta\psi$ (membrane potential) and a highly reduced complex I.

The data that exists on mitochondrial energetics and ROS production with age have come from studies on isolated mitochondria, using specific substrates and inhibitors for different complexes of the ETC, to localize presumptive *in vivo* sites of oxygen radical generation. Uncertainty exists about the degree to which this data, generated from mitochondria under nonphysiological conditions, reflects the real rate of ROS production and mitochondrial function with aging *in vivo*.

Studies using electron microscopy have shown the presence with aging of larger mitochondria exhibiting several morphological changes in a number of tissues. Due to selective fragility and loss during the isolation procedure, this increased mitochondrial heterogeneity observed with age may not be reflected in the purified preparations.[81] The use of intact cells to study oxidative stress with age avoids this problem and allows ROS generations and oxidative damage to be studied throughout the cell and not just in mitochondria.

Attempts have been made, using flow-cytometry techniques, to confirm in intact hepatocytes the observations on ROS production with age obtained from isolated mitochondria. To my knowledge, no studies have been reported using cytometric flow analysis with cells isolated from CR animals. Rhodamine-123 has been used to determine membrane potential in intact rat hepatocytes as a function of aging.[82] This data was confirmed using the hydrophobic cation [³H]tetraphenylphosphonium bromide. The formation of oxidants within cells was determined using 2′,7′-dichlorofluorescin diacetate. Hepatocytes isolated from fully fed animals showed an impairment with age of the mitochondrial membrane potential and an increase in intracellular peroxide levels and mitochondrial size.[75,82] Electron transport in mitochondria within a significant fraction of older cells appears less efficiently coupled to ATP synthesis, which may lead to increased superoxide generation.

This approach has been developed further by Harper et al.[83] to apply metabolic control analysis, specifically top-down elasticity analysis as developed by Brand and Brown,[84] to the effect of age on mitochondrial proton leak and ATP turnover reactions in isolated mouse hepatocytes. Such an approach allows not only the important sites of control to be identified within a metabolic pathway, but top-down elasticity analysis permits differences in pathway regulation to be determined in comparison of different metabolic states, that is, under hormonal stimulation. A comparison of hepatocytes isolated from mice of 3 and 30 months revealed an effect of age on the overall kinetics of the mitochondrial proton leak. A significant decrease in the amount of oxygen used to support the synthesis and use of ATP was observed in old compared to young cells, and top-down metabolic analysis showed a shift in control away from substrate oxidation reactions towards increased control by the proton leak and by ATP turnover reactions. Harper et al.[83] interpret the age-related increase in proton leak as a possible effect of oxidative damage to the lipids of the mitochondrial inner membrane. Documented age-related changes in the membrane lipid composition include increased levels of polyunsaturated fatty acids such as 22:4 and 22:5 and a decrease in 18:2, 18:1, and 16:1, increasing the likelihood of peroxidation. This change in the unsaturation of the polyunsaturated lipid component is the reverse of the adaptation induced by CR. The increased rate of oxygen consumption required to balance the increased proton leak with aging could result in increased ROS production as the proton pumps attempt to maintain the protonmotive force. Similar data from cells isolated from CR animals has not been published, nor the effect of CR on ATP concentrations and turnover. When applied to cells isolated from CR animals, top-down control analysis clearly has great potential to identify important adaptive responses in oxidative phosphorylation and metabolic control to the CR feeding regime.

SUMMARY

A general consensus has been reached that CR feeding regimes reduce the rate of accrual of oxidative damage as measured by lipid peroxidation, genomic and mtDNA damage, and protein carbonyl formation. The demonstration that this reduced rate of accrual of oxidative damage is sufficient to explain the retarded appearance of the aging phenotype in the CR animal awaits further clarification. It is unclear whether CR protects against random oxidative damage *per se* or is protective of those vulnerable proteins in key pathways, that is, those containing iron-sulfur centers of the ETC or DNA-binding signaling proteins. The question remains open whether oxidative damage in genomic and mtDNA with aging is purely random or whether there are susceptible locations to ROS which may vary with specific tissue expression patterns or cell cycle status. The answer to these questions will clearly affect the magnitude and importance of the functional consequences resulting from ROS-induced damage.

Age-related accrual of ROS-induced damage is a balance between generation, defence enzyme activities, repair, and turnover. It has been demonstrated that there is at specific ages and in certain tissues, an upregulation in the activity of these enzymes, particularly catalase, but the overall response is specific to the species, tissue, and enzyme under study; and it is not possible to conclude that CR always induces an upregulation in the activities of those enzymes that dismutate and reduce ROS species.

With the exception of limited studies on liver microsomes, studies on ROS production in CR animals with aging have used mitochondrial preparations isolated from a variety of tissues. It is established that under *in vitro* assay conditions mitochondria from CR-maintained animals produce fewer total ROS per nanomol of O during state 4 respiration. Why this should be so is not fully understood, although recent work on the ETC complexes has suggested changes in the K_m of complex III and a retention of high-affinity binding sites for complex IV, which may reduce ROS formation.[77] The signaling pathway leading to these adaptive responses is unknown. It is possible to speculate that the observed changes in the unsaturated lipid composition of the mitochondrial membrane not only may protect against ROS-induced lipid peroxidation, but also may influence the binding characteristics of the ETC complexes embedded in the membrane and its transport properties. It is recognized that the sensitivity to circulating concentrations of T_3 and insulin is enhanced in CR animals. Basal metabolic or 24-hour metabolic-rates are not significantly different in age-matched comparisons between control and CR animals in spite of a reduction of 60% and 30%, respectively, in the circulating concentrations of these hormones. Whether the changes in membrane lipid composition are also under endocrine control is unknown.

The measurement of ROS production in isolated mitochondria *in vitro* suffers from a number of technical limitations such as the use of nonphysiological substrate concentrations, high O_2 tensions, and the presence of inhibitors of specific complexes of the ETC. Until comparable data is available on ROS production under more physiological conditions, such as that provided by studies with isolated cells, it may be prudent to interpret the current data with some caution.

The effect of chronic calorie restriction on the rate of oxidative phosphorylation within individual mitochondria is unknown. Detailed studies of mitochondrial respiration kinetics and protonmotive force for mitochondria from CR animals is lacking. Similarly, the effect of chronic calorie restriction on mitochondrial number and turnover rates in tissues from animals exhibiting retarded aging is not available. Detailed knowledge of the age-related tissue heterogeneity of mitochondria are lacking. Characterization, therefore, of the adaptive response of mitochondrial respiration kinetics and the production of superoxide to chronic calorie-restriction feeding regimes is still very incomplete, as indeed is our knowledge of aging on mitochondrial energetics and ROS generation.

REFERENCES

1. McCAY, C.M. *et al.* 1939. Retarded growth, life span, ultimate body size and age changes in the albino rat after feeding diets restricted in calories. J. Nutr. **18:** 1–13.
2. McCAY, C.M. *et al.* 1935. The effect of retarded growth upon the length of the life span and ultimate body size. J. Nutr. **10:** 63–79.
3. WEINDRUCH, R. & R.L. WALFORD. 1988. The retardation of aging and disease by dietary restriction. Charles C Thomas. Springfield, IL.
4. MERRY, B.J. 1991. Effect of dietary restriction on life spans. Rev. Clin. Gerontol. **1:** 203–213.
5. HOLEHAN, A.M. & B.J. MERRY. 1986. The experimental manipulation of ageing by diet. Biol. Rev. **61:** 329–368.
6. YU, B.P. 1996. Aging and oxidative stress: modulation by dietary restriction. Free Radic. Biol. Med. **21:** 651–668.
7. YU, B.P. *et al.* 1985. Nutritional influences on aging of Fischer 344 rats: I. Physical, metabolic and longevity characteristics. J. Gerontol. **40:** 657–670.
8. MERRY, B.J. 1987. Food restriction and the aging process. *In* Biological Age and Aging Risk Factors. A. Ruiz-Torres, Ed.: 259–272. Tecnipublicaciones. Madrid.
9. YU, B.P. 1994. How diet influences the aging process of the rat. Proc. Soc. Exp. Biol. Med. **205:** 97–105.
10. BECKMAN, K.B. & B.N. AMES. 1998. Mitochondrial aging: open questions. Ann. N.Y. Acad. Sci. **854:** 118–127.
11. DE, A.K. *et al.* 1983. Some biochemical parameters of ageing in relation to dietary protein. Mech. Ageing Dev. **21:** 37–48.
12. CHIPALKATTI, S. *et al.* 1983. Effect of diet restriction on some biochemical parameters related to aging in mice. J. Nutr. **113:** 944–950.
13. MATSUO, M. *et al.* 1993. Food restriction suppresses an age-dependent increase in the exhalation rate of pentane from rats—a longitudinal study. J. Gerontol. **48:** B133–B138.
14. KOIZUMI, A. *et al.* 1987. Influences of dietary restriction and age on liver enzyme activities and lipid peroxidation in mice. J. Nutr. **117:** 361–367.
15. LAGANIERE, S. & B.P. YU. 1987. Anti-lipoperoxidation action of food restriction. Biochem. Biophys. Res. Commun. **145:** 1185–1191.
16. LAGANIERE, S. & B.P. YU. 1989. Effect of chronic food restriction in aging rats II. Liver cytosolic antioxidants and related enzymes. Mech. Ageing Dev. **48:** 221–230.
17. LAGANIERE, S. & B.P. YU. 1989. Effect of chronic food restriction in aging rats I. Liver subcellular membranes. Mech. Ageing Dev. **48:** 207–219.
18. TIAN, L.Q. *et al.* 1995. Effects of caloric restriction on age-related oxidative modifications of macromolecules and lymphocyte-proliferation in rats. Free Radic. Biol. Med. **19:** 859–865.
19. BAILEY, J.W. *et al.* 1993. Fatty acid composition of adipose tissue in aged rats—effects of dietary restriction and exercise. Exp. Gerontol. **28:** 233–247.
20. LEE, J. *et al.* 1999. Modulation of cardiac mitochondrial membrane fluidity by age and calorie intake. Free Radic. Biol. Med. **26:** 260–265.

21. VORBECK, M.L. *et al.* 1982. Aging-dependent modification of lipid composition and lipid structural order parameter of hepatic mitochondria. Arch. Biochem. Biophys. **217:** 351–361.
22. DOBRESTOV, G.E. *et al.* 1977. The increase of phospholipid bilayer rigidity after lipid peroxidation. FEBS Lett. **84:** 125–128.
23. YU, B.P. *et al.* 1992. Effect of age-related lipid peroxidation on membrane fluidity and phospholipase-A$_2$—modulation by dietary restriction. Mech. Ageing Dev. **65:** 17–33.
24. CHOE, M. *et al.* 1995. Lipid peroxidation contributes to age-related membrane rigidity. Free Rads. Biol. Med. **18:** 977–984.
25. KIM, J.D. *et al.* 1996. Influence of age, exercise, and dietary restriction on oxidative stress in rats. Aging Clin. Exp. Res. **8:** 123–129.
26. CHOI, J.H. & B.P. YU. 1995. Brain synaptosomal aging: free radicals and membrane fluidity. Free Radic. Biol. Med. **18:** 133–139.
27. GABBITA, S.P. *et al.* 1997. Aging and caloric restriction affect mitochondrial respiration and lipid membrane status: an electron paramagnetic resonance investigation. Free Radic. Biol. Med. **23:** 191–201.
28. DAX, E.M. *et al.* 1989. Food restriction prevents an age-associated increase in rat liver β-adrenergic receptors. J. Gerontol. **44:** B72–76.
29. GAO, E.H. *et al.* 1998. Age-related decline in β-adrenergic and adenosine A(1) receptor function in the heart are attenuated by dietary restriction. J. Pharmacol. Exp. Ther. **285:** 186–192.
30. CHEN, G. *et al.* 1997. Effects of aging and food restriction on α(1)-adrenergic receptors and G-protein in the rat parotid gland. J. Gerontol. A Biol. Sci. Med. Sci. **52:** B103–B110.
31. BERTLETT, B.S. & E.R. STADTMAN. 1997. Protein oxidation in aging, disease and oxidative stress. J. Biol. Chem. **272:** 20313–20316.
32. AGARWAL, S. & R.S. SOHAL. 1994. Aging and proteolysis of oxidized proteins. Arch. Biochem. Biophys. **309:** 24–28.
33. DUBEY, A. *et al.* 1996. Effect of age and caloric intake on protein oxidation in different brain regions and on behavioral functions of the mouse. Arch. Biochem. Biophys. **333:** 189–197.
34. PLESHAKOVA, O.V. *et al.* 1998. Study of protein carbonyls in subcellular fractions isolated from liver and spleen of old and gamma-irradiated rats. Mech. Ageing Dev. **103:** 45–55.
35. OLIVER, C.N. *et al.* 1987. Age-related changes in oxidized proteins. Biol. Chem. **262:** 5488–5491.
36. PALLARDO, F.V. *et al.* 1998. Late onset administration of oral antioxidants prevents age-related loss of motor co-ordination and brain mitochondrial DNA damage. Free Radic. Res. **29:** 617–623.
37. CARNEY, J.M. *et al.* 1991. Reversal of age-related increase in brain protein oxidation, decrease in enzyme activity, and loss in temporal and spatial memory by chronic administration of the spin-trapping compound *N-tert*-butyl-α-phenylnitrone. Proc. Natl. Acad. Sci. USA **88:** 3633–3636.
38. SOHAL, R.S. & A. DUBEY. 1994. Mitochondrial oxidative damage, hydrogen peroxide release, and aging. Free Radic. Biol. Med. **16:** 621–626.
39. FORSTER, M.J. *et al.* 1996. Age-related losses of cognitive function and motor skills in mice are associated with oxidative protein damage in the brain. Proc. Natl. Acad. Sci. USA **93:** 4765–4769.
40. MASORO, E.J. *et al.* 1982. Action of food restriction in delaying the aging process. Proc. Natl. Acad. Sci. USA **79:** 4239–4241.
41. LASS, A. *et al.* 1998. Caloric restriction prevents age-associated accrual of oxidative damage to mouse skeletal muscle mitochondria. Free Radic. Biol. Med. **25:** 1089–1097.
42. LEEUWENBURGH, C. *et al.* 1997. Caloric restriction attenuates cross-linking of cardiac and skeletal muscle proteins in aging mice. Arch. Biochem. Biophys. **346:** 74–80.
43. SHIGENAGA, M.K. *et al.* 1994. Oxidative damage and mitochondrial decay in aging. Proc. Natl. Acad. Sci. USA **91:** 10771–10778.

44. YAN, L.J. *et al.* 1997. Oxidative damage during aging targets mitochondrial aconitase. Proc. Natl. Acad. Sci. USA **94:** 11168–11172.
45. BECKMAN, K.B. & B.N. AMES. 1998. The free radical theory of aging matures. Physiol. Rev. **78:** 547–581.
46. FRAGA, C.G. *et al.* 1990. Oxidative damage to DNA during aging: 8-hydroxy-2'-deoxyguanosine in rat organ DNA and urine. Proc. Natl. Acad. Sci. USA **87:** 4533–4537.
47. CHUNG, M.H. *et al.* 1990. Protection of nuclear DNA damage by dietary restriction. Age **13:** 101.
48. CHUNG, M.H. *et al.* 1992. Protection of DNA damage by dietary restriction. Free Radic. Biol. Med. **12:** 523–525.
49. KANEKO, T. *et al.* 1997. Retarding effect of dietary restriction on the accumulation of 8-hydroxy-2'-deoxyguanosine in organs of Fischer 344 rats during aging. Free Radic. Biol. Med. **23:** 76–81.
50. CHEN, J.J. & B.P. YU. 1994. Alterations in mitochondrial membrane fluidity by lipid-peroxidation products. Free Radic. Biol. Med. **17:** 411–418.
51. ZHANG, Y. *et al.* 1990. The oxidative inactivation of mitochondrial electron-transport chain components and ATPase. J. Biol. Chem. **265:** 16330–16336.
52. KRISTAL, B.S. *et al.* 1994. Sensitivity of mitochondrial transcription to different free-radical species. Free Radic. Biol. Med. **16:** 323–329.
53. RICHTER, C. 1992. Reactive oxygen and DNA damage in mitochondria. Mutat. Res. **275:** 249–255.
54. RICHTER, C. 1995. Oxidative damage to mitochondrial DNA and its relationship to ageing. Int. J. Biochem. Cell Biol. **27:** 647–653.
55. GADALETA, M.N. *et al.* 1992. Mitochondrial DNA copy number and mitochondrial DNA deletion in adult and senescent rats. Mutat. Res. **275:** 181–193.
56. GADALETA, M.N. *et al.* 1998. Aging and mitochondria. Biochimie **80:** 863–870.
57. LEE, C.M. *et al.* 1993. Multiple mitochondrial DNA deletions associated with age in skeletal muscle of *Rhesus* monkeys. J. Gerontol. **48:** B201–B205.
58. KANG, C.M. *et al.* 1997. Age-related mitochondrial DNA deletions: effect of dietary restriction. Free Radic. Biol. Med. **24:** 148–154.
59. LUHTALA, T.A. *et al.* 1994. Dietary restriction attenuates age-related increases in rat skeletal muscle antioxidant enzyme activities. J. Gerontol. **49:** B231–B238.
60. SOHAL, R.S. *et al.* 1994. Oxidative damage, mitochondrial oxidant generation and antioxidant defenses during aging and in response to food restriction in the mouse. Mech. Ageing Dev. **74:** 121–133.
61. RAO, G. *et al.* 1990. Effect of dietary restriction on the age-dependent changes in the expression of antioxidant enzymes in rat liver. J. Nutr. **120:** 602–609.
62. GOMI, F. & M. MATSUO. 1998. Effects of aging and food restriction on the antioxidant enzyme activity of rat livers. J. Gerontol. A Biol. Sci. Med. Sci. **53:** B161–B167.
63. MCCARTER, R.J. & J. PALMER. 1992. Energy metabolism and aging: a lifelong study of Fischer 344 rats. Am. J. Physiol. **263:** E448–452.
64. MASORO, E.J. *et al.* 1992. Dietary restriction alters characteristics of glucose fuel use. J. Gerontol. **47:** B202–208.
65. MERRY, B.J. & A.M. HOLEHAN. 1985. The endocrine response to dietary restriction in the rat. *In* The Molecular Biology of Aging. A.D. Woodhead, A.D. Blackett & A. Hollaender, Eds.: 117–137. Plenum Press. New York.
66. CASTILHO, R.F. *et al.* 1998. 3,5,3'-triiodothyronine induces mitochondrial permeability transition mediated by reactive oxygen species and membrane protein thiol oxidation. Arch. Biochem. Biophys. **354:** 151–157.
67. HAFNER, R.P. *et al.* 1988. Altered relationship between protonmotive force and respiration rate in non-phosphorylating liver mitochondria from rats of different thyroid status. Eur. J. Biochem. **178:** 511–518.
68. BARJA, G. *et al.* 1994. Low mitochondrial free-radical production per unit O_2 can explain the simultaneous presence of high longevity and aerobic metabolic rate in birds. Free Radic. Res. **21:** 317–327.
69. PAMPLONA, R. *et al.* 1996. Low fatty-acid unsaturation protects against lipid peroxidation in liver mitochondria from long-lived species—the pigeon and human case. Mech. Ageing Dev. **86:** 53–66.

70. BOVERIS, A. *et al.* 1999. Regulation of mitochondrial respiration by adenosine diphosphate, oxygen, and nitric oxide. Methods Enzymol. **301:** 188–198.
71. BOVERIS, A. & B. CHANCE. 1973. The mitochondrial generation of hydrogen peroxide. Biochem. J. **134:** 707–716.
72. HANSFORD, R.G. *et al.* 1997. Dependence of H_2O_2 formation by rat heart mitochondria on substrate availability and donor age. J. Bioenerg. Biomembr. **29:** 89–95.
73. CORBISIER, P. & J. REMACLE. 1990. Involvement of mitochondria in cell degeneration. Eur. J. Cell Biol. **51:** 173–182.
74. TAKEYAMA, N. *et al.* 1993. Oxidative damage to mitochondria is mediated by the Ca^{2+}-dependent inner membrane permeability transition. Biochem. J. **294:** 719–725.
75. SASTRE, J. *et al.* 1996. Aging of the liver: age-associated mitochondrial damage in intact hepatocytes. Hepatology **24:** 1199.
76. KRISTAL, B.S. & B.P. YU. 1998. Dietary restriction augments protection against induction of the mitochondrial permeability transition. Free Radic. Biol. Med. **24:** 1269–1277.
77. FEUERS, R.J. 1998. The effects of dietary restriction on mitochondrial dysfunction in aging. Ann. N.Y. Acad. Sci. **854:** 192–201.
78. MERRY, B.J. & A.M. HOLEHAN. 1991. Effect of age and restricted feeding on polypeptide chain assembly kinetics in liver protein synthesis in vivo. Mech. Ageing Dev. **58:** 139–150.
79. SOHAL, R.S. 1993. Aging, cytochrome oxidase activity, and hydrogen peroxide release by mitochondria. Free Radic. Biol. Med. **14:** 583–588.
80. CHAPPELL, J.B. & R.G. HANSFORD. 1972. Preparation of mitochondria from animal tissues and yeast. *In* Subcellular Components: Preparation and Fractionation. G.D. Birnie, Ed.: 77–91. Butterworth. London.
81. WILSON, P.D. & L.M. FRANKS. 1975. The effect of age on mitochondrial ultrastructure and enzymes. Adv. Exp. Med. Biol. **53:** 171–183.
82. HAGEN, T.M. *et al.* 1997. Mitochondrial decay in hepatocytes from old rats: membrane potential declines, heterogeneity and oxidants increase. Proc. Natl. Acad. Sci. USA **94:** 3064–3069.
83. HARPER, M.E. *et al.* 1998. Age-related increase in mitochondrial proton leak and decrease in ATP turnover reactions in mouse hepatocytes. Am. J. Physiol. Endocrinol. Metab. **275:** E197–E206.
84. BRAND, M.D. & G.C. BROWN. 1994. The experimental application of control analysis to metabolic systems. *In* Biothermokinetics. H.V. Westerhoff, Ed.: 22–35. Intercept. Andover.

Role of Mitochondrial DNA Mutations in Disease and Aging

D.A. COTTRELL,[a] E.L. BLAKELY, G.M. BORTHWICK, M.A. JOHNSON,
G.A. TAYLOR, E.J. BRIERLEY, P.G. INCE, AND D.M. TURNBULL

*Department of Neurology, The Medical School, University of Newcastle upon Tyne,
Newcastle upon Tyne, NE2 4HH, UK*

ABSTRACT: Since Harman in 1972 first proposed a role in the process of aging
for the mitochondrial genome, a wealth of evidence has been accumulated to
support this theory. We discuss the hereditary mitochondrial DNA disorders,
which we believe may give insight into both normal aging and neurodegenerative conditions. We then review the evidence for the role of mitochondrial DNA
mutations in both aging and age-related disorders and also discuss new approaches for investigating the mitochondrial genome at a single cell level, by
observing the activity of the mitochondrial enzyme cytochrome c oxidase.

INTRODUCTION

In addition to the nuclear genome, each human cell contains multiple copies of a
small (16.5-kb) double-stranded genome within each mitochondria, the mitochondrial genome (mtDNA). Although mtDNA constitutes less than 1% of the total cellular nucleic acid, it is still essential for the normal survival of mitochondria and
hence the cell. This small genome is highly efficient in terms of expressed DNA,
with virtually no introns except for a small portion known as the D-loop. It encodes
for 37 genes, all of which are involved in synthesizing subunits of the respiratory
chain complex. Thirteen genes encode for polypeptide components of the respiratory
chain. The remaining 25 genes encode for 22 transfer RNAs and two ribosomal
RNAs essential for mitochondrial protein synthesis (FIG. 1).

Human cells contain a few hundred to more than one thousand mitochondria;
each mitochondrion in turn has 2–10 copies of mtDNA.[1] Thus, several thousand copies of the genome can be present within a single cell. Mutated and wild-type (normal) mtDNA can coexist in any proportion, a situation termed heteroplasmy. Levels
of mutation can vary considerably between mitochondria, cells, and even tissues
within the same individual.

[a]Address for correspondence: Dr. D.A. Cottrell, Department of Neurology, The Medical
School, University of Newcastle upon Tyne, Newcastle upon Tyne, NE2 4HH, UK. Phone: 44-
191-2225101; fax: 44-191-2228553.
d.a.cottrell@ncl.ac.uk

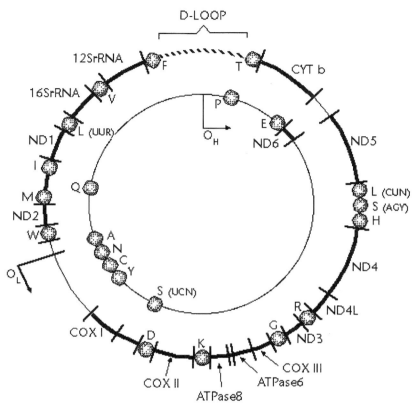

FIGURE 1. Human mitochondrial DNA labeled to show the locations of the 37 genes. Each sphere indicates the position of a single tRNA.

HEREDITARY MITOCHONDRIAL DISORDERS

It is now well established that mtDNA defects are an important cause of disease. Since the first description, a decade ago, of mutations of mtDNA, more than 100 pathologic mtDNA defects have been characterized.[2] Point mutations may involve either the RNA or protein-encoding genes, and rearrangements may take the form of deletions or duplications. We believe that an essential factor in our understanding of the role of defects of mitochondrial function in aging is knowledge of the pathogenic mechanisms involved in mtDNA defects and disease. In the presence of heteroplasmy there is a critical ratio of mutant to wild-type mitochondrial genomes that is necessary before the disease becomes both biochemically and clinically apparent. While this threshold level is dependent upon both the nature of the mutation and the tissue affected, the vast majority of pathologic mtDNA tRNA mutations investigated are extremely recessive, and biochemical dysfunction is only apparent when levels of wild-type mtDNA fall below 30%.

As might be expected, mtDNA disorders are phenotypically diverse given the ubiquitous presence of mitochondria and the variation in the levels of heteroplasmy throughout the body. Yet many have predominant neurologic and muscular symptoms, including dementia, seizures, ataxic syndromes, peripheral neuropathies, and myopathy which progress temporally. Higher levels of mutant are found in these postmitotic tissues, perhaps because they have no opportunity for cells with high mutant mtDNA concentrations to be selected against. CNS imaging of patients with mtDNA disorders often reveals moderate degrees of cerebral or cerebellar atrophy that are consistent with neurodegeneration and are often comparable with senescent brains or those with dementia. Limited neuropathologic studies have confirmed massive neuronal cell loss and demyelination indicating neurodegeneration.[3,4]

THE MITOCHONDRIAL THEORY OF AGING

Harman[5] in 1972 was among the first to propose that mitochondria may have a central role in the process of aging. Mitochondria and mtDNA are essential for survival at the cellular level, yet these organisms are particularly vulnerable to damage. Their high consumption of metabolic fuels in the pursuit of ATP production via the inner membrane-incorporated respiratory chain releases the toxic by-products known as free radicals.[6–8] These highly unstable molecules, which are normally disposed of by free radical scavenging enzymes such as superoxide dismutase and catalase, can escape these defense mechanisms and increase the levels of oxidative stress.

During aging some of the free radical scavenging systems are decreased,[9–11] so that more free radicals escape, increasing the level of oxidative stress within the organelle. These free radicals can oxidize membrane lipids in close proximity to the respiratory chain, decreasing the fluidity and increasing the permeability of the inner mitochondrial membrane.[7] Oxidation of mitochondrial proteins and DNA also occurs. Oxidatively damaged proteins are known to increase markedly with age.[12] The role of oxidative damage to proteins has long been questioned. It was assumed that any damaged protein would not accumulate, as cellular proteins are continuously turned over and damaged proteins are removed more rapidly than normal. However, functionally inactive forms of enzymes have been shown to accumulate with age and are thought to be a result of oxygen-free radical–mediated damage. An age-related increase in the amount of oxidized protein is possibly the result of an accumulation of damage to DNA which affects the factors responsible for protein oxidation and the degradation of oxidized protein.[12] An accumulation of oxidized protein may also result in crosslinking to other proteins, which would alter their biochemical and physiologic function in mitochondria.

Levels of the oxidized nucleotide 8-hydroxy-deoxyguanosine (8-OH-dG), a biomarker of DNA damage, have also been shown to accumulate with aging.[13–15] In several tissues, including the CNS and muscle, levels of 8-OH-dG in mtDNA exceed nuclear DNA (nDNA) some 16-fold.[16] Several studies indicate that 8-OH-dG most frequently base pairs correctly with cytosine, but also that it has the ability to mispair with adenine approximately 1% of the time.[17–20] It also has the ability to cause misreading at adjacent residues. Hayakawa *et al.*[13] found that increased levels of 8-

OH-dG in human heart correlated with increases in levels of a 7.4-kb deletion.[13] Mecocoi *et al.*[15] found a similar correlation in human brain.

Ultrastructural changes were also reported to occur in mitochondria with age. They become larger and less numerous with vacuolization, cristae rupture, and the accumulation of intramitochondrial paracrystalline inclusions.[21,22] Cardiolipin, an acidic phospholipid that occurs only in mitochondria, was also shown to decrease with age.[23–25] This inner membrane lipid is known to have excellent electrical insulating properties and is thought to contribute significantly to the transmembrane potential that drives the formation of ATP via ATP synthase, the terminal complex of the respiratory chain. Indeed, studies have shown a decrease in membrane potential in mitochondria from older animals.[26,27]

ACCUMULATION OF mtDNA MUTATIONS WITH AGING

Linnane[28] in 1989 proposed that the accumulation of mtDNA mutations during life is a major cause of age-related disease. The compact and efficient mitochondrial genome is a particularly vulnerable piece of DNA. Not only does it lack introns, but also it has a 10-fold higher mutation rate than does nuclear DNA.[16,29,30] There are several reasons for this high mutagenic propensity. Firstly, mitochondria have limited nucleotide excision and recombination DNA repair mechanisms.[16] Secondly, mtDNA lack the structurally protective histone proteins. Thirdly, mtDNA reside and replicate close to the inner mitochondrial membrane and hence are exposed to the enriched free radical milieu produced by both the respiratory chain and the monoamine oxidases located within the membrane.

A large body of evidence indicates that mtDNA mutations increase temporally, with the highest levels seen in CNS and muscle. To date over 20 different types of deletions have been shown to accumulate in aging human tissues. The first published report of an age-related increase in an mtDNA deletion was the so-called common deletion, originally found in elderly brain tissue and patients with Parkinson's disease.[31] The common deletion occurs between two 13-bp sequence repeats beginning at nucleotides 8470 and 13447, removing almost a 5-kB region of mtDNA between ATPase 8 and the ND5 genes. The deletion is thought to occur during replication of the mtDNA. This absent arc encodes for six essential polypeptides of the respiratory chain and five tRNAs. It has been associated with several different clinical entities, including chronic progressive external ophthalmoplegia (CPEO) and Kearns Sayre syndrome (KSS). The common deletion was also shown to increase with age in skeletal muscle,[32–36] cardiac muscle,[33,37–39] diaphragm,[39] retina,[40] skin,[41,42] ovary,[43] and sperm,[44] but it is undetectable in fetal tissues. This accumulation suggests that mutant compared with wild-type mtDNA has a replicative advantage. Another mutation widely reported to increase with age in skeletal and cardiac muscle, brain, and skin is a 7.4-kb deletion between np 8649 and np 16084. This deletion is also found in the muscle of patients with CPEO and KSS.

The levels of individual mutations observed, however, are low (<1%) even in tissues from very elderly subjects, and by analogy to the situation in patients with mtDNA diseases it is difficult to see how individual mutations could affect mitochondrial function. The low levels of mtDNA mutation observed have all been from homogenate samples of tissue. Khrapko *et al.*[45] recently used long polymerase chain reaction (PCR) techniques in single cell cardiomyocytes from old patients. Their results indicate that various different deletions occur within homogenized cardiac tissue, but that clonal expansion of a single mutation occurs within individual cells.

INCREASED LEVELS OF mtDNA MUTATIONS IN AGE-RELATED DISORDERS

Several age-related disorders have been shown to harbor higher levels of mtDNA mutations than their age-matched counterparts. In one study the common deletion in cardiac muscle from patients with ischemic heart disease contained up to 240 times the normal cardiac muscle deletion levels.[38] In another study by Pang *et al.*,[41] skin that was sun exposed in one individual had 30 times the level of common deletion than did non-sun–exposed skin. In the CNS, Ikebe *et al.*[31] found 17 times the level of the common deletion in the striatum of patients with Parkinson's disease compared with age-matched controls.[31] Evidence also shows that levels of the common deletion are higher in patients with Alzheimer's disease.[46] Levels of the oxidized nucleotide 8-OH-dG are also higher in the brain of patients with Alzheimer's disease.[47]

AN AGE-RELATED INCREASE IN CELLS DEFICIENT IN CYTOCHROME C OXIDASE

Another method to examine the presence of mtDNA dysfunction is to measure activities of respiratory chain complexes. Complex IV of the respiratory chain, cytochrome c oxidase, is essential for respiratory chain function. Cytochrome c oxidase (COX) is composed of 13 subunits, which are partly encoded by the mitochondrial genome and partly by nuclear DNA. The larger subunits, I, II, and III, are encoded by mitochondrial DNA and are synthesized within the mitochondria; the remaining 10 subunits are encoded by the nuclear genome. Therefore, mtDNA integrity is essential for the successful synthesis of active COX. A major feature of mitochondrial DNA disease in humans is the presence of cells with low COX activity. Previous studies from our group have shown that the mechanism of these changes is likely to be clonal expansion of individual mtDNA deletions within individual cells.[48]

Many studies have reported the finding of cytochrome c oxidase deficiency at a single cell level. Muller-Hocker *et al.*[49] first demonstrated cytochrome c oxidase-deficient cardiomyocytes in human heart that increased with age. Cox-deficient cardiomyocytes were regularly present from the sixth decade of life, only occurring sporadically prior to this. It was observed that the loss of enzyme activity was always confined to single randomly distributed cardiomyocytes. The density of COX-deficient cardiomyocytes increased from an average in the third decade of life of 3 defects/cm^2 to 50 defects/cm^2 in heart aged over 70 years.

Histochemical analysis of cytochrome c oxidase activity in various extraocular muscles also revealed randomly distributed COX-deficient fibers.[50] Defects were observed in the second decade of life in some subjects and were consistently apparent from the third decade onwards. In limb muscle and diaphragm, an almost 10-fold increase in the incidence of defective fibers was noted for those in the eighth and ninth decade (54–60 defects/cm^2) compared to those between the third and sixth decades (5–7). The affected isolated muscle fibers showed normal SDH activity.[49] These observations were confirmed by Byrne et al.,[51] who reported an accumulation of COX-deficient fibers in the diaphragm from the fourth decade onwards, with an exponential increase in later life.

Our recent studies have concentrated on COX activity within the CNS. This, like muscle, consists predominantly of postmitotic cells, and previous studies have shown low levels of mtDNA deletions in aged brains. We have already shown that COX-negative neurons exist in abundance in the CNS of a patient with a mitochondrial disorder.[4] In addition, we are completing a comprehensive study exploring whether cytochrome c oxidase negative neurons exist in specific brain regions in the normal brain and the possible mechanisms for these changes.

In COX-negative skeletal muscle fibers from elderly patients, our group, using competitive three primer PCR techniques, has shown that clonal expansion of one particular mutation occurs within individual cells.[48] Because many different mutations could be "expanding" in different aging cells, the overall quantity of a single mutation would be expected to be low.

CONCLUSIONS

We have come a long way since Harman first proposed the mitochondrial theory of aging. In the past decade, numerous mutations and indicators of oxidative damage have been shown to increase with age in mitochondria throughout the body in both animal and human studies. What effects these changes within the mitochondria have on the cells they support remains to be elucidated. In cybrid studies the ATPase-dependent mitochondrial transmembrane potential falls at thresholds for mtDNA mutation.[52] It may be expected that when the lack of ATP synthesis reaches a threshold, the cell ceases to function and die. Yet, in our morphologic studies, COX-deficient muscle fibers and neurons look identical to COX-positive cells. Further studies are needed at the cellular level to look for evidence of neurodegeneration within these COX-deficient cells, in particular, whether these cells exhibit evidence of apoptosis, in which mitochondria and their contents appear to play such a pivotal role.[53] Evidence that mtDNA damage occurs and accumulates with age is substantial, but the consequences are still elusive.

REFERENCES

1. ROBIN, E.D. & R. WONG. 1988. Mitochondrial DNA molecules and virtual number of mitochondria per cell in mammalian cells. J. Cell. Physiol. **136:** 507–513.
2. CHINNERY, P.F. & D.M. TURNBULL. 1999. Mitochondrial DNA and disease. Lancet **354:** 17–21.
3. SPARACO, M., E. BONILLA, S. DIMAURO et al. 1993. Neuropathology of mitochondrial encephalomyopathies due to mitochondrial DNA defects. J. Neuropathol. Exp. Neurol. **52:** 1–10.
4. COTTRELL, D.A., P. INCE, T.M. WARDELL et al. 1999. Detection of cytochrome oxidase deficient neurons in a patient with multiple mitochondrial DNA deletions. Ann. Neurol. Submitted.

5. HARMAN, D. 1972. Free radical theory of ageing: dietary implications. Am. J. Clin. Nutr. **25:** 839–843.
6. SOHAL, R.S. & B.H. SOHAL. 1991. Hydrogen peroxide release by mitochondria increases during ageing. Mech. Ageing Dev. **57:** 187–202.
7. HRUSZKEWYCZ, A.M. 1992. Lipid peroxidation and mtDNA degeneration. A hypothesis. Mutat. Res. **275:** 243–248.
8. SOHAL, R.S, H.H. KU, S. AGARWAL *et al.* 1994. Oxidative damage, mitochondrial oxidant generation and antioxidant defences during ageing and in response to food restriction in the mouse. Mech. Ageing Dev. **74:** 121–133.
9. SOHAL, R.S., L.A. ARNOLD & B.H. SOHAL. 1990. Age-related changes in antioxidant enzymes and prooxidant generation in tissues of the rat with special reference to parameters in two insect species. Free Rad. Biol. Med. **9:** 495–500.
10. SOHAL, R.S., L. ARNOLD & W.C. ORR. 1990. Effect of age on superoxide dismutase, catalase, glutathione reductase, inorganic peroxides, TBA-reactive material, GSH/GSSG, NADPH/NADP+ and NADH/NAD+ in *Drosophila melanogaster.* Mech. Ageing Dev. **56:** 223–235.
11. SEMSEI, I., G. RAO & A. RICHARDSON. 1991. Expression of superoxide dismutase and catalase in rat brain as a function of age. Mech. Ageing Dev. **58:** 13–19.
12. STADTMAN, E.R. 1992. Protein oxidation and ageing. Science **257:** 1220–1224.
13. HAYAKAWA, M., K. TORII, S. SUGIYAMA *et al.* 1991. Age-associated accumulation of 8-hydroxydeoxyguanosine in mitochondrial DNA of human diaphragm. Biochem. Biophys. Res. Commun. **179:** 1023–1029.
14. HAYAKAWA M., K. HATTORI, S. SUGIYAMA *et al.* 1992. Age-associated oxygen damage and mutations in mitochondrial DNA in human hearts. Biochem. Biophys. Res. Commun. **189:** 979–985.
15. MECOCCI, P., U. MACGARVEY, A.E. KAUFMAN *et al.* 1993. Oxidative damage to mitochondrial DNA shows marked age-dependent increases in human brain. Ann. Neurol. **34:** 609–616.
16. RICHTER, C., J.W. PARK & B.N. AMES. 1988. Normal oxidative damage to mitochondrial and nuclear DNA is extensive. Proc. Natl. Acad. Sci. USA **85:** 6465–6467.
17. KUCHINO, Y., F. MORI, H. KASAI *et al.* 1987. Misreading of DNA templates containing 8-hydroxydeoxyguanosine at the modified base and at adjacent residues. Nature **327:** 77–79.
18. SHIBUTANI, S., M. TAKESHITA & A.P. GROLLMAN. 1991. Insertion of specific bases during DNA synthesis past the oxidation-damaged base 8-OH-dG. Nature **349:** 431–434.
19. CHENG, K.C., D.S. CAHILL, H. KASAI *et al.* 1992. 8-Hydroxyguanine, an abundant form of oxidative DNA damage, causes G----T and A----C substitutions. J. Biol. Chem. **267:** 166–172.
20. WOOD, M.L., M. DIZDAROGLU, E. GAJEWSKI *et al.* 1990. Mechanistic studies of ionising radiation and oxidative mutagenesis: genetic effects of a single 8-hydroxyguanine (7-hydro-8-oxoguanine) residue inserted at a unique site in a viral genome. Biochemistry **29:** 7024–7032.
21. FELDMAN, D., R.L. SWARM & J. BECKER. 1981. Ultrastructural study of rat liver and liver neoplasms after long-term treatment with phenobarbital. Cancer Res. **41:** 2151–2162.
22. FRENZEL, H. & J. FEIMANN. 1984. Age-dependent structural changes in the myocardium of rats. A quantitative light- and electron-microscopic study on the right and left chamber wall. Mech. Ageing Dev. **27:** 29–41.
23. PARADIES, G. & F.M. RUGGIERO. 1990. Age-related changes in the activity of the pyruvate carrier and in the lipid composition in rat-heart mitochondria. Biochem. Biophys. Acta **1016:** 207–212.
24. PARADIES, G. & F.M. RUGGIERO. 1991. Effect of ageing on the activity of the phosphate carrier and on the lipid composition in rat liver mitochondria. Arch. Biochem. Biophys. **284:** 332–337.
25. RUGGIERO, F.M., F. CAFAGNA, V. PETRUZZELLA *et al.* 1992. Lipid composition in synaptic and nonsynaptic mitochondria from rat brains and effect of ageing. J. Neurochem. **59:** 487–491.

26. LINNANE, A.W., M. DEGLI ESPOSTI, M. GENEROWICZ et al. 1995. The universality of bioenergetic disease and amelioration with redox therapy. Biochem. Biophys. Acta **1271:** 191–194.

27. HAGEN, T.M., D.L. YOWE, J.C. BARTHOLOMEW et al. 1997. Mitochondrial decay in hepatocytes from old rats: membrane potential declines, heterogeneity and oxidants increase. Proc. Natl. Acad. Sci. USA **94:** 3064–3069.

28. LINNANE A.W., S. MARZUKI, T. OZAWA et al. 1989. Mitochondrial DNA mutations as an important contributor to ageing and degenerative diseases. Lancet **1:** 642–645.

29. MERRIWETHER, D.A., A.G. CLARK, S.W. BALLINGER et al. 1991. The structure of human mitochondrial DNA variation. J. Molec. Evol. **33:** 543–555.

30. FRAGA, C.G., M.K. SHIGENAGA, J.W. PARK et al. 1990. Oxidative damage to DNA during ageing: 8-hydroxy-2'-deoxyguanosine in rat organ DNA and urine. Proc. Natl. Acad. Sci. USA **87:** 4533–4537.

31. IKEBE, S., M. TANAKA, K. OHNO et al. 1990. Increase of deleted mitochondrial DNA in the striatum in Parkinson's disease and senescence. Biochem. Biophys. Res. Commun. **170:** 1044–1048.

32. COOPER, J.M., V.M. MANN & A.H. SCHAPIRA. 1992. Analyses of mitochondrial respiratory chain function and mitochondrial DNA deletion in human skeletal muscle: effect of ageing. J. Neurol. Sci. **113:** 91–98.

33. SIMONETTI, S., X. CHEN, S. DIMAURO et al. 1992. Accumulation of deletions in human mitochondrial DNA during normal ageing: analysis by quantitative PCR. Biochem. Biophys. Acta **1180:** 113–122.

34. DIDONATO, S., M. ZEVIANI, P. GIOVANNINI et al. 1993. Respiratory chain and mitochondrial DNA in muscle and brain in Parkinson's disease patients. Neurology **43:** 2262–2268.

35. LEE, H.C., C.Y PANG, H.S. HSU et al. 1994. Differential accumulations of 4,977 bp deletion in mitochondrial DNA of various tissues in human ageing. Biochem. Biophys. Acta **1226:** 37–43.

36. LEZZA, A.M., D. BOFFOLI, S. SCACCO et al. 1994. Correlation between mitochondrial DNA 4977-bp deletion and respiratory chain enzyme activities in ageing human skeletal muscles. Biochem. Biophys. Res. Commun. **205:** 772–779.

37. CORTOPASSI, G.A. & N. ARNHEIM. 1990. Detection of a specific mitochondrial DNA deletion in tissues of older humans. Nucleic Acids Res. **18:** 6927–6933.

38. CORRAL-DEBRINSKI, M., G. STEPIEN, J.M. SHOFFNER et al. 1991. Hypoxemia is associated with mitochondrial DNA damage and gene induction. Implications for cardiac disease [see comments]. JAMA **266:** 1812–1816.

39. CORTOPASSI, G.A., D. SHIBATA, N.W. SOONG et al 1992. A pattern of accumulation of a somatic deletion of mitochondrial DNA in ageing human tissues. Proc. Natl. Acad. Sci. USA **89:** 7370–7374.

40. BARREAU, E., J.Y. BROSSAS, Y. COURTOIS et al. 1996. Accumulation of mitochondrial DNA deletions in human retina during ageing. Invest. Ophthalmol. Vis. Sci. **37:** 384–391.

41. PANG, C.Y., H.C. LEE, J.H. YANG et al. 1994. Human skin mitochondrial DNA deletions associated with light exposure. Arch. Biochem. Biophys. **312:** 534–538.

42. YANG, J.H., H.C. LEE, K.J. LIN et al. 1994. A specific 4977-bp deletion of mitochondrial DNA in human ageing skin. Arch. Dermatol. Res. **286:** 386–390.

43. KITAGAWA, T., N. SUGANUMA, A. NAWA et al. 1993. Rapid accumulation of deleted mitochondrial deoxyribonucleic acid in postmenopausal ovaries. Biol. Reprod. **49:** 730–736.

44. KAO, S.H., H.T. CHAO & Y.H. WEI. 1995. Mitochondrial deoxyribonucleic acid 4977-bp deletion is associated with diminished fertility and motility of human sperm. Biol. Reprod. **52:** 729–736.

45. KHRAPKO, K., N. BODYAK, W.G. THILLY et al. 1999. Cell-by-cell scanning of whole mitochondrial genomes in aged human heart reveals a significant fraction of myocytes with clonally expanded deletions. Nucleic Acids Res. **27:** 2434–2441.

46. CORRAL-DEBRINSKI, M., T. HORTON, M.T. LOTT et al. 1994. Marked changes in mitochondrial DNA deletion levels in Alzheimer brains. Genomics **23:** 471–476.

47. MECOCCI, P., U. MACGARVEY & M.F. BEAL. 1994. Oxidative damage to mitochondrial DNA is increased in Alzheimer's disease. Ann. Neurol. **36:** 747–751.

48. BRIERLEY, E.J., M.A. JOHNSON, R.N. LIGHTOWLERS *et al.* 1998. Role of mitochondrial DNA mutations in human ageing: implications for the central nervous system and muscle. Ann. Neurol. **43:** 217–223.
49. MULLER-HOCKER, J. 1990. Cytochrome c oxidase deficient fibres in the limb muscle and diaphragm of man without muscular disease: an age-related alteration. J. Neurol. Sci. **100:** 14–21.
50. MULLER-HOCKER, J., P. SEIBEL, K. SCHNEIDERBANGER *et al.* 1993. Different in situ hybridisation patterns of mitochondrial DNA in cytochrome c oxidase-deficient extraocular muscle fibres in the elderly. Virchows Arch. A, Pathol. Anat. Histopathol. **422:** 7–15.
51. BYRNE, E. & X. DENNETT. 1992. Respiratory chain failure in adult muscle fibres: relationship with ageing and possible implications for the neuronal pool. Mutat. Res. **275:** 125–131.
52. PORTEOUS, W.K., A.M. JAMES, P.W. SHEARD *et al* 1998. Bioenergetic consequences of acccumulating the common 4977-bp mitochondrial DNA deletion. Eur. J. Biochem. **257:** 192–201.
53. GREEN, D.R. & J.C. REED. 1998. Mitochondria and apoptosis. Science **281:** 1309–1312.

Inherited Variability of the Mitochondrial Genome and Successful Aging in Humans

G. DE BENEDICTIS,[a] G. CARRIERI,[a] O. VARCASIA,[a] M. BONAFÈ,[b] AND C. FRANCESCHI[b,c]

[a]*Department of Cell Biology, University of Calabria, Calabria, Italy*

[b]*Department of Experimental Pathology, University of Bologna, Bologna, Italy*

[c]*Department of Gerontology, INRCA, Ancona, Italy*

ABSTRACT: Increasing data indicate that polymorphic variants of nuclear loci can affect rate and quality of aging in humans. However, the mitochondrial genome is another good candidate, because of the central role played by mitochondrial genes in oxidative phosphorylation (OXPHOS) and cell metabolism. A characteristic of the mitochondrial genome (mtDNA) is the high level of interindividual variability that ensues from high mutation rate and unilinear inheritance. Related groups of germline/inherited mtDNA polymorphisms (haplogroups) have been identified as continent-specific sets of stable/ancient/ associated restriction fragment length polymorphisms in the mtDNA coding region, representing markers capable of exactly depicting the mtDNA pool of a specific population. The hypothesis can be put forward that mtDNA variants included in a haplogroup may have similar OXPHOS efficiency and therefore act as genetic factors predisposing to individual successful or unsuccessful aging. This idea can be explored by sampling groups of individuals of different ages from a well-defined population and comparing the pools of mtDNA haplogroups between samples. The results obtained by screening mtDNA haplogroups in about 800 Italians of different ages, including more that 200 centenarians, agree with the hypothesis that the inherited variability of the mitochondrial genome is associated with the chance of successful aging and longevity in humans.

GENETICS OF AGING AND LONGEVITY

Only recently have successful aging and longevity become an interesting topic of investigation, probably as a consequence of the recent, rapid increase in mean life span, particularly of the "proliferation" of the oldest-old. At the same time, old tenets in gerontology, such as the idea that the aging process is a simple deterioration of every function and parameter, have become obsolete. Indeed, recent data on healthy elderly and centenarians suggest that physiologic aging and longevity are characterized by a remodeling and a profound adaptation of every cell and organ of the body in order to cope with the continuous attrition caused by internal and external damaging agents.[1] Another oversimplification is vanishing, that is, the idea that the aging process is homogeneous and that old people are similar. On the contrary, elderly people and centenarians are quite heterogeneous.[2] This phenotypic heterogeneity represents a serious confounding variable, together with many others such as gender, geographic origin, ethnicity, lifestyle, personality, and nutrition. Despite the

Aging Affecting Genes		
Mutations	Polymorphisms	Mutations
Life-span shortening	Adaptive capability Frail alleles Robust alleles	Life-span lengthening
Epistatic suppressors	Epistatic suppressors	Hypostatic genes

FIGURE 1. A hypothetical model of aging-affecting genes. These genes may have rare variants (mutations) shortening or lengthening individual life span. The former ones would be epistatic over all the other genes, acting before their phenotypic expression; the latter ones would be hypostatic with respect to all the other genes, acting very late in the life and therefore requiring a favorable gene network to reveal themselves. The majority of these putative aging-affecting genes could carry polymorphisms (common variants) that control the adaptive capability of the organism to cope with age-related environmental stresses (frail alleles, robust alleles).[12]

complexity of human aging and longevity, data on the characteristics of centenarians are emerging.[3–5] In particular, data on the genetics of longevity are accumulating[6–10] despite the low proportion of variance attributable to genetic factors in the longevity trait.[11] Most of the difficulties encountered in studies aimed to disentangle the environmental, genetic, and stochastic components of human longevity were overcome by studying simpler creatures. A variety of longevity genes have been identified in classical model systems, such as *Saccharomyces cerevisiae, Caenorhabditis elegans,* and *Drosophila melanogaster.* The lesson from these studies in lower organisms, together with the first available data in mammalians, such as mice and humans, inspired us to propose a classification of aging-affecting genes,[12] as illustrated in FIGURE 1. According to this conceptualization, most of the genetic variance in the longevity phenotype could be attributed to polymorphic variants at several loci (tens?, hundreds?, thousands?) and their interactions. Moreover, we hypothesized that frail and robust alleles exist (ultimately responsible for population heterogeneity) and that their biologic role can change in opposite directions (positive or negative for survival) during the life span as a consequence of the adaptation and remodeling process.

SUCCESSFUL AGING AND MITOCHONDRIAL GENOME

Is there a role for the mitochondrial genome (mtDNA) in successful aging and longevity? Several arguments favor a positive answer to this question.

The first rationale regards the central role of mitochondria in energy production by oxidative phosphorylation (OXPHOS). For this reason, mitochondria have attracted the attention of scientists interested in unraveling the complex changes associated with aging and age-related diseases.[13] It is known that the OXPHOS capacity declines with aging; such a decline seems to parallel the age-related increase in mtDNA somatic damages, probably as a consequence of the high production of reactive oxygen species (ROS) that occurs within mitochondria.[14] More than four decades ago Harman[15] proposed the free radical theory of aging, indicating that the accumulation of molecular damages caused by free radicals is a crucial factor in aging. The conclusions of the intensive research developed in subsequent years clearly indicated that: (1) oxidative damage to macromolecules, including mtDNA, occurs during aging;[16,17] (2) postmitotic tissues become a bioenergetic mosaic during the process of normal aging;[18] and (3) the instability of the mitochondrial genome leads to mitochondrial dysfunction and increased oxidative stress.[19] However, the low number of mutant molecules as well as the mosaic distribution of mutant mtDNAs in each cell or tissue does not support the hypothesis that mtDNA somatic mutations are *per se* the cause of the observed age-related OXPHOS decline.[20] Indeed, the direct causal linkage between age-related tissue dysfunction and mtDNA mutant load is controversial.[21,22]

The second rationale suggesting a possible central role of mtDNA in successful aging and longevity regards the complex cross-talk between mitochondrial genome and nuclear genome. Mitochondria are under the control of two distinct genomes, their own (mtDNA) and that of the nucleus. For instance, it was recently demonstrated that a single, nucleus-encoded gene product, mitochondrial transcription factor A (mtTFA), has a direct bearing on mtDNA copy number, and cells that have been artificially depleted of mtDNA exhibit concomitantly low levels of mtTFA.[23] From a genetic point of view the mitochondrial phenotype should therefore be regarded as a complex trait that is controlled by about 1,000 genes, a small percentage of which (13 structural, 2 rRNA, and 22 tRNA mitochondrial genes)[24] have peculiar genetic features: maternal inheritance, lack of recombination, high mutation rate, replicative segregation in cell divisions, possible heteroplasmy within cells and tissues, and last but not least, some code signals for amino acids that are different from those of nuclear DNA.[25] A correct cross-talk between these genomes is expected to be crucial for efficient mitochondrial working and, in general, for efficient cell metabolism. In any case, the complexity of the mitochondrial system as well as the complexity of the aging process[26] should be kept well in mind when a possible role of the mitochondrial genome in successful aging and longevity is explored.

INHERITED VARIABILITY OF THE MITOCHONDRIAL GENOME

The third rationale suggesting an important role of mtDNA in successful aging and longevity regards its interindividual and intraindividual sequence variability. Population genetics studies tell us that a characteristic feature of the mitochondrial genome is the high level of interindividual variability; indeed, the probability that unrelated individuals share the same mtDNA sequence is very low. Moreover, it is known that each cell contains more than 1,000 copies of mtDNA. Therefore, we can ask whether the mtDNA molecules in each cell and tissue of the same individual are all identical.

mtDNA INTRAINDIVIDUAL VARIABILITY

Heteroplasmy (more than one type of mtDNA genome in one individual) is observed in stable tissues, although at a low level, in both degenerative diseases and aging. Recent data confirm that heteroplasmy is significantly present in the noncoding region of mtDNA from brain tissue, whereas it is virtually absent in leukocytes.[27] These tissue-specific results raise the question of whether heteroplasmy is present in germline cells; such a question is important because the presence of heteroplasmy in this line has theoretic implications relevant to origin, transmission, and fixation of mtDNA mutations.[28] It has been proposed that female germline mitochondria have a repressed bioenergetic function.[29] As a consequence, they would escape mutagenesis due to free radicals produced by OXPHOS and replicate their genome with minimal damage. On the contrary, male germline mitochondria would pay the cost of their active (and mutagenic) OXPHOS by accumulating mutations in their mtDNA. According to this theory, anisogamy and maternal inheritance of mtDNA are the evolutionary solution to the conflict between mitochondrial respiration (and therefore free radical production) and transmission of high quality mitochondrial genome across generations. However, as a matter of fact, mutation rate is higher in mitochondrial than in nuclear genes[30] and recombination occurs rarely, if at all. Thus, theoretically, heteroplasmy in female germ cells cannot be ruled out. However, when a cell replicates, mitochondria are distributed randomly between the daughter cells (replicative segregation). If the mother cell is heteroplasmic, the daughter cells can become homoplasmic (all wild-type or all mutant) just by chance. Consequently, over repeated divisions, the mitochondrial genotypes can progressively drift towards either pure mutant or pure wild-type mtDNAs (homoplasmy). Segregation of mtDNA genotypes occurs early in oogenesis,[31] between the developmental stages of the primordial germ cells and primary oocytes, before amplification of the mtDNA copy number which is observed in maturation of oocytes. Although in principle this process may produce cells carrying low quality mitochondrial genomes, these cells may be eliminated before fertilization. An attractive paper by Krakauer and Mira[32] proposes the intriguing hypothesis that atresia (death of most female germ cells before fertilization) plays an important role in the elimination of germ cells carrying deleterious mitochondrial genomes. Selective atresia would be chiefly important in species producing a small number of offspring, where high quality mitochondria (and thereby high quality offspring) are required to ensure survival and viability of the descendants. This death of germ cells could be the developmental answer to the accumulation of deleterious mutations that is expected in the absence of recombination. After fertilization, the copy number of mtDNA decreases dramatically, and a sort of "bottleneck" arises from this partitioning event. The stochastic distribution of hidden mutant mtDNAs during cell division may complicate the patterns of familial transmission[33] and contribute to modulating the penetrance of other mutations (possibly present in the nuclear genome) that could affect multifactorial phenotypes, such as complex diseases and rate and quality of aging.

INTERINDIVIDUAL VARIABILITY AND mtDNA HAPLOGROUPS

If mtDNA intraindividual variability is low, interindividual variability is very high because of the high mutation rate and strict maternal inheritance. The unilinear inheritance has important consequences for mtDNA diversity and evolution. At a given time, many types of mtDNA, which differ because of mutations that have arisen in their ancestry, are present in the population. By reconstructing the genealogy of these types, we can always find a common ancestor. This does not mean that the entire population arises from one female, but that mitochondria from all the other women who lived at that time and had other types of mtDNA are extinct; at every generation, some of the original types are lost by random drift and new types are produced by mutations. If the total size of the population remains constant, the number of different mtDNA types remains constant too (equilibrium). Although a single common ancestor can be found for a unilinearly transmitted set of markers, mtDNA types that are descendant from the original type will differ among themselves, because of the new mutations that they will have accumulated. There is therefore a proportion between mtDNA diversity and mtDNA evolution time. According to this model, the mitochondrial genome evolves by simple accumulation of neutral, or near neutral, mutations that irradiate sequentially along female lineages, and analysis of mtDNA polymorphisms in human populations provides a powerful tool for reconstructing ancient human migrations.[34] Starting from pioneer studies on mtDNA restriction fragment length polymorphisms (RFLP) carried out by Southern blotting analysis,[35] the general picture of human mtDNA evolution has been outlined; the modern human mtDNA tree likely originated in Africa (150 ky according to a certain calibration of the mitochondrial clock). As humans migrated out of Africa towards new lands, additional neutral mutations were established by random genetic drift, resulting in continent-specific mtDNA polymorphisms.

The introduction of PCR-based procedures[36,37] allowed high resolution RFLP studies and sequence analyses of the mtDNA control region to be performed. With this approach, related groups of mtDNAs (haplogroups) were identified as continent-specific sets of stable/ancient/associated RFLPs in the mtDNA coding region. Thus, each haplogroup includes a number of evolutionary-related different types of mtDNA.[38,39] The associated RFLPs that define mtDNA haplogroups are germline/inherited markers capable of exactly depicting the mtDNA pool of a specific population. Although the level of mtDNA variability can be high within a haplogroup and an enormous number of individual mtDNA variants exist in the population, most mtDNA polymorphisms observed in individuals belonging to modern human populations have occurred on preexisting haplogroups. Therefore, analysis of haplogroup-specific RFLPs overcomes the methodologic problems due to the high level of interindividual variability of mtDNA, and recently it was exploited to investigate the possible contribution of mitochondrial genome to complex traits.[40]

mtDNA HAPLOGROUPS IN SUCCESSFUL AGING AND LONGEVITY

A maternal component of life expectancy has been shown by epidemiologic studies,[41] suggesting the possible involvement of mtDNA-inherited variants in successful aging and longevity.[42] However, because of the complexity of the longevity trait, the contribution to the variance of the trait by a single locus, such as the mitochondrial genome, may be difficult to identify. In this case, a crucial role is played by the methodologic approach, its power and appropriateness. A comparison between mtDNA genotypic pools of several hundred long-lived individuals such as centenarians (cases) and younger people (controls), extracted from the same population, is apparently powerful enough, but the choice of the mtDNA markers as well as the matching between cases and controls is crucial for the reliability of the conclusions. In this regard, two papers were recently published, indicating that indeed mtDNA variants can affect longevity. However, in our opinion both studies suffer from methodologic problems. In French centenarians the frequency of the HaeII morph-2 was greater in centenarians than in controls (13.7% vs 6.6%; p <0.01). [43] From this finding the authors infer that this morph may be significantly associated with longevity. However, their conclusion is biased and weakly supported by experimental data. Indeed, the number of morphs generated by the six endonucleases used in the screening is high in Caucasians.[44] Accordingly, multiple comparisons are possible, but they were not taken into account (Bonferroni's correction) when statistical significance was estimated. We think it unlikely that independent analysis of single RFLPs (morphs), without considering their association in mtDNA types, is a reliable approach to revealing differences between cases and controls. More appropriate than analysis of single restriction morphs is the sequencing of entire mtDNA molecules, followed by a comparison of these sequences between centenarians and controls. Using this approach, Tanaka *et al.*[45] found three linked variants (Mt3010A, Mt5178A, and Mt8414T) whose frequency was greater in centenarians than in controls, again indicating that the mitochondrial genome has a role in longevity. However, in this case too, the high level of interindividual variability increases the number of multiple comparisons and reduces the level of statistical significance.

Analysis of mtDNA haplogroups overcomes these problems and offers the opportunity to evaluate the changes in the mitochocondrial genotypic pool, while the population ages and survival selection occurs. In fact, if an mtDNA type favorable (or unfavorable) to longevity is included in a certain haplogroup, the frequency of that haplogroup is expected to increase (or decrease) in centenarians in comparison with controls. We applied this idea to the study of successful aging and longevity by typing the set of associated RFLPs that define the nine European haplogroups (H, I, J, K, T, U, V, W, and X; ref. 46) in individuals of different ages, paying attention to their health status, sex, ethnicity, and the geographic origin of the samples.[47] We found that the population pool of mtDNA haplogroups differed between centenarians and younger individuals. This difference was found only in males and only in a well-defined geographic area (Northern Italy). Geographic effects on genes affecting aging are not unexpected, taking into account that longevity is the result of a prolonged interaction among geographic-specific factors, both genetic and environmental. Likewise, different results in males and females can be explained, considering that longevity is a multifactorial trait, heavily affected by the physiologic scenario in which a gene is expressed. Indeed, sex-specific findings have been observed in genetic studies on centenarians.[48,49] It can be speculated that males more than females require additional protective mechanisms, such as particular mitochondrial genomes, in order to attain longevity. In this regard, a theoretic basis for male-female asymmetry in the phenotypic expression of complex traits affected by the mitochondrial genome has been proposed by Frank and Hurst.[50]

Interestingly, the main difference that we found between male centenarians and younger people involved the J haplogroup (20% vs about 2%). Increasing evidence indicates that the genetic background of the J haplogroup affects the penetrance of primary mutations and therefore the phenotypic expression of complex traits.[51,52] This phenomenon became evident when the penetrance of primary mtDNA mutations responsible for specific diseases was investigated. Thus, it was implicitly assumed that mtDNA haplogroups could act as genetic backgrounds modulating the phenotypic expression of deleterious gene variants. Our finding that the J haplogroup is more frequent in centenarians than in younger people suggests that the J-specific genetic background can also affect the penetrance of advantageous mutations. At present, sequencing analysis of specific mtDNA haplogroups are performed in our laboratory to identify and characterize these putative beneficial mutations.

In previous studies, we showed that genotypic frequencies of longevity-associated polymorphisms can follow nonlinear age-related trajectories.[53] Such trajectories are expected as a consequence of the crossing of mortality curves occurring in subgroups of individuals carrying different genotypes. (See Appendix in ref. 53.) The same considerations apply to population studies regarding people selected for healthy status, when crossing of the susceptibility to age-related diseases is considered. This hypothesis has been verified by extending the screening of mtDNA haplogroups to a sample of about 800 individuals aged 20 to more than 100 years.[54] All subjects were clinically healthy and had normal hematologic and functional parameters. The main results we obtained are shown in FIGURES 2, 3, and 4 where the maximum likelihood curves fitting the observed haplogroup frequencies in several age-classes (including centenarians) are shown. The frequency of the J haplogroup

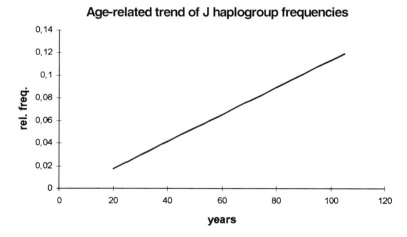

FIGURE 2. Age-related trend of J haplogroup frequencies. Maximum likelihood regression line fitting the observed frequencies of J haplogroup in age-classes. The hypothesis that frequencies are not different in age-classes is rejected with $p < 0.0001$ by the bootstrap method. Statistical analysis shows that trajectories with lower positive slope also are consistent with the data.

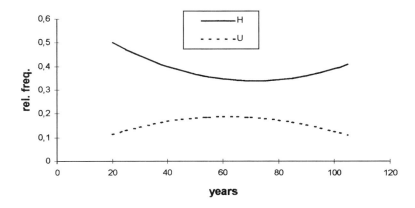

FIGURE 3. Age-related trends of H and U haplogroup frequencies. Maximum likelihood regression curves fitting the observed frequencies of H and U haplogroups in age-classes $(0.05 < p < 0.06)$.

(FIG. 2) increases with age, according to a linear trajectory ($p < 0.0001$). It must be noted that the slope of the regression line of FIGURE 2 would require a very high mortality rate in younger people, which is not compatible with current demographic data. However, statistical analysis (the bootstrap method) shows that trajectories with a lower positive slope are also consistent with the data, because of their large standard errors. The frequencies of H and U haplogroups show opposite nonlinear trajectories (FIG. 3) that are borderline statistically significant ($0.05 < p < 0.06$ for both H and U trajectories). Other haplogroups, such those considered in FIGURE 4, do not show any significant age-related trajectory ($p > 0.30$ for both K and T). On the whole, these results show that some mtDNA haplogroups are associated with successful aging and longevity, although the pattern of association is not simple and is likely affected by the physiologic age-related remodeling.

The pattern of FIGURE 2 enables us to classify, at least provisionally, the J haplogroup as an allele that increases the chance of successful aging, unless deleterious mutations do not occur on it. This result is confirmed by a completely different analytic approach, applied to a subset of the same data, which integrates demographic and genetic data.[55] The method is based on evaluation of relative risks for individuals carrying candidate alleles and is described in detail elsewhere.[56,57] In agreement with the trend shown in FIGURE 2, we found that in males the presence of the J haplogroup reduces the relative risks from 1 to 0.711 ($p = 0.012$). This finding confirms that in males the J haplogroup is associated with increased chance of survival. The novel approach proposed by Yashin *et al.*[56,57] overcomes the gene-frequency method currently used in gene/longevity association studies and probably represents the future of research in this field.

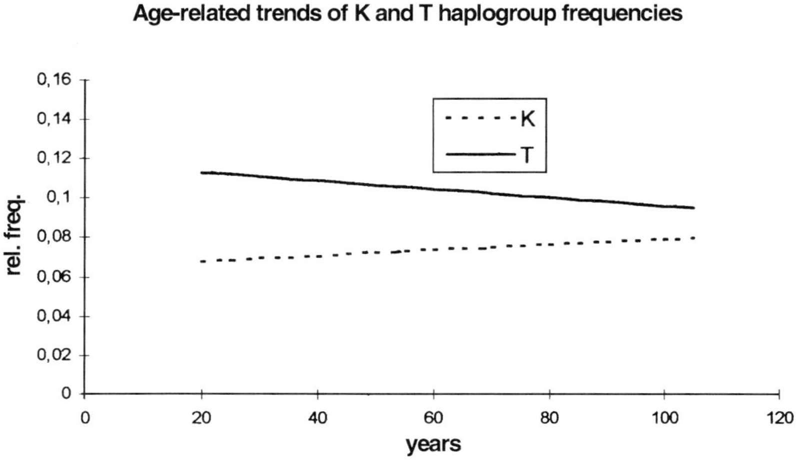

Age-related trends of K and T haplogroup frequencies

FIGURE 4. Age-related trends of K and T haplogroup frequencies. Maximum likelihood regression lines fitting the observed frequencies of K and T haplogroups in age-classes ($p > 0.30$).

ACKNOWLEDGMENTS

This work was partially supported by MURST Project "Genetic Determinants of Human Longevity" (1998–2000).

REFERENCES

1. RESNICK, N.M. & E.R. MARCANTONIO. 1997. How should clinical care of the aged differ? Lancet **350:** 1157–1158.
2. FORETTE, B. 1997. Centenarians: health and frailty. *In* Longevity: to the Limits and Beyond. J.M. Robine *et al.*, Eds. :105–112. Springer-Verlag. Berlin.
3. FRANCESCHI, C. *et al.* 1995. The immunology of exceptional individuals: the lesson of centenarians. Immunol. Today **16:** 12–15.
4. MARI, D. *et al.* 1995. hypercoagulability in centenarians: the paradox of successful aging. Blood **85:** 3144–3149.
5. FORETTE, B. 1999. Are common risk factors relevant in the eldest old? *In* The Paradoxes of Longevity. J.M. Robine *et al.*, Eds.: 73–79. Springer-Verlag. Berlin.
6. SCHACHTER, F. *et al.* 1994. Genetic associations with human longevity at the APOE and ACE loci. Nature Genet. **6:** 29–32.
7. MANNUCCI, P.M. *et al.* 1997. Gene polymorphisms predicting high plasma levels of coagulation and fibrinolysis proteins. A study in centenarians. Arterios. Thromb. Vasc. Biol. **17:** 755–759.
8. FAURE-DELANEF, L. *et al.* 1997. Methylenetetrahydrofolate reductase thermolabile variant and human longevity. Am. J. Hum. Genet. **60:** 999–1001.
9. DE BENEDICTIS, G. *et al.* 1997. DNA multiallelic systems reveal gene/longevity associations not detected by diallelic systems. Hum. Genet. **99:** 312–318.
10. BONAFE', M. *et al.* 1999. Variants predisposing to cancer are present in healthy centenarians. Am. J. Hum. Genet. **64:** 292–295.

11. HERSKIND, A.M. *et al.* 1996. The heritability of human longevity: a population-based study on 2872 Danish twin pairs born 1870–1900. Hum. Genet. **97**: 319–323.
12. DE BENEDICTIS, G. & C. FRANCESCHI. 1998. The genetics of successful aging. Aging Clin. Exp. Res. **10**: 147–148.
13. WALLACE, D.C 1999. Mitochondrial diseases in man and mouse. Science **283**: 1482–1488.
14. CORTOPASSI, G.A. & A. WONG. 1999. Mitochondria in organismal aging and degeneration. Biochim. Biophys. Acta **1410**: 183–193.
15. HARMAN, D. 1956. Ageing: theory based on free radical and radiation chemistry. J. Gerontol. **11**: 298–300.
16. MARTIN, G.M. *et al.* 1996. Genetic analysis of ageing: role of oxidative damage and environmental stresses. Nature Genet. **13**: 25–34.
17. KHRAPKO, K. *et al.* 1999. Cell-by-cell scanning of whole mitochondrial genomes in aged human heart reveals a significant fraction of myocytes with clonally expanded deletions. Nucleic Acids Res. **27**: 2434–2441.
18. KOPSIDAS, G. *et al.* 1998. An age-associated correlation between cellular bioenergy decline and mtDNA rearrangements in human skeletal muscle. Mutat. Res. **421**: 27–36.
19. OSIEWACZ, H.D. 1997. Genetic regulation of aging. J. Mol. Med. **75**: 715–727.
20. GADALETA, M.N. *et al.* 1999. Aged-linked changes in the genotype and phenotype of mitochondria. *In* Frontiers in Cellular Bioenergetics. S. Papa *et al.*, Eds. : 693–727. Kluwer Academic/Plenum Publishers. New York.
21. WEI, Y.H. *et al.* 1998. Oxidative damage and mutation to mitochondrial DNA and age-dependent decline of mitochondrial respiratory function. Ann. N.Y. Acad. Sci. **854**: 155–170.
22. LIGHTOWLERS, R.N. *et al.* 1999. Mitochondrial DNA- all things bad? TIG **15**: 91–93.
23. LARSSON, N.G. *et al.* 1998. Mitochondrial transcription factor A is necessary for mtDNA maintenance and embryogenesis in mice. Nature Genet. **18**: 231–236.
24. MITOMAPDATABASE. http://websvr.mips.biochem.mpg.de/proj/medgen/mitop/
25. LIGHTOWLERS, R.N. *et al.* 1997. Mammalian mitochondrial genetics: heredity, heteroplasmy and disease. TIG **13**: 450–455.
26. KIRKWOOD, T.B.L. 1996. Human senescence. BioEssays **18**: 1009–1016.
27. JAZIN, E.E. *et al.* 1996. Human brain contains high level of heteroplasmy in the noncoding regions of mitochondrial DNA. Proc. Natl. Acad. Sci. USA **93**: 12382–12387.
28. GIBBONS, A. 1998. Calibrating the mitochondrial clock. Science **279**: 28–29.
29. ALLEN, J.F. 1996. Separate sexes and the mitochondrial theory of ageing. J. Theor. Biol. **180**: 135–140.
30. LYNCH, M. 1996. Mutation accumulation in transfer RNAs: molecular evidence for Muller's ratchet in mitochondrial genomes. Mol. Biol. Evol. **13**: 209–220.
31. JENUTH, J.P. *et al.* 1996. Random genetic drift in the female germline explains the rapid segregation of mammalian mitochondrial DNA. Nature Genet. **14**: 146–151.
32. KRAKAUER, D.C. & A. MIRA. 1999. Mitochondria and germ-cell death. Nature **400**: 125–126.
33. KOEHLER, C.M. *et al.* 1991. Replacement of bovine mitochondrial DNA by a sequence variant within one generation. Genetics **129**: 247–255.
34. CAVALLI-SFORZA, L.L. *et al.* 1993. The History and Geography of Human Genes. : 83–88. Princeton University Press. Princeton, NJ.
35. JOHNSON, M.J. *et al.* 1983. Radiation of human mitochondria DNA types analyzed by restriction endonuclease cleavage patterns. J. Mol. Evol. **19**: 255–271.
36. BALLINGER, S.W. *et al.* 1992. Southeast Asian mitochondrial DNA analysis reveals genetic continuity of ancient mongoloid migrations. Genetics **130**: 139–152.
37. TORRONI, A. *et al.* 1992. Native American mitochondrial DNA analysis indicates that the Amerind and the Nadene populations were founded by two independent migrations. Genetics **130**: 153–162.
38. GRAVEN, L.G. *et al.* 1995. Evolutionary correlation between control region sequence and restriction polymorphisms in the mitochondrial genome of a large Senegalese Mandeka population. Mol. Biol. Evol. **12**: 334–345.
39. TORRONI, A. *et al.* 1993. Asian affinities and continental radiation of the four founding native American mtDNAs. Am. J. Hum. Genet. **53**: 563–590.

40. TORRONI, A. & D.C. WALLACE. 1994. Mitochondrial DNA variation in human populations and implication for detection of mitochondrial DNA mutations of pathological significance. J. Bioenerg. Biomembr. **26:** 261–271.
41. BRAND, F.N. *et al.* 1992. Family patterns of coronary heart disease mortality: the Framingham longevity study. J. Clin. Epidemiol. **45:** 169–174.
42. SONT, J.K & J.P. VANDENBROUCKE. 1993. Life expectancy and mitochondrial DNA. J. Clin. Epidemiol. **46:** 199–201.
43. IVANOVA R. *et al.* 1998. Mitochondrial genotype associated with French Caucasian Centenarians. Gerontology **44:** 349.
44. DE BENEDICTIS, G. *et al.* 1989. Restriction fragment length polymorphism of human mitochondrial DNA in a sample population from Apulia (Southern Italy). Ann. Hum. Genet. **53:** 311–318.
45. TANAKA, M. *et al.* 1998. Mitochondrial genotype associated with longevity. Lancet **351:** 185–186.
46. TORRONI, A. *et al.* 1996. Classification of European mtDNAs from analysis of three European population. Genetics **144:** 1835–1850.
47. DE BENEDICTIS, G. *et al.* 1999. Mitochondrial DNA inherited variants are associated with successful aging and longevity in humans. FASEB J. **13:** 1532–1535.
48. IVANOVA, R. *et al.* 1998. HLA-DR alleles display sex-dependent effects on survival and discriminate between individual and family longevity. Hum. Mol. Genet. **7:** 187–194.
49. DE BENEDICTIS, G. *et al.* 1998. Gene/longevity association studies at four autosomal loci (REN, THO, PARP, SOD2). Eur. J. Hum. Genet. **6:** 534–541.
50. FRANK, S.A. & L.D. HURST. 1996. Mitochondria and male disease. Nature **383:** 224.
51. TORRONI, A. *et al.* 1997. Haplotype and phylogenetic analyses suggest that one European-specific mtDNA background plays a role in the expression of Leber hereditary optic neuropathy by increasing the penetrance of the primary mutations 11778 and 14484. Am. J. Hum. Genet. **60:** 1107–1121.
52. REYNIER, P. *et al.* 1999. MtDNA haplogroup J: a contributing factor of optic neuritis. Eur. J. Hum. Genet. **7:** 404–406.
53. DE BENEDICTIS, G. *et al.* 1998. Age-related changes of the 3'APOB-VNTR genotype pool in ageing cohorts. Ann. Hum. Genet. **62:** 115–122.
54. DE BENEDICTIS, G. *et al.* 1999. Age-related changes of the population pool of mtDNA haplogroups in aging cohorts. Manuscript in preparation.
55. YASHIN, A.I. *et al.* 2000. Genes and longevity: lessons from centenarians studies. J. Gerontol. In press.
56. YASHIN, A.I. *et al.* 1998. Combining genetic and demographic information in population studies of aging and longevity. J. Epidemiol. Biostat. **3:** 289–294.
57. YASHIN, A.I. *et al.* 1999. Genes, demography and life span: the contribution of demographic data in genetic studies of aging and longevity. Am. J. Hum. Genet. **65:** 1178–1193.

Mitochondrial Oxidative Stress

Physiologic Consequences and Potential for a Role in Aging

SIMON MELOV[a]

Buck Center for Research in Aging, PO Box 638, Novato, California 94948–0638, USA

ABSTRACT: During the last 10 years, the theory known as the "free radical theory of aging" has achieved prominence as one of the most compelling explanations for many of the degenerative changes associated with aging. Although its appeal derives from a long-standing body of supporting correlative data, the theory was only recently more rigorously tested. Ongoing researches in the study of free radical biochemistry and the genetics of aging have been at the forefront of this work. First, transgenic approaches in invertebrate models with candidate genes such as superoxide dismutase (SOD) involved in the detoxification of reactive oxygen species (ROS) have shown that the endogenous production of ROS due to normal physiologic processes is a major limiter of life span.[1,2] Genes involved in ROS detoxification are highly conserved among eukaryotes; hence, the physiologic processes that limit life span in invertebrates are likely to be similar in higher eukaryotes. Secondly, transgenic mice deficient in the antioxidant enzyme mitochondrial superoxide dismutase (SOD2) die within their first week of life, demonstrating the importance of limiting endogenous mitochondrial free radicals in mammals. Together, data from studies using transgenic invertebrates and those using *sod2* mutant mice demonstrate that modulation of metabolic ROS can have a profound effect on life span. We show here that the effects of mitochondrial ROS can be modulated through appropriate catalytic antioxidant intervention. These catalytic antioxidants are discussed in the context of mitochondrial oxidative stress and their potential role in intervening in mitochondrial oxidative stress and aging.

OXIDATIVE STRESS AND MITOCHONDRIA

During the course of normal oxidative phosphorylation (OXPHOS), between 0.4 and 4% of all oxygen consumed is converted into the superoxide free radical ($O_2^{\cdot-}$).[3–6] Free radicals have the capacity to oxidize proteins, lipids, DNA, and RNA.[7–9] Therefore, superoxide is potentially damaging within the mitochondria, and eukaryotes have evolved defenses against the free radical by-products of OXPHOS. The primary defense is the mitochondrial form of superoxide dismutase (*sod2*) whose role is to detoxify superoxide into hydrogen peroxide. The less reactive H_2O_2 can then either diffuse out of the mitochondrion and be converted to water by cytosolic catalase or, alternatively, if the H_2O_2 is confined within the mitochondrion, it is converted to water by mitochondrial glutathione peroxidase.[10] The rate of production of reactive oxygen species (ROS) by the mitochondria is essentially

[a]Phone: 415 899 1800; fax: 415 209 2231.
smelov@buckcenter.org

determined by the metabolic rate, although ROS production can clearly be altered by respiratory chain architecture and antioxidant concentration.

Mitochondrial OXPHOS is essential for ATP generation. When the production of ATP is disrupted through mutations of proteins involved in OXPHOS, tissues such as the brain and heart, which are dependent on high levels of ATP, are the most affected. In human brain, this can be phenotypically observed as ataxia, tremor, mental retardation, epilepsy, and stroke-like episodes and progressive dementia.[11] It has often been speculated that the production of free radicals within the mitochondria is a potentially deleterious process and that in mitochondrial disease, mutations of subunits involved in OXPHOS can result in an increased propensity for the respiratory chain to produce free radicals and hence overwhelm the mitochondria's ability to protect itself from endogenous oxidative stress.[12] Currently, the clearest example of a disease in which there are hallmarks of mitochondrial oxidative stress is Friedreich's ataxia.[13,14]

OXIDATIVE STRESS AND LIFE SPAN

The most compelling arguments that oxidative stresses are a major limiter of life span have primarily come from genetic studies using invertebrates and the mouse.[1,2,15–21] Overexpression of genes that are responsible for neutralizing the superoxide free radical have increased the mean and maximum life span of *Drosophila melanogaster*, indicating that endogenous superoxide production limits life span in the fly.[1,2] Conversely, inactivation of the mitochondrial form of superoxide dismutase via homologous recombination in the mouse is neonatal lethal, demonstrating the serious consequences of endogenous unregulated mitochondrial superoxide production in a mammal.[18,20,21]

In addition to direct cause and effect type experiments such as those just mentioned, much work has been done correlating the differential production of ROS from mitochondria with the lifespans of different species.[22–30] Moreover, there are broad correlations in metabolism related to maximum and mean life span, and a number of notable exceptions to this generalization.[31,32] Perhaps the best known exception is the comparison of non-passerine birds to rats. Pigeons and rats have high, approximately equivalent metabolic rates, yet pigeons live about 3–5 times longer than the rat. An apparent contradiction if the simplistic assumption of metabolic rate being equivalent to free radical production were to be valid. However, mitochondria isolated from each species and compared for ROS production do not show equivalence as might be expected based on metabolic rate.[32] Instead, isolated mitochondria from a variety of tissues of the pigeon have approximately 2–4-fold lower levels of ROS production than do those from the rat.[32] Hence, ROS production and its effects, in addition to metabolic rate, must be taken into account in evaluating the potential for free radicals to be a major contributor in limiting life span.

ANTIOXIDANTS AND THEIR EFFICACY

A variety of studies on the amelioration or attenuation of aging or the progression of disease by chronic or acute antioxidant treatments have been carried out.[33–41]

Such studies have at best yielded equivocal or mildly beneficial results in slowing disease progression or improving mean or maximal life span.[40] One possible reason for the apparent preciousness in obtaining robust results in extending life span is that the antioxidants used are not particularly efficacious in mitochondrial oxidative stress. The antioxidants utilized in the studies just listed (vitamin C, E, etc.) can be considered "one shot" antioxidants, that is, unlike enzymes, they are not catalytically active within the cell. However, synthetic classes of antioxidants now exist that are catalytically active[42–44] and have demonstrable antioxidant activity within mitochondria,[45] the source of most superoxide within the cell.

Eukarion (Bedford, MA) has developed a class of small molecular weight molecules that are catalytic scavengers of ROS. These molecules are salen-manganese complexes, which act as SOD and catalase "mimics" catalytically destroying both superoxide and H_2O_2.[46, 47] The EUK (from Eukarion) class of compounds is effective in a broad variety of experimental paradigms that involve oxidative stress. EUK-8 effectively inhibits lipid peroxidation compared to vitamin E in a brain microsome/ROS system.[47] Furthermore, EUK-8 has been effective in two mouse models of Parkinson's disease, protecting nigrostriatal dopaminergic neurons from damage induced by MPTP or 6-hydroxydopamine.[47] EUK-8 has also shown efficacy in a cell culture model of Alzheimer's disease, protecting hippocampal slices from the cytotoxicity of β-amyloid peptide.[48] EUK-134, an analog of EUK-8, has increased catalase activity and equal SOD activity compared to EUK-8.[49] It demonstrated a high degree of protection in a rat stroke model, in an MPTP mouse model, and in kidney ischemia/reperfusion-induced injury.[49,50] As both EUK-8 and EUK-134 readily cross the blood/brain barrier in pharmacologically active amounts, this makes them attractive candidate molecules for testing their efficacy in inhibiting mitochondrial superoxide in the brain in addition to peripheral organ systems.

TREATMENT OF *SOD2* MUTANT MICE WITH CATALYTIC ANTIOXIDANTS

The lack of mitochondrial SOD is particularly deleterious to organismal physiology, as revealed in studies of mice that have had SOD inactivated through homologous recombination. There are three isoforms of SOD: *sod1*, a cytosolic Cu/Zn SOD; *sod2*, a mitochondrial MnSOD; and *sod3*, an extracellular Cu/Zn SOD. Inactivation of *sod1* or *sod3* via homologous recombination in mice results in mild nonlethal phenotypes.[51,52] In marked contrast, inactivation of the mitochondrial SOD results in a neonatal lethal phenotype characterized by a dilated cardiomyopathy and fibrosis, neurodegeneration, metabolic acidosis, hepatic fat accumulation, DNA oxidative damage, tissue-specific mitochondrial respiratory chain abnormalities, and abnormalities in TCA cycle enzymes.[18,21,53] This emphasizes the importance of mitochondrial SOD and mitochondrially generated ROS and demonstrates the potential for heterogeneous pathologies due to mitochondrial oxidative stress.

We wished to determine if it was possible to make up for the lack of endogenous mitochondrial *sod* by chronically treating mice lacking SOD2 with a variety of synthetic catalytic antioxidants. The first such antioxidant we tested was the SOD mimetic Manganese 5,10,15,20-tetrakis (4-benzoic acid) porphyrin (MnTBAP). By chronically treating the mutant mice with MnTBAP through daily intraperitoneal injection at a dosage of 5 mg/kg, we were able to increase the life span of *sod* mutant mice from a mean of 8 days to 16 days of age, with a concomitant dramatic increase in survival kinetics.[53] Not only was MnTBAP treatment of the mutant mice able to increase the survival of the animals substantially, but also it rescued the dilated cardiomyopathy and the hepatic lipid accumulation.[53] However, MnTBAP is a large molecule and does not penetrate the blood/brain barrier. This means that by preventing the early neonatal death of the *sod2* mutant mice through effective treatment of the heart failure, we allowed the brain sufficient time to develop its own free radical mediated pathology due to the endogenous production of mitochondrial free radicals via metabolism in the absence of SOD2.

This is reflected as an age-related brain disorder in the MnTBAP-treated *sod2* mutant mice. As the MnTBAP-treated *sod2* mutant mice age, they develop a profound movement disorder starting at about 12 days of age, which is characterized by a mild tremor accompanied by circling, falling, and barrel-like rolls.[53] By 3 weeks of age the MnTBAP-treated animals are severely behaviorally compromised and develop an inability to feed due to their overt neurologic problems; hence, at this point they are sacrificed.[53] Neuropathologic analysis of the brains of MnTBAP-treated *sod2* mutant mice showed that they develop a severe spongiform encephalopathy in the frontal cortex and focally within specific regions of the brainstem. Most animals showed spongiform changes in the motor nucleus of cranial nerves V and VII, the reticulotegmental nucleus of the pons, and the superior and medioventral periolivary nuclei. Hence, these results demonstrate the potential for *sod2* mutant mice to be used to screen effective antioxidants *in vivo* in relation to protecting different organ systems.

To definitively test the hypothesis that mitochondrial free radicals are a major limiter of life span in the mammal, it must be demonstrated that (1) endogenous production of free radicals can cause a wide range of pathologies, (2) such pathologies can be attenuated or prevented through exogenous antioxidant treatment; and (3) an increase in mean and maximal life span can be achieved through treatment of normal animals with such antioxidants.

We believe that we have demonstrated that the first two criteria are satisfied and are currently attempting to falsify the third criterion. If the mean and maximum life span is not increased in organisms treated with antioxidants (which are effective by criterion 2), it is unlikely that the endogenous production of mitochondrial free radicals is a major limiter of life span.

REFERENCES

1. PARKES, T.L., A.J. ELIA, D. DICKINSON *et al.* 1998. Extension of Drosophila lifespan by overexpression of human SOD1 in motorneurons. Nature Genet. **19:** 171–174.
2. SUN, J. & J. TOWER. 1999. FLP recombinase-mediated induction of Cu/Zn-superoxide dismutase transgene expression can extend the life span of adult *Drosophila melanogaster* flies [In Process Citation]. Mol. Cell Biol. **19:** 216–228.
3. BOVERIS, A. 1984. Determination of the production of superoxide radicals and hydrogen peroxide in mitochondria. Methods Enzymol. **105:** 429–435.
4. CHANCE, B., H. SIES & A. BOVERIS. 1979. Hydroperoxide metabolism in mammalian organs. Physiol. Rev. **59:** 527–605.

5. TURRENS, J.F. & A. BOVERIS. 1980. Generation of superoxide anion by the NADH dehydrogenase of bovine heart mitochondria. Biochem. J. **191:** 421–427.
6. HANSFORD, R.G., B.A. HOGUE & V. MILDAZIENE. 1997. Dependence of H_2O_2 formation by rat heart mitochondria on substrate availability and donor age. J. Bioenerg. Biomembr. **29:** 89–95.
7. BECKMAN, K.B. & B.N. AMES. 1999. Endogenous oxidative damage of mtDNA. Mutat. Res. **424:** 51–58.
8. HALLIWELL, B. & S. CHIRICO. 1993. Lipid peroxidation: its mechanism, measurement, and significance. Am. J. Clin. Nutr. **57:** 715S–724S; discussion 724S–725S.
9. STADTMAN, E.R. 1995. Role of oxidized amino acids in protein breakdown and stability. Methods Enzymol. **258:** 379–393.
10. PANFILI, E., G. SANDRI & L. ERNSTER. 1991. Distribution of glutathione peroxidases and glutathione reductase in rat brain mitochondria. FEBS Lett. **290:** 35–37.
11. SCHAPIRA, A.H.V. 1998. Mitochondrial dysfunction in neurodegenerative disorders. Biochim. Biophys. Acta **1366:** 225–233.
12. ROBINSON, B.H. 1998. Human complex I deficiency: clinical spectrum and involvement of oxygen free radicals in the pathogenicity of the defect. Biochim. Biophys. Acta **1364:** 271–286.
13. PANDOLFO, M. 1998. Molecular genetics and pathogenesis of Friedreich ataxia [In Process Citation]. Neuromusc. Disord. **8:** 409–415.
14. ROTIG, A., P. DE LONLAY, D. CHRETIEN *et al.* 1997. Aconitase and mitochondrial iron-sulphur protein deficiency in Friedreich ataxia. Nature Genet. **17:** 215–217.
15. MARTIN, G.M., S.N. AUSTAD & T.E. JOHNSON. 1996. Genetic analysis of ageing: role of oxidative damage and environmental stresses. Nature Genet. **13:** 25–34.
16. LARSEN, P.L. 1993. Aging and resistance to oxidative damage in *Caenorhabditis elegans*. Proc. Natl. Acad. Sci. USA **90:** 8905–8909.
17. VANFLETEREN, J.R. 1993. Oxidative stress and ageing in *Caenorhabditis elegans*. Biochem. J. **292:** 605–608.
18. MELOV, S., P. COSKUN, M. PATEL *et al.* 1999. Mitochondrial disease in superoxide dismutase 2 mutant mice. Proc. Natl. Acad. Sci. USA **96:** 846–851.
19. MELOV, S., P.E. COSKUN & D.C. WALLACE. 1999. Mouse models of mitochondrial disease, oxidative stress, and senescence. Mutat. Res. **434:** 233–242.
20. LI, Y., T.T. HUANG, E.J. CARLSON *et al.* 1995. Dilated cardiomyopathy and neonatal lethality in mutant mice lacking manganese superoxide dismutase. Nature Genet. **11:** 376–381.
21. LEBOVITZ, R.M., H. ZHANG, H. VOGEL *et al.* 1996. Neurodegeneration, myocardial injury, and perinatal death in mitochondrial superoxide dismutase-deficient mice. Proc. Natl. Acad. Sci. USA **93:** 9782–9787.
22. SOHAL, R.S., S. AGARWAL, M. CANDAS *et al.* 1994. Effect of age and caloric restriction on DNA oxidative damage in different tissues of C57BL/6 mice. Mech. Ageing Dev. **76:** 215–224.
23. SOHAL, R.S., S. AGARWAL & B.H. SOHAL. 1995. Oxidative stress and aging in the Mongolian gerbil (*Meriones unguiculatus*). Mech. Ageing Dev. **81:** 15–25.
24. SOHAL, R.S., H.H. KU, S. AGARWAL *et al.* 1994. Oxidative damage, mitochondrial oxidant generation and antioxidant defenses during aging and in response to food restriction in the mouse. Mech. Ageing Dev. **74:** 121–133.
25. SOHAL, R.S. & A. DUBEY. 1994. Mitochondrial oxidative damage, hydrogen peroxide release, and aging. Free Rad. Biol. Med. **16:** 621–626.
26. SOHAL, R.S., H.H. KU & S. AGARWAL. 1993. Biochemical correlates of longevity in two closely related rodent species. Biochem. Biophys. Res. Commun. **196:** 7–11.
27. SOHAL, R.S. & W.C. ORR. 1992. Relationship between antioxidants, prooxidants, and the aging process. Ann. N.Y. Acad. Sci. **663:** 74–84.
28. SOHAL, R.S., I. SVENSSON & U.T. BRUNK. 1990. Hydrogen peroxide production by liver mitochondria in different species. Mech. Ageing Dev. **53:** 209–215.
29. MIQUEL, J., R. BINNARD & J.E. FLEMING. 1983. Role of metabolic rate and DNA-repair in Drosophila aging: implications for the mitochondrial mutation theory of aging. Exp. Gerontol. **18:** 167–171.

30. PEREZ-CAMPO, R., M. LOPEZ-TORRES, S. CADENAS *et al.* 1998. The rate of free radical production as a determinant of the rate of aging: evidence from the comparative approach. J. Comp. Physiol. [B]. **168:** 149–158.
31. BARJA, G., S. CADENAS, C. ROJAS *et al.* 1994. Low mitochondrial free radical production per unit O_2 consumption can explain the simultaneous presence of high longevity and high aerobic metabolic rate in birds. Free Rad. Res. **21:** 317–327.
32. KU, H.H. & R.S. SOHAL. 1993. Comparison of mitochondrial pro-oxidant generation and anti-oxidant defenses between rat and pigeon: possible basis of variation in longevity and metabolic potential. Mech. Ageing Dev. **72:** 67–76.
33. BEHL, C. 1999. Alzheimer's disease and oxidative stress: implications for novel therapeutic approaches. Prog. Neurobiol. **57:** 301–323.
34. GEY, K.F. 1995. Cardiovascular disease and vitamins. Concurrent correction of 'suboptimal' plasma antioxidant levels may, as important part of 'optimal' nutrition, help to prevent early stages of cardiovascular disease and cancer, respectively. Bibl. Nutr. Diet. **52:** 75–91.
35. HALLER, J., R.M. WEGGEMANS, M. FERRY & Y. GUIGOZ. 1996. Mental health: minimental state examination and geriatric depression score of elderly Europeans in the SENECA study of 1993. Eur. J. Clin. Nutr. **50:** S112–116.
36. PERRIG, W.J., P. PERRIG & H.B. STAHELIN. 1997. The relation between antioxidants and memory performance in the old and very old. J. Am. Geriatr. Soc. **45:** 718–724.
37. GALE, C.R., C.N. MARTYN & C. COOPER. 1996. Cognitive impairment and mortality in a cohort of elderly people. Br. Med. J. **312:** 608–611.
38. MIQUEL, J., J. FLEMING & A.C. ECONOMOS. 1982. Antioxidants, metabolic rate and aging in Drosophila. Arch. Gerontol. Geriatr. **1:** 159–165.
39. HARRINGTON, L.A. & C.B. HARLEY. 1988. Effect of vitamin E on lifespan and reproduction in *Caenorhabditis elegans.* Mech. Ageing Dev. **43:** 71–78.
40. LIPMAN, R.D., R.T. BRONSON, D. WU *et al.* 1998. Disease incidence and longevity are unaltered by dietary antioxidant supplementation initiated during middle age in C57BL/6 mice. Mech. Ageing Dev. **103:** 269–284.
41. JAMA, J.W., L.J. LAUNER, J.C. WITTEMAN *et al.* 1996. Dietary antioxidants and cognitive function in a population-based sample of older persons. The Rotterdam Study. Am. J. Epidemiol. **144:** 275–280.
42. FAULKNER, K.M., S.I. LIOCHEV & I. FRIDOVICH. 1994. Stable Mn(III) porphyrins mimic superoxide dismutase in vitro and substitute for it in vivo. J. Biol. Chem. **269:** 23471–23476.
43. DAY, B.J. & J.D. CRAPO. 1996. A metalloporphyrin superoxide dismutase mimetic protects against paraquat-induced lung injury in vivo. Toxicol. Appl. Pharmacol. **140:** 94–100.
44. PATEL, M., B.J. DAY, J.D. CRAPO *et al.* 1996. Requirement for superoxide in excitotoxic cell death. Neuron **16:** 345–355.
45. SZABO, C., B.J. DAY & A.L. SALZMAN. 1996. Evaluation of the relative contribution of nitric oxide and peroxynitrite to the suppression of mitochondrial respiration in immunostimulated macrophages using a manganese mesoporphyrin superoxide dismutase mimetic and peroxynitrite scavenger. FEBS Lett. **381:** 82–86.
46. BAUDRY, M., S. ETIENNE, A. BRUCE *et al.* 1993. Salen-manganese complexes are superoxide dismutase-mimics. Biochem. Biophys. Res. Commun. **192:** 964–968.
47. DOCTROW, S.R., K. HUFFMAN, C.B. MARCUS *et al.* 1996. Salen-manganese complexes: combined superoxide dismutase/catalase mimics with broad pharmacological efficacy. Adv. Pharmacol. **38:** 247–269.
48. BRUCE, A.J., B. MALFROY & M. BAUDRY. 1996. Beta-amyloid toxicity in organotypic hippocampal cultures: protection by EUK-8, a synthetic catalytic free radical scavenger. Proc. Natl. Acad. Sci. USA **93:** 2312–2316.
49. BAKER, K., C.B. MARCUS, K. HUFFMAN *et al.* 1998. Synthetic combined superoxide dismutase/catalase mimetics are protective as a delayed treatment in a rat stroke model: a key role for reactive oxygen species in ischemic brain injury. J. Pharmacol. Exp. Ther. **284:** 215–221.
50. GIANELLO, P., A. SALIEZ, X. BUFKENS *et al.* 1996. EUK-134, a synthetic superoxide dismutase and catalase mimetic, protects rat kidneys from ischemia-reperfusion-induced damage. Transplantation **62:** 1664–1666.

51. REAUME, A.G., J.L. ELLIOTT, E.K. HOFFMAN *et al.* 1996. Motor neurons in Cu/Zn superoxide dismutase-deficient mice develop normally but exhibit enhanced cell death after axonal injury. Nature Genet. **13:** 43–47.

52. CARLSSON, L.M., J. JONSSON, T. EDLUND & S.L. MARKLUND. 1995. Mice lacking extracellular superoxide dismutase are more sensitive to hyperoxia. Proc. Natl. Acad. Sci. USA **92:** 6264–6268.

53. MELOV, S., J.A. SCHNEIDER, B.J. DAY *et al.* 1998. A novel neurological phenotype in mice lacking mitochondrial manganese superoxide dismutase [see comments]. Nature Genet. **18:** 159–163.

Tissue Mitochondrial DNA Changes

A Stochastic System

GEORGE KOPSIDAS, SERGEY A. KOVALENKO, DAMIEN R. HEFFERNAN, NATALIA YAROVAYA, LUDMILLA KRAMAROVA, DIANE STOJANOVSKI, JUDY BORG, MOHAMMED M. ISLAM, APHRODITE CARAGOUNIS, AND ANTHONY W. LINNANE[a]

Centre for Molecular Biology & Medicine, Epworth Medical Centre, Richmond, Melbourne, Victoria 3121, Australia

ABSTRACT: Several lines of evidence support the view that the bioenergetic function of the mitochondria in postmitotic tissue deteriorates during normal aging. Skeletal muscle is one such tissue that undergoes age-related fiber loss and atrophy and an age-associated rise in the number of cytochrome *c* oxidase (COX) deficient fibers. With such metabolic pressure placed on skeletal muscle it would be an obvious advantage to supplement the cellular requirement for energy by up-regulating glycolysis, an alternative pathway for energy synthesis. Analysis of rat skeletal muscle utilizing antibodies directed against key enzymes involved in glycolysis has provided evidence of an age-associated increase in the enzymes involved in glycolysis. Fructose-6-phosphate kinase, aldolase, glyceraldehyde-3-phosphate dehydrogenase, and pyruvate kinase protein levels appeared to increase in the soleus, gracilis, and quadriceps muscle from aged rats. The increase in the level of these proteins appeared to correlate to a corresponding decrease in the amount of cytochrome *c* oxidase protein measured in the same tissue. Together these results are interpreted to represent a general upregulation of glycolysis that occurs in response to the age-associated decrease in mitochondrial energy capacity. Mitochondrial DNA (mtDNA) damage and mutations may accumulate with advancing age until they reach a threshold level were they impinge on the bioenergy capacity of the cell or tissue. Evidence indicates that mtDNA from the skeletal muscle of both aged rats and humans not only undergoes changes at the nucleotide sequence level (mutations and DNA damage), but also undergoes modifications at the tertiary level to generate unique age-related conformational mtDNA species. One particular age-related conformational form was only detected in aged rat tissues with high demands on respiration, specifically in heart, kidney, soleus muscle, and, to a lesser extent, the quadriceps muscle. The age-related form was not detected in gracilis muscle which is predominantly dependent upon glycolysis with regard to its energy requirements. Finally, a comprehensive hypothesis is presented that features the stochastic nature of the mitochondrial system. The basis of the hypothesis is that a dynamic relationship exists between endogenous mutagen production, DNA repair, mtDNA turnover, and nuclear control of mtDNA copy number and that age-associated changes in the dynamics of this relationship lead to a loss of functional full-length mtDNA that eventually leads to bioenergy decline.

[a]Address for correspondence: Centre for Molecular Biology and Medicine, Epworth Medical Centre, 185-187 Hoddle Street, Richmond, Melbourne, Victoria 3121, Australia. Phone: 61-3-9426 4200; fax: 61-3-9426 4201.

tlinnane@cmbm.com.au

INTRODUCTION: EVOLUTION OF A CONCEPT

In 1989, Linnane and colleagues[1] proposed that the occurrence of mtDNA mutations and their accumulation during life would make a significant contribution to the aging process. This proposal was based on the observation of numerous similarities between the characteristics of aging and overt mitochondrial diseases in terms of mitochondrial bioenergy deficits and the associated mtDNA mutations.[2,3] The essence of the hypothesis stated that random mtDNA mutations occur in somatic tissues and progressively accumulate throughout life with the distinctive metabolism of cells located within different types of tissue influencing the mutation rate of the mtDNA. It was conceptualized that mtDNA mutations would not be evenly distributed among the cells of a given tissue by reason of the random nature of mutations, the stochastic occurrence of mutations in dividing and nondividing cells, and the random segregation of mitochondrial genomes that occurs in dividing cells. A mtDNA mosaic among the cells of a tissue would ultimately occur that would represent the uneven distribution of wild-type and various mutant mtDNA molecules. This in turn would result in a cellular bioenergy mosaic. As the proportion of mtDNA-damaged cells accumulates above a certain threshold, generally achieved by late age, a progressive decline would occur in the physiological and biochemical performance of individual tissues and organs, making a significant contribution to the aging process.

The original proposal has since gained substantial support. Data from different laboratories including our own provides compelling evidence that mtDNA mutations accumulate with age in postmitotic tissues.[4–9] For instance, we initially showed in 1990 that a particular large mtDNA deletion of 4977 bp (known as the 5 kb or "common deletion" because of its frequent occurrence in certain classes of mitochondrial disease patients) occurs in an age-related manner in a wide variety of human tissues,[4] a finding that has been repeatedly confirmed (references compiled in Refs. 10 and 11). Shortly afterwards in the further development of our own studies, we identified at least 10 different mtDNA deletions in tissues from a single 69-year-old woman.[12] Most of these multiple mtDNA deletions occurred at breakpoints involving short, direct-repeat mtDNA sequences. For example, the 5-kb deletion occurred between a pair of 13-bp direct-repeat sequences in mtDNA. Many thousands of short direct repeat sequences are located along the human mitochondrial sequence, ranging from 4–13 bp,[13] illustrating the potential for multiple and varying deletions in human mtDNA.[14]

The potential to generate multiple deletions is without doubt a predisposing facet of the mitochondrial genome. Multiple deletions, however, have been difficult or laborious to demonstrate because analysis of mutations was always limited to procedures based on the small DNA fragments that could be amplified with the polymerase chain reaction (PCR) or to relatively insensitive non-PCR methods such as Southern blot analysis or high-resolution restriction enzyme analysis of purified mtDNA genomes. Although these approaches give a localized view of one or a few mtDNA mutations, they do not give an indication of the overall genetic integrity of the mitochondrial genome. An alternative approach is to apply extra-long PCR (XL-PCR), a procedure utilizing a mixture of thermostable DNA polymerases (generally *Taq* DNA polymerse and a second DNA polymerase with a proofreading function such as P*wo*) to overcome the limitations of conventional PCR for amplifying long

DNA fragments. By applying XL-PCR to amplify the entire human 16,569-bp mtD-NA sequence, a collective appraisal of any mtDNA mutations (localized between the primer pair) that can be characterized by a change in overall genome size becomes possible. This type of approach also determines the amount of full-length mtDNA that is essentially undamaged and in a form that is available for amplification by PCR. Consequently, a mtDNA profile can be generated by determining the relative concentration of full-length mtDNA that can be amplified by XL-PCR (presumably full-length mtDNA plus any point mutations) with respect to the relative proportions of any mtDNA rearrangements.

Our laboratory has applied XL-PCR to analyze the mtDNA genome extracted from a series of human tissues, including deltoid muscle[15] and quadriceps muscle[16] taken from different aged individuals, in addition to several other tissues (liver and heart) sampled from the same individual.[17] The most definitive data were obtained with skeletal muscle. The results showed that the amount of full-length mtDNA that could be amplified by XL-PCR (referred to here as amplifiable full-length mtDNA) decreased markedly with age and that multiple heterogeneous mtDNA rearrangements also occurred. By simultaneously monitoring COX activity in the same tissue (an enzyme dependent on the contribution of particular mitochondrial genes for its assembly and function), a correlation became apparent that linked a decrease in the bioenergy capacity of skeletal muscle to a concurrent decrease in the amount of amplifiable full-length mtDNA and an accumulation of mtDNA rearrangements. The phenomenon was to a lesser extent essentially reproduced in cardiac tissue, but was not observed in liver.

Considering the bioenergetic mosaicism within tissues such as skeletal muscle, any definitive resolution relating mtDNA mutations to an age-associated decline in cellular bioenergy would need to correlate mtDNA mutations to the bioenergetic capacity of a *single* cell of established COX activity. This was accomplished by isolating single human skeletal muscle fibers of predetermined COX activity and applying XL-PCR to generate a mtDNA mutation profile.[18] COX positive muscle fibers dissected from individuals of various ages contained amplifiable full-length mtDNA together with a limited number of mtDNA rearrangements. By contrast, COX negative fibers taken from the same individuals did not contain detectable full-length mtD-NA, but they did contain a heterogeneous mixture of rearranged mtDNA species with the frequency and occurrence of each deletion varying considerably from fiber to fiber. These data lead to the conclusion that amplifiable full-length mtDNA and COX activity are linked. It was proposed that the amount of amplifiable full-length mtDNA equates to the level of functional mtDNA and is an exemplifier of the bioenergetic capacity of a tissue.[18]

Paradoxically, when total cellular DNA extracts were analyzed by Southern blot, the total amount of full-length mtDNA detected in the skeletal muscle of young and old subjects was similar.[16,19] The reduced levels of full-length mtDNA that could be amplified by XL-PCR that were observed in aged muscle were not reflected by a significant reduction in the total amount of full-length mtDNA extracted from the same tissue. We recently published a short communication that in part seemed to resolve the paradox.[16] The resolution was based on the finding that DNA polymerases fail to insert a base opposite a variety of DNA base adducts, which leads to a detectable reduction in PCR amplification efficiency.[20,21] We proposed that mtDNA from aged skeletal muscle is not only significantly mutated but also extensively damaged. Persistent mtDNA damage would account for the inability of the XL-PCR reaction to amplify full-length mtDNA from aged tissues. Damaged mtDNA templates that could not be amplified by XL-PCR *in vitro* would presumably not be replicated or transcribed *in vivo*. Consequently, the level of mtDNA that can be amplified with XL-PCR was postulated to reflect only that portion of the mtDNA pool that is fully functional and would not contribute to the age-associated loss of functional mitochondrial genomes. Thus, a major change in the amount of total cellular mtDNA need not occur.

In this report, we present several recent observations that have contributed to the continued evolution of our original hypothesis. The dynamic equilibrium between mtDNA damage, DNA repair, and mtDNA turnover in the mitochondria is discussed with respect to the stochastic nature of the mitochondrial system.

AGE-ASSOCIATED CONFORMATIONAL CHANGES TO mtDNA

The first mtDNA mutation to be associated with a disease phenotype was a duplication of mtDNA found in chronic myelogenous leukemia.[22] Since this early discovery, complex conformational forms of mtDNA including catenates and multiple monomeric species (duplications, partial duplications, and duplication-deletion combinations) have been described in aged and cancerous tissues and have been associated with overt mitochondrial diseases such as Kears-Sayre syndrome.[23–25]

Our own current studies on human and rat skeletal muscle, which have used field inversion gel electrophoresis (FIGE) to perform high-resolution separation of intact high molecular weight mtDNA molecules, have confirmed age-associated conformational differences in mtDNA. Southern blot analysis of FIGE-separated mtDNA from different rat tissues identified the numerous conformational forms of mtDNA that are normally detected (open circular, supercoiled, and linear fractions). Analysis also noted the appearance of at least one extra age-associated conformational form or "age-related band" in rat, as previously reported by other workers in mice and humans.[26,27] The age-related band (ARB) was observed in aged rat quadriceps muscle, heart, and kidney, but it was not detected in adrenal gland or liver (FIG. 1 and TABLE 1). The ARB was also detected in human quadriceps muscle (FIG. 2). Interestingly, the adrenal gland displayed only two conformational forms of mtDNA, presumably linear and open circular mtDNA, and did not show any of the other conformational forms generally noted in other tissues. This observation suggests that many of the conformational forms seen in muscle, heart, and kidney are not artefacts of the extraction process.

An important aspect of these muscle studies is the recognition that skeletal muscle consists of two major fiber groups, type I and type II. Type I muscle fibers have a greater oxidative capacity and contain more mitochondria per fiber. Type II muscle fibers which can be classified into two subgroups, IIa and IIb, are more glycolytic than type I fibers with respect to their energy requirements. Accordingly, mtDNA changes were investigated in different types of rat muscle that contain distinct proportions of type I and type II muscle fibers, the soleus muscle, which is predominantly comprised of type I muscle fibers (95%), the gracilis muscle, which is principally

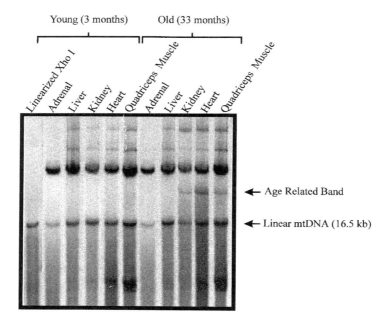

FIGURE 1. Southern blot of unrestricted total DNA extracted from various organs from a young and aged rat. Total DNA (5 μg) was loaded onto a 0.8% TAE agarose gel and separated with FIGE Electrophoresis (60 V, 0.5 sec forward, and 60 V, 0.2 sec reverse) for 72 hours. Agarose gels containing DNA of interest were neutralized with HCl (0.25 M for 30 min), then capillary blotted and fixed onto nylon Hybond-N+ membranes (Amersham) using 20X SSC (3 M NaCl, 0.3 M $Na_3Citrate$, pH 7). Full-length mtDNA probes were synthesized by incorporating DIG-11-dUTP into the DNA during XL-PCR. The template used was genomic DNA extracted from white bloods cells (Roche). Labeling reactions were separated on an ethidium bromide agarose gel and the appropriate DNA band extracted from the gel and purified (Qiaex II agarose extraction, Qiagen, Germany). Hybridization and wash conditions were carried out using DIG Easy Hyb under highly stringent conditions as outlined by the manufacturer (Boehringer-Mannheim). DIG-labeled DNA bound to Hybond-N+ membrane was subsequently detected by enzyme-linked immunoassay using anti-DIG antibodies conjugated to alkaline phosphatase (Boehringer-Mannheim). An enzyme-catalyzed color reaction with 5-bromo-4-chloro-3-indolyl phosphate and nitroblue tetrazolium salt was used to visualize amplified sequences via a blue precipitate.

type II muscle fibers (95%), and the quadriceps muscle, which is a mixture of both fiber types in proportions that vary throughout the muscle. FIGE-Southern blot analysis of the three rat muscle types revealed that the ARB appeared to be absent from both young and aged gracilis muscle and was identified in 23% of aged quadriceps muscle specimens and in 60% of aged soleus muscle samples (FIG. 3 and TABLE 1). These results suggest that the ARB occurs in tissues that have a high aerobic requirement, such as soleus muscle, and not in gracilis muscle, which is more glycolytic with regard to its bioenergy requirements.

TABLE 1. Percentage occurrence of the age-related band in various tissues from young and aged rats[a]

Tissue	Young rats (<5 mo.)	Aged rats (>29 mo.)
Heart	0% ($n = 2$)	100% ($n = 4$)
Kidney	0% ($n = 4$)	93% ($n = 14$)
Adrenals	0% ($n = 4$)	0% ($n = 5$)
Liver	0% ($n = 2$)	0% ($n = 10$)
Vastus lateralis (quadriceps)	0% ($n = 6$)	24% ($n = 30$)
Soleus	0% ($n = 3$)	60% ($n = 11$)
Gracilis caudalis	0% ($n = 3$)	0% ($n = 6$)

[a]Number of animals used for each determination is shown in parentheses. Young animals were aged less than 5 months with an average age of 3 months. Aged animals were generally 30–35 months of age.

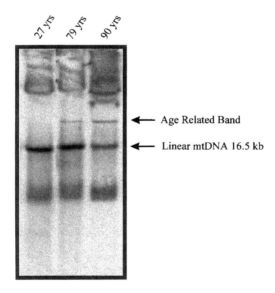

FIGURE 2. Southern blot of unrestricted total human DNA extracted from the quadriceps muscle of various aged individuals. Total DNA (2.5 mg) extracted from quadriceps muscle was loaded onto 0.8% TAE agarose gel, separated with FIGE Electrophoresis, blotted, and probed as outlined in FIGURE 1.

The ARB was not observed in DNA extracts from liver. Liver is a mitotic tissue that undergoes rapid cell turnover and apoptosis, thereby potentially removing any age-associated mtDNA changes. Indeed, previous studies on human liver have indicated that this organ does not demonstrate many of the age-associated mutations observed with postmitotic tissues,[17] suggesting that liver may be an inappropriate organ for aging studies of this type.

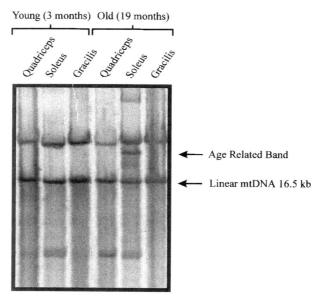

FIGURE 3. Southern blot of unrestricted total DNA extracted from various skeletal muscles from a young and aged rat. Total DNA (5 mg) extracted from soleus, gracilis caudalis, and vastus lateralis (quadriceps) muscle was loaded onto 0.8% TAE agarose gel, separated with FIGE Electrophoresis, blotted, and probed as outlined in FIGURE 1.

Evidence is accumulating that free radical generation is significantly higher in the skeletal muscle of aged rats.[28–30] The generation of reactive oxygen species has been reduced in the mitochondria isolated from calorically restricted animals compared to ad libitum fed animals.[31] Also, the occurrence of the ARB in mice has been shown to be reduced following caloric restriction.[27] Together these results suggest that oxidative damage may be an important parameter in the formation of the ARB. Perhaps the simplest interpretation that accounts for the data and also provides a mechanism for the generation of the ARB is that the band represents a damaged pool of mtDNA or a damage-repair intermediate that is partially unwound as a consequence of DNA repair. Preliminary evidence obtained with human cell lines indicates that a damage-repair intermediate accumulates following an *in vivo* mutagenic assault on the mtDNA with bleomycin.[32] FIGE-Southern blot analysis shows that this bleomycin-induced conformational form appears to correspond to the ARB, as they both run with a similar mobility on agarose gels. The bleomycin-induced band became less abundant when bleomycin was removed and the cells were allowed several hours to repair any mtDNA damage.

The preliminary data just outlined highlights the importance of DNA repair in the mitochondria. Until recently, the redundancy of mtDNA in mammalian cells was thought to preclude the requirement for extensive DNA repair of the type seen in the nucleus. The initial report by Clayton and colleagues[33] showing that cyclobutane pyrimidine dimers were not removed from mammalian mtDNA generated a perception that DNA repair in mitochondria was completely absent or severely limited. For instance, determinations of oxidative damage, generally of 8-oxo-deoxyguanosine (oxo-[8]dG) levels, led to the assumption that mtDNA is particularly prone to oxidative damage that is orders of magnitude greater than that observed in the nucleus. However, an increasing number of mammalian enzyme systems are being uncovered that are capable of repairing mtDNA oxidative damage, including base excision repair, uracil DNA glycosylase, AP endonuclease, some recombinational repair in mammalian cells and repair of certain UV-induced photoproducts, and subsequent induction of mtDNA mutations (for reviews see refs. 34 and 35). Furthermore, the mitochondrial capacity to repair oxo-[8]dG damage increases with age, suggesting an age-associated upregulation of at least one repair pathway in rats.[36]

Other explanations of the ARB also need to be considered. An alternative plausible explanation is that the ARB may represent a segment of the mtDNA population that is generated as a result of aborted replication events. A unique feature of mtDNA replication is the observation that most of the newly synthesized H-strands terminate prematurely, resulting in the formation of displacement loop (D-loop) strands.[37] Only the H-strand, which successfully proceeds past the premature termination site, will be replicated into a circular mtDNA molecule. This innate control mechanism has been postulated to be partially responsible for the regulation of mtDNA copy number.[38] Although the precise physiological significance of the D-loop formation is still unclear, premature termination may indicate that replication has initiated on a damaged or inappropriate template and is subsequently aborted. An increase in the frequency of incomplete replication events would be predicted to eventually form a distinct, detectable pool of mtDNA in aged tissue.

AS MITOCHONDRIAL BIOENERGY DECREASES, GLYCOLYSIS IS ENHANCED

A fundamental property of any biological system is the ability to adjust metabolic pathways to embrace adaptations and allow continued cell function. It would be an apparent advantage to augment the requirement for cellular bioenergy in skeletal muscle when there is an age-associated decline in mitochondrial function by upregulating alternative pathways for energy synthesis such as glycolysis. However, the effects of aging on the relation between oxidative phosphorylation and glycolysis in skeletal muscle is unclear, leading to considerable debate in the literature on whether or not glycolysis is reduced or is increased as a consequence of the aging process.[39–42] Undoubtedly, skeletal muscle undergoes age-associated fiber loss, which coupled with an overall decline in bioenergy capacity as exemplified by a global reduction in COX activity,[4,43] places significant metabolic pressure on aged skeletal muscle.

Our group has initiated an investigation of the relationship between oxidative phosphorylation and glycolysis during the course of aging. Rat skeletal muscle sections probed with fluorescently labeled antibodies has provided evidence that indicates a coordinated increase in key enzymes involved in glycolysis that appeared to

Young   Aged

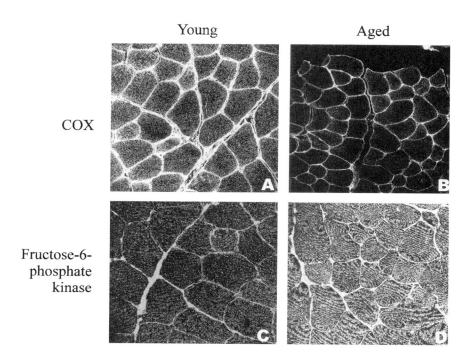

COX

Fructose-6-
phosphate
kinase

FIGURE 4. Antibody labeling of cytochrome *c* oxidase and a glycolytic enzyme in rat gracilis muscle. Immunofluorescence was performed on young (4-month) and old (34-month) Sprague-Dawley rat gracilis muscle using antibodies against cytochrome c oxidase subunit I (**A** = young and **B** = aged, magnification × 150) and fructose-6-phophate kinase (**C** = young and **D** = aged, magnification × 200). An increase in the labeling intensity of the cytoplasm suggests an increase in protein levels within the muscle fiber. Note that the labeling between fibers is nonspecific secondary antibody binding. Images were collected using a BioRad MRC 1024ES laser scanning confocal microscope equipped with a Krypton/Argon laser using the 488 nm laser line.

parallel a corresponding decrease in the amount of COX protein in the same tissue. FIGURE 4 illustrates immunofluorescence images captured with a confocal laser microscope of young and senescent rat gracilis muscle stained with antibodies directed against COX protein subunit I and fructose-6-phosphate kinase. The images clearly demonstrate an age-associated decrease in the level of antibody labeling of the COX protein (panels A and B). The same tissue displayed an age-associated increase in the immunolabeling of fructose-6-phosphate kinase (panels C and D). A similar increase was recorded with aldolase, glyceraldehyde-3-phosphate dehydrogenase, pyruvate kinase, but not with hexokinase type II (TABLE 2). This general phenomenon (i.e., an age-related decrease in COX protein level and corresponding increase in the immunolabeling of glycolytic enzymes) was also observed in soleus and quadriceps muscle. Experiments are in progress to determine the empirical relation between the intensity of immunolabeling and glycolytic enzyme concentration.

TABLE 2. Antibody labeling comparison of five glycolytic enzymes in gracilis, soleus, and quadriceps muscles from young and old rats[a]

Antibody specificity	Young	Old
Cytochrome *c* oxidase subunit I	++	+
Aldolase	+	++
Fructose-6-phosphate kinase	+	++
Glyceraldehyde-3-phosphate dehydrogenase	+	++
Pyruvate kinase	+	++
Hexokinase type II	+	+

[a]A decrease in labeling intensity of cytochrome *c* oxidase subunit I was observed in old tissue compared to young tissue. Antibodies against aldolase, fructose-6-phosphate kinase, glyceraldehyde-3-phosphate kinase, and pyruvate kinase showed an increase in labeling intensity of old tissue compared to young tissue. Hexokinase type II demonstrated no detectable difference between young and old tissue. Labeling intensities: intermediate (+) and high (++).

A DYNAMIC VIEW OF mtDNA MUTATIONS AND DAMAGE: A HYPOTHESIS

The aging process is postulated to involve an age-associated accumulation of mtDNA damage and mutations that are retained and accumulated throughout life until the damage/mutations reach a point were they significantly compromise bioenergy production. However, many of the current views on mitochondria and mtDNA damage fail to take into account the dynamic nature of the cell and of the organelle itself. For instance, mitochondria form extensive networks within the cell that are continually undergoing morphometric change. Furthermore, mtDNA is actively turned over in both mitotic and postmitotic tissue every 7–31 days, depending on the

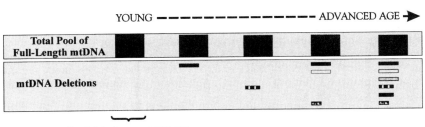

FIGURE 5. Turnover of deleted mtDNA. Refer to text for detailed explanation. Deletions and other mutations continuously occur throughout life; however, they are minimized or removed during each mitochondrial turnover cycle. The number of deletions generated within each mitochondrial turnover period increases with age, possibly due to an age-associated increase in the production of endogenous mutagens. Consequently, due to turnover the observed occurrence of deletions at any one time is limited and the total amount of full-length mtDNA undergoes little age-associated change.

cell type and tissue.[44] We now refine our earlier hypothesis by considering the dynamic nature of the mitochondria, the stochastic characteristics of the system, and the eventual effect on mtDNA and cellular bioenergy. Our hypothesis now incorporates the dynamic relationship between endogenous mutagen production, DNA repair, mtDNA turnover, and nuclear control of mtDNA copy number. These concepts are summarized in FIGURES 5 and 6, the features of which will be discussed.

Mitochondrial DNA Repair

It is apparent that throughout the life span of an organism, the total amount of full-length mtDNA for a given surviving postmitotic cell undergoes little change.[19] However, deletions, point mutations, and mtDNA damage occur continuously throughout life, the total extent dependent on age. In the young, it is proposed that DNA repair activity is adequate, so that intrinsic mtDNA damage is kept to a low level. Consequently, the total amount of functional mtDNA that is available for transcription and replication is essentially unchanged in the mitochondria of young subjects and equates to the amount of full-length mtDNA that can be amplified with XL-PCR. In the mitochondria of aged cells, we propose that there is an increase in the amount of mtDNA damage relative to the rate of repair. As a function of age, the equilibrium between mtDNA damage and repair shifts towards increased damage due to an age-associated increase in the generation of endogenous mutagens in combination with either inadequate, overwhelmed, or reduced repair capacity. The amount of functional full-length mtDNA available for transcription and replication at any one time in aged cells is reduced and leads to bioenergy decline. Because of the stochastic nature of the process, some cells will contain little functional full-length mtDNA and carry extensive mtDNA damage, a situation that will lead to cell loss (see section on A Stochastic System).

DNA Turnover Minimizes mtDNA Damage and Mutations

Our hypothesis envisages that mtDNA molecules damaged with persistent DNA lesions (DNA adducts that cannot be repaired) are eliminated during the continual mtDNA turnover process. The remaining undamaged pool of mtDNA (represented by full-length mtDNA that can be amplified with XL-PCR) would be expanded under the control of the nucleus[45,46] to largely restore the total pool of functional mtDNA. Restoring the pool of functional mtDNA is possibly based on the faithful replication of the remaining undamaged templates (see below) and the limited number of mtDNA membrane attachment sites.

In addition to removing damaged mtDNA, DNA turnover would also eventually eliminate many deletions and other mutations. Accordingly, an important feature of the hypothesis is that it predicts that deletion mutations would be generated within the mtDNA turnover period (which may be as short as 7 days, for example, in rat heart muscle[44]) and would only persist for a relatively brief period before being minimized during subsequent turnover events. Consequently, deletions involving short, direct-repeat sequences would be predicted to occur more frequently following turnover events (ie. would have a greater potential to be generated within the turnover period) than deletions that occur independently of repeat sequences. As mtDNA deletions are thought to be precipitated through replication or repair of damaged DNA,

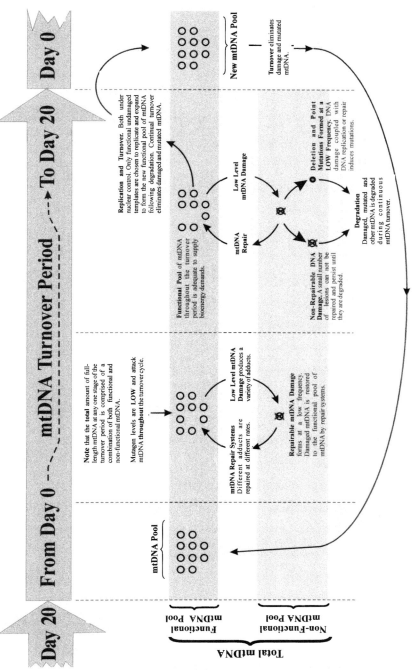

FIGURE 6A. *See legend on page 239.*

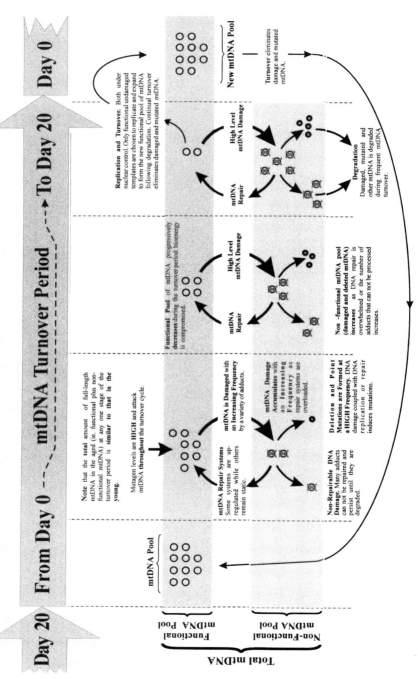

FIGURE 6B. *See legend on page 239.*

it also follows that deletions would be formed more frequently in aging cells due to increasing amounts of mtDNA damage and overwhelmed repair systems as outlined earlier.

The proposed model also allows for the elimination of certain point mutations during turnover events. This elimination is predicted to be based on the requirement for conserved nucleotide sequences at the sites of mtDNA replication and transcription.[47] As point mutations would be expected to occur randomly throughout the mitochondrial genome, a proportional amount of point mutations would be expected to occur at positions critical for mtDNA replication or transcription or at control elements for either of these processes. A significant number of point (and other) mutations could be predicted to affect DNA unwinding, bending, or the binding of protein elements. Our model predicts that the selection of appropriate templates for replication or transcription may be partially based on the integrity of the sequences located

---

FIGURE 6. Outline of hypothesis: A stochastic system. Mitochondrial DNA is in a dynamic relationship that is formed between endogenous mutagen damage, DNA repair, organelle and mtDNA turn-over, and nuclear control of mtDNA copy number. (**A**) Representation of mtDNA found in postmitotic tissue mitochondria of a young subject. In the mitochondria of young subjects, repair and damage are in equilibrium, so that the majority of damage that occurs is effectively repaired. Some types of DNA damage cannot be repaired and will persist or may be fixed by replication or repair to generate low levels of mtDNA mutations. However, mtDNA turnover minimizes the continual occurrence of damaged and mutated templates (see also FIG. 5), in effect, eliminating the nonfunctional pool of mtDNA. Only undamaged full-length mtDNA (represented experimentally by the portion of mtDNA that can be amplified with XL-PCR) can be replicated to reestablish the new pool of functional mtDNA. As nuclear genes control mtDNA copy number and maintain the total amount of cellular mtDNA, the amount of total cellular mtDNA overall undergoes little change. Note that mtDNA degradation may not be sequential and occur at the end of a turnover cycle, as illustrated here. Degradation may be ongoing throughout the turnover period. The turnover period chosen for the purpose of this diagram has been arbitrarily allocated a value of 20 days. The half-life of mtDNA is species, tissue, and cell dependent and, for example, can range in rats from 9–31 days. (**B**) Representation of mtDNA found in postmitotic tissue mitochondria of an aged subject. In the mitochondria of aged subjects, there is an increase in endogenous mutagen production with a corresponding increase in DNA damage. This leads to some mtDNA repair pathways being overwhelmed, whereas other repair pathways may be induced. Regardless of any upregulation, repair pathways are insufficient to accommodate all DNA lesions. This in turn leads to an accumulation of mtDNA templates that are damaged and hence nonfunctional. As in the young, mitochondrial turnover continually eliminates the majority of this damaged mtDNA with the remaining undamaged full-length mtDNA forming the mtDNA pool that is replicated and expanded under nuclear gene control and selection. Note that in aged tissue, damaged full-length mtDNA is formed earlier and more frequently prior to turnover. Consequently, at any one time in aged tissue there is a reduced number of functional mtDNA molecules available for transcription relative to young tissue, a situation that leads to age-associated bioenergy decline. It is important to note that the total amount of full-length mtDNA is basically comprised of both functional and nonfunctional mtDNA, so that the total amount of mtDNA in the aged is similar to the total amount of mtDNA in the young. The relative proportions of functional and nonfunctional mtDNA between young and aged are however considerably different.

Legend: O = Functional Full-Length mtDNA; ☒ = Damaged Full-Length mtDNA: Non- Functional; ● = Deletion Mutation: Non-Functional.

within these control regions. Mitochondrial DNA containing certain point mutations would not serve as templates for replication. Accumulating evidence in the literature suggests that the majority of replication events that are initiated are subsequently aborted with only a small percentage of replication events proceeding to completion.[37,38] Aborted replication events may be indicative of replication attempts on damaged or altered mtDNA templates. This model also predicts that a number of point mutations would not affect the way DNA behaves or interacts with proteins and would not infringe on the selection of that particular template for replication. Indeed, some mutations may even incite a replication advantage. Consequently, a small percentage of mutations that would have been generated in between turnover events may be conserved and may possibly appear at a higher frequency during subsequent turnover cycles.

A Stochastic System

An important aspect of the hypothesis is that mtDNA turnover is asynchronous and that mitochondrial changes within the cell are stochastic. Consequently, any multicellular analysis of the whole tissue represents an instant in time of each individual cell. Single cell analysis, therefore, provides insight into the individual status of the cells that collectively specify the nature of the whole tissue. A very small proportion of cells will contain little or no functional full-length mtDNA, whereas other cells will contain varying amounts of functional full-length mtDNA that may range from intermediate to completely functional. In addition, a heterogeneous population of individual mutations would also be predicted. Accordingly, adjacent cells within the same tissue, or the same tissue from different similarly aged individuals, would be anticipated to demonstrate a diverse variety and number of mutations, ranging from a few individual mutations to an extensive array comprised of numerous mutations.

Turnover and repair are not absolute. There will be a small proportion of mitochondria in which damaged mtDNA escapes repair or turnover and deleterious mutations are fixed. Eventually, over a long period of time, mitochondria in postmitotic tissue will progressively move towards bioenergetic dysfunction until a point is reached where the cell is lost from the organ. The process of cell loss is slow and is measured over the life span of the organism. In postmitotic tissue such as human skeletal muscle, the loss of cells to age 40 years is minimal; however, it progressively accelerates and increases about 5% per decade thereafter.[48–50] Conversely, in mitotic tissues such as liver, these cells are being continually replaced so that the function of the organ is maintained.

Hypothesis Summary

The essence of the proposal states that mtDNA damage and mutations do not accumulate with the advancing age of the organism *per se*, but rather are generated within the shorter turnover cycle of the mitochondrial organelle itself. Continuous mitochondrial and mtDNA turnover eliminates most of the damage and mutations to produce and constantly form new functional mtDNA. However, damaged and mutated mtDNA molecules continue to be generated at a faster rate in aged subjects (compared to young) due to overwhelmed or reduced repair and enhanced mtDNA assault. The eventual outcome is that over the life span of the organism, even though the total pool of mtDNA is undergoing continual turnover and remains relatively constant, there are nonetheless ongoing stochastic changes that maintain an mtDNA and bioenergy cellular mosaic and ultimately results in attrition of the number of cells in postmitotic tissues as exemplified by skeletal muscle. Mitotic tissues function to replace defective cells, thereby maintaining tissue cell number.

REFERENCES

1. LINNANE, A.W., S. MARZUKI, T. OZAWA & M. TANAKA.1989. Mitochondrial DNA mutations as an important contributor to ageing and degenerative diseases. Lancet **1:** 642–645.
2. LINNANE, A.W., C. ZHANG, A. BAUMER & P. NAGLEY. 1992. Mitochondrial DNA mutation and the ageing process: bioenergy and pharmacological intervention. Mutat. Res. **275:** 195–208.
3. NAGLEY, P., I.R. MACKAY, A. BAUMER *et al.* 1992. Mitochondrial DNA mutation associated with aging and degenerative disease. Ann. N.Y. Acad. Sci. **673:** 92–102.
4. LINNANE, A.W., A. BAUMER, R.J. MAXWELL *et al.* 1990. Mitochondrial gene mutation: the ageing process and degenerative diseases. Biochem. Int. **22:** 1067–1076.
5. ZHANG, C., A.W. LINNANE & P. NAGLEY. 1993. Occurrence of a particular base substitution (3243 A to G) in mitochondrial DNA of tissues of ageing humans. Biochem. Biophys. Res. Commun. **195:** 1104–1110.
6. BAUMER, A., C. ZHANG, A.W. LINNANE & P. NAGLEY. 1994. Age-related human mitochondrial DNA deletions: a heterogeneous set of deletions arising at a single pair of directly repeated sequences. Am. J. Hum. Genet. **54:** 618–630.
7. LUFT, R. 1994. The development of mitochondrial medicine. Proc. Natl. Acad. Sci. USA **91:** 8731–8738.
8. OZAWA, T. 1994. Mitochondrial cardiomyopathy. Herz **19:** 105–118.
9. WALLACE, D.C., V.A. BOHR, G. CORTOPASSI *et al.* 1995. Group report: The role of bioenergetics and mitochondrial DNA mutations in aging and age-related diseases. *In* Molecular Aspects of Aging. K. Esser & G.M. Martin, Eds. :199–225. John Wiley & Sons. New York.
10. WALLACE, D.C. 1992. Mitochondrial genetics: a paradigm for aging and degenerative diseases? Science **256:** 628–632.
11. WEI, Y.H. 1992. Mitochondrial DNA alterations as ageing-associated molecular events. Mutat. Res. **275:** 145–155.
12. ZHANG, C., A. BAUMER, R.J. MAXWELL *et al.* 1992. Multiple mitochondrial DNA deletions in an elderly human individual. FEBS Lett. **297:** 34–38.
13. NAGLEY, P., C. ZHANG, R.D. MARTINUS *et al.* 1993. Mitochondrial DNA mutation and human aging: molecular biology,, bioenergetics and redox therapy. *In* Mitochondrial DNA in Human Pathology. S. DiMauro & D.C. Wallace, Eds. :137–157. Raven Press. New York.
14. SOONG, N.W., G. HINTON, G. CORTOPASSI & N. ARNHEIM. 1992. Mosaicism for a specific somatic mitochondrial DNA mutation in adult human brain. Nature Genet. **2:** 318–323.
15. KOVALENKO, S.A., G. KOPSIDAS, J.M. KELSO & A.W. LINNANE. 1997. Deltoid human muscle mtDNA is extensively rearranged in old age subjects. Biochem. Biophys. Res. Commun. **232:** 147–152.
16. KOVALENKO, S.A., G. KOPSIDAS, M. ISLAM *et al.* 1998. The age-associated decrease in the amount of amplifiable full-length mitochondrial DNA in human skeletal muscle. Biochem. Mol. Biol. Int. **46:** 1233–1241.
17. KOVALENKO, S.A., G. KOPSIDAS, J. KELSO *et al.* 1998. Tissue-specific distribution of multiple mitochondrial DNA rearrangements during human aging. Ann. N.Y. Acad. Sci. USA **854:** 171–181.

18. KOPSIDAS, G., S.A. KOVALENKO, J.M. KELSO & A.W. LINNANE. 1998. An age-associated correlation between cellular bioenergy decline and extensive mtDNA rearrangements in human skeletal muscle. Mutat. Res. **421:** 27–36.
19. GADALETA, M.N., V. PETRUZZELLA, L. DADDABBO *et al.* 1994. Mitochondrial DNA transcription and translation in aged rat. Effect of acetyl-L-carnitine. Ann. N.Y. Acad. Sci. USA **717:** 150–160.
20. SALAZAR, J.J. & B. VAN HOUTEN. 1997. Preferential mitochondrial DNA injury caused by glucose oxidase as a steady generator of hydrogen peroxide in human fibroblasts. Mutat. Res. **385:** 139–149.
21. YAKES, F.M. & B. VAN HOUTEN. 1997. Mitochondrial DNA damage is more extensive and persists longer than nuclear DNA damage in human cells following oxidative stress. Proc. Natl. Acad. Sci. USA **94:** 514–519.
22. CLAYTON, D.A. & J. VINOGRAD. 1967. Circular dimer and catenate forms of mitochondrial DNA in human leukaemic leucocytes. J. Pers. **35:** 652–657.
23. BULPITT, K.J. & L. PIKO. 1984. Variation in the frequency of complex forms of mitochondrial DNA in different brain regions of senescent mice. Brain Res. **300:** 41–48.
24. PIKO, L. 1992. Accumulation of mtDNA defects and changes in mtDNA content in mouse and rat tissues with aging. Ann. N. Y. Acad. Sci. USA **663:** 450–452.
25. POULTON, J., K.J. MORTEN, D. MARCHINGTON *et al.* 1995. Duplications of mitochondrial DNA in Kearns-Sayre syndrome. Muscle Nerve 3: S154–158.
26. MELOV, S., J.M. SHOFFNER, A. KAUFMAN & D.C. WALLACE. 1995. Marked increase in the number and variety of mitochondrial DNA rearrangements in aging human skeletal muscle. Nucleic Acids. Res. **23:** 4122–4126.
27. MELOV, S., D. HINERFELD, L. ESPOSITO & D.C. WALLACE. 1997. Multi-organ characterization of mitochondrial genomic rearrangements in ad libitum and caloric restricted mice show striking somatic mitochondrial DNA rearrangements with age. Nucleic Acids Res. **25:** 974–982.
28. JI, L.L., C. LEEUWENBURGH, S. LEICHTWEIS *et al.* 1998. Oxidative stress and aging. Role of exercise and its influences on antioxidant systems Ann. N.Y. Acad. Sci. **54:** 102–117.
29. BEJMA, J. & L.L. JI. 1999. Aging and acute exercise enhance free radical generation in rat skeletal muscle. J. Appl. Physiol. **87:** 465–470.
30. WEI, Y.H., C.Y. LU, H.C. LEE *et al.* 1998. Oxidative damage and mutation to mitochondrial DNA and age-dependent decline of mitochondrial respiratory function. Ann. N.Y. Acad. Sci. **854:** 155–170.
31. SOHAL, R.S., H.H. KU, S. AGARWAL *et al.* 1994. Oxidative damage, mitochondrial oxidant generation and antioxidant defenses during aging and in response to food restriction in the mouse. Mech. Ageing Dev. **74:** 121–133.
32. KOPSIDAS, G., D.R. HEFFERNAN, S.A. KOVALENKO & A.W. LINNANE. In preparation.
33. CLAYTON, D.A., J.N. DODA & E.C. FRIEDBERG. 1974. The absence of a pyrimidine dimer repair mechanism in mammalian mitochondria. Proc. Natl. Acad. Sci. USA **71:** 2777–2781.
34. LINN, S. 1995. DNA repair in mitochondria: how is it limited? What is its function? *In* Molecular Aspects of Aging. K. Esser & G.M. Martin, Eds. :191–196. John Wiley & Sons. New York.
35. CROTEAU, D.L., R.H. STIERUM & V.A. BOHR. 1999. Mitochondrial DNA repair pathways. Mutat. Res. **434:** 137–148.
36. SOUZA-PINTO, N.C., D.L. CROTEAU, E.K. HUDSON *et al.* 1999. Age-associated increase in 8-oxo-deoxyguanosine glycosylase/AP lyase activity in rat mitochondria. Nucleic Acids. Res. **27:** 1935–1942.
37. Kai, Y., K. Miyako, T. Muta *et al.* 1999. Mitochondrial DNA replication in human T lymphocytes is regulated primarily at the H-strand termination site. Biochim. Biophys. Acta **1446:** 126–134.
38. ROBERTI, M., C. MUSICCO, P.L. POLOSA *et al.* 1998. Multiple protein-binding sites in the TAS-region of human and rat mitochondrial DNA. Biochem. Biophys. Res. Commun. **243:** 36–40.
39. BANDY, B. & A.J. DAVISON. 1990. Mitochondrial mutations may increase oxidative stress: implications for carcinogenesis and aging? Free. Rad. Biol. Med. **8:** 523–539.

40. DHAHBI, J.M., P.L. MOTE, J. WINGO *et al.* 1999. Calories and aging alter gene expression for gluconeogenic, glycolytic, and nitrogen-metabolizing enzymes. Am. J. Physiol. **277:** 352–360.
41. TAKEKURA, H., N. KASUGA & T. YOSHIOKA. 1994. Differences in ultrastructural and metabolic profiles within the same type of fibres in various muscles of young and adult rats. Acta Physiol. Scand. **150:** 335–344.
42. KIRKENDALL, D.T. & W.E. GARRETT, JR. 1998. The effects of aging and training on skeletal muscle. Am. J. Sports Med. **26:** 598–602.
43. CORTOPASSI, G.A. & A. WONG. 1999. Mitochondria in organismal aging and degeneration. Biochim. Biophys. Acta **1410:** 183–193.
44. GROSS, N.J., G.S. GETZ & M. RABINOWITZ. 1969. Apparent turnover of mitochondrial deoxyribonucleic acid and mitochondrial phospholipids in the tissues of the rat. J. Biol. Chem. **244:** 1552–1562.
45. HOLT, I.J., D.R. DUNBAR & H.T. JACOBS. 1997. Behaviour of a population of partially duplicated mitochondrial DNA molecules in cell culture: segregation, maintenance and recombination dependent upon nuclear background. Hum. Mol. Genet. **6:** 1251–1260.
46. CHINNER, Y.P.F. & D.C. SAMUELS. 1999. Relaxed replication of mtDNA: a model with implications for the expression of disease. Am. J. Hum. Genet. **64:** 1158–1165.
47. SHADEL, G.S. & D.A. CLAYTON. 1997. Mitochondrial DNA maintenance in vertebrates. Annu. Rev. Biochem. **66:** 409–435.
48. AOYAGI, Y. & R.J. SHEPHARD. 1992. Aging and muscle function. Sports Med. **14:** 376–396.
49. BROOKS, S.V. & J.A. FAULKNER. 1994. Skeletal muscle weakness in old age: underlying mechanisms. Med. Sci. Sports Exerc. **26:** 432–439.
50. FRISCHKNECHT, R. 1998. Effect of training on muscle strength and motor function in the elderly. Reprod. Nutr. Dev. **38:** 167–174.

Inflamm-aging

An Evolutionary Perspective on Immunosenescence

CLAUDIO FRANCESCHI,[a,b,e] MASSIMILIANO BONAFÈ,[a] SILVANA VALENSIN,[a] FABIOLA OLIVIERI,[b] MARIA DE LUCA,[d] ENZO OTTAVIANI,[c] AND GIOVANNA DE BENEDICTIS[d]

[a]Department of Experimental Pathology, University of Bologna, Bologna, Italy

[b]Department of Gerontological Research, Italian National Research Center on Aging (INRCA), Ancona, Italy

[c]Department of Animal Biology, University of Modena and Reggio Emilia, Modena, Italy

[d]Department of Cell Biology, University of Calabria, Calabria, Italy

ABSTRACT: In this paper we extend the "network theory of aging," and we argue that a global reduction in the capacity to cope with a variety of stressors and a concomitant progressive increase in proinflammatory status are major characteristics of the aging process. This phenomenon, which we will refer to as "inflamm-aging," is provoked by a continuous antigenic load and stress. On the basis of evolutionary studies, we also argue that the immune and the stress responses are equivalent and that antigens are nothing other than particular types of stressors. We also propose to return macrophage to its rightful place as central actor not only in the inflammatory response and immunity, but also in the stress response. The rate of reaching the threshold of proinflammatory status over which diseases/disabilities ensue and the individual capacity to cope with and adapt to stressors are assumed to be complex traits with a genetic component. Finally, we argue that the persistence of inflammatory stimuli over time represents the biologic background (first hit) favoring the susceptibility to age-related diseases/disabilities. A second hit (absence of *robust* gene variants and/or presence of *frail* gene variants) is likely necessary to develop overt organ-specific age-related diseases having an inflammatory pathogenesis, such as atherosclerosis, Alzheimer's disease, osteoporosis, and diabetes. Following this perspective, several paradoxes of healthy centenarians (increase of plasma levels of inflammatory cytokines, acute phase proteins, and coagulation factors) are illustrated and explained. In conclusion, the beneficial effects of inflammation devoted to the neutralization of dangerous/harmful agents early in life and in adulthood become detrimental late in life in a period largely not foreseen by evolution, according to the antagonistic pleiotropy theory of aging.

INTRODUCTION: THE NETWORK HYPOTHESIS OF AGING

According to the "network theory,"[1,2] the aging process is indirectly controlled by a variety of defense functions or anti-stress responses, globally acting as anti-aging mechanisms (FIG. 1). The stressors are very diverse and include different phys-

[e]Address for correspondence: Claudio Franceschi, Department of Experimental Pathology, University of Bologna, Via s. Giacomo 12, 40126 Bologna, Italy. Phone: 0039 051 209 4730 or 0039 051 209 4747.

clafra@alma.unibo.it

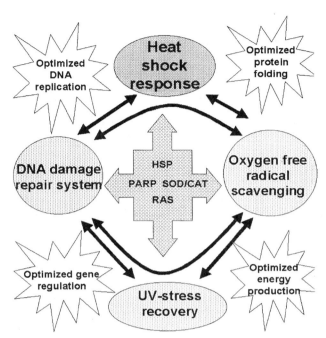

FIGURE 1. The "anti-aging" cellular defense network (inspired by Kirkwood and Franceschi, 1992, with modifications). Each part of the system is controlled by a variety of genes, which can serve multiple pathways. Only a few relevant examples of genes affecting life span throughout evolution are reported in the figure. The activity exerted by each specific defense mechanism is expected to contribute to the overall individual longevity. However, the capacity to reach the extreme limits of human life span in good shape is likely to depend on the global optimization of this network of defense mechanisms, which in turn allows for the age-related persistence of appropriate efficient integrated functions (DNA replication, gene regulation, protein folding, and energy production).

ical (UV and gamma radiation, heat), chemical (components of the body and products of metabolism such as oxygen-free radicals and reducing sugars), and biological (viruses, bacteria) agents.[1] The efficiency of this network, which includes DNA repair enzymes and poly(ADP-ribosyl)polymerase (PARP), antioxidants (superoxide dismutase, SOD; catalase, CAT), heat shock proteins (HSP) and other stress proteins (H-Ras, RAS), and the immune system, among others, is genetically controlled and differs in various species and individuals, accounting for the observed differences in life span.[2] On average, the longevity assured by each specific defense function is expected to be similar within a species. However, the genetic variability within the population is expected to result in variation in the individual extent to which the organism is predisposed to age and die from specific causes. For example, some individuals are likely to be less well protected against oxygen radicals than others, and these individuals will therefore experience a greater toll of oxidative damage. Instances of extreme longevity, such as human centenarians, are of special interest, for they are likely to be endowed with unusually high levels of each of the important ingredients of the anti-aging network. [2] In this paper we extend the network theory of aging, and we argue that a global reduction in the capability to cope with a variety of stressors and a concomitant progressive increase in the proinflammatory status, which we will call "inflamm-aging," is a major characteristic of the aging process and that it is provoked by a continuous antigenic load and stress. We also argue that the immune and the stress responses are equivalent and that antigens are nothing other than particular types of stressors, on the basis of evolutionary studies strongly supporting this hypothesis. Following this perspective, several paradoxes of healthy centenarians are illustrated and explained. Finally, current studies on the genetics of human longevity (centenarians) are starting to identify the possible variants of genes, which lifelong modulate and optimize the immune/stress response.

THE CAPABILITY TO COPE WITH STRESS AND THE EXTENSION OF LONGEVITY IN LOWER EUKARYOTES: EVIDENCE OF A CONSERVED ANTI-AGING NETWORK

In recent years a number of lower eukaryotic organisms have been identified as useful models for studying the genetic mechanisms involved in aging and longevity. In particular, studies on yeast (*Saccharomyces cerevisiae*), worms (*Caenorhabditis elegans*), and insects (*Drosophila melanogaster*) indicate that molecular variants of the genes involved in the response to oxidative, radiation-induced, and thermal stress, among others, markedly influence organismal longevity.[3–5] Surprisingly, most of the molecular pathways acting in these organisms are similar to those settling the stress defense network of mammals, including humans.[3–5] Some of the most interesting examples are briefly summarized: (1) the response to oxidative stress requires the induction of superoxide dismutase and catalase (free radical scavenger enzymes) to protect *D. melanogaster* as well as human cells, a superinduction of these enzymes resulting in extended longevity;[5] (2) heat shock induces the transcription of heat shock proteins (HSP) in all eukaryotes, in which they are responsible for the refolding of heat-denatured proteins;[4] (3) PARP (a DNA repair gene) has a specific enzymatic activity proportional to life span in a variety of species, from insects to primates;[6] (4) H-Ras is an intracellular signaling molecule involved in the recovery from radiation-induced stress (UV) of *S. cerevisiae*, very similar to what occurs in human fibroblasts. Moreover, the human H-Ras gene can modulate yeast longevity.[3] In conclusion, a number of ancestral, evolutionary-conserved molecular pathways responsible for resistance to a variety of stressors are involved in longevity from yeast to humans (FIG.1).

AN ADDITIONAL, POWERFUL ANTI-STRESS MECHANISM: THE IMMUNE SYSTEM

As the evolutionary key to understanding the aging process has been successfully identified in the mechanisms of stress resistance, we propose to go further in the equation *stress resistance = ability to survive*. In particular, we refer to the acquisition of specialized functions for the defense of the organism, which occurred in invertebrates, where some cells (immunocytes, responsible for chemotaxis and

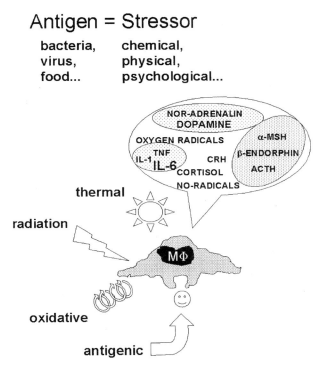

FIGURE 2. The equivalence between antigens and stressors (inspired by Ottaviani *et al.*, 1991, with modifications). From invertebrates to humans, the cellular response to a number of stressors appears to be highly maintained and involves the upregulation of a variety of evolutionary-conserved mediators, such as oxygen-free radicals, nitric oxide (NO), proinflammatory cytokines (IL-1, IL-6, TNFα), propiomelanocortin-derived peptides (ACTH, β-endorphin, α-MSH), steroids (cortisol), biogenic amines (noradrenaline, adrenaline, dopamine), and neuropeptides (CRH). From this evolutionary perspective the equivalence between stressors and antigens emerges, and the stimuli that activate the immune system are nothing other than a particular type of stressor.

phagocytosis) became able to fight against a variety of external stimuli and gave rise to a complex, stereotyped, and molecularly conserved set of mechanisms, termed "prototype stress response" (FIG. 2).[7,8] Thus, the response to stress became more complex in invertebrates and involved other mechanisms, such as inflammation and innate immunity, sharing ancestral molecular pathways with the response to non-antigenic stressors.[9,10] A peculiar cell type, the macrophage, emerged with a central role in this scenario.[11,12] As illustrated in FIGURE 2, macrophages, that is, cells with phagocytic activity, can be stimulated and/or activated by bacterial products, neuropeptides, neurohormones, cytokines, and other stimuli and can release large amounts of proinflammatory cytokines, nitric oxide, biogenic amines, neuropeptides, hormones, among others.[13] This occurs in both invertebrates and vertebrates. Thus, we propose to return macrophage to its rightful place as central actor, at the

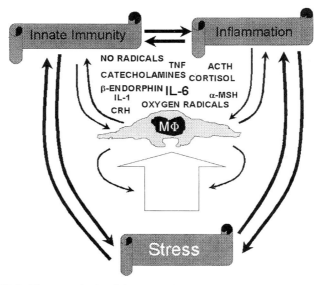

FIGURE 3. The central role of the macrophage (inspired by Ottaviani and Franceschi, 1997, with modifications) in innate immunity, stress response, and inflammation is illustrated. In invertebrates, a single cell is responsible for all these integrated responses, collectively called "prototype stress response." In higher vertebrates, such as humans, a complex "immuno-neuro-endocrine suprasystem" is the player for this role.

cellular level, not only of the inflammatory response and immunity, as originally suggested by Metchnikoff, but also of the stress response.[11] The march of complexity of the immune system reaches greater sophistication in vertebrates, particularly mammals, with the emergence of clonotypic immunity. It is conceivable that the immune system (or better the immuno-neuro-endocrine system) of humans is the ultimate result of the selection for fitness impinging upon an integrated network of mediators and cells, which allows the individual to respond to environmental changes and stressors.[14] In this perspective, most of the biomolecular features of aging can be considered the result of an adaptive capacity at the organismal level, due to a cooperating mixture of old, conserved (macrophage-centered) and new, evolutionary recent (lymphocyte-centered) immune and anti-stress mechanisms (FIG. 3).[14]

PROINFLAMMATORY STATUS OF AGING (INFLAMM-AGING) AS A RESULT OF CHRONIC ACTIVATION OF THE MACROPHAGE WITH AGE (MACROPH-AGING)

Current data are compatible with the conceptualization of aging as the result of chronic stress impinging upon the macrophage as one of the major target cells in this process.[14,15] This hypothesis, illustrated in FIGURE 4, suggests the existence of a direct relation between age and macrophage activation, mostly responsible for the presence of a subclinical chronic inflammatory process in the elderly. This phenom-

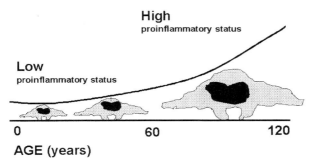

FIGURE 4. Inflamm-aging as a consequence of macroph-aging. The increase in proinflammatory status at an organismal level, caused by chronic age-related stimulation of the macrophage, called "macroph-aging," is referred to as "inflamm-aging."

enon is only part of the whole spectrum of change characteristic of immunosenescence, and indeed the macrophage is not the only cell involved in the aging process. Lymphocytes are also largely affected during immunosenescence, and the continuous age-related, largely inescapable, antigenic stress provokes a variety of changes even in the most evolutionary recent, clonotypical immune system. The results were illustrated elsewhere[14] and include the expansion of memory cells, the decrease and even the exhaustion of naive cells, the shrinkage of the T-cell repertoire, and the global reduction of the "immunological space."[14] On the whole, data on immunosenescence indicate that changes occurring over time might be considered the result of global reshaping, where the immune system continuously looks for possible stable points for optimal functioning.[16] Conceivably, this phenomenon is more general, likely involving the stress response in every tissue and organ. To explain a variety of experimental data concerning the pathophysiology of immunosenescence, we propose an oversimplified but heuristically powerful framework, according to which the effects of stress (including antigenic stress) accumulate, the capability of reshaping decreases, and the inflammatory status becomes pervasive over time.

As Hans Selye had suggested since the beginning of his pioneering research, the biological effects of stress, defined as a general adaptation syndrome without any connotative (positive or negative) value, depend on the intensity of the stressor and the individual capacity for coping.[10] Usually, the effects of continuous exposure of living organisms to minimal stress have been neglected, whereas the effects of strong stress have been implicitly assumed as the only ones deserving attention and having biological consequences. We can therefore understand the unexpected finding that stress, particularly minimal stress, can increase survival in a variety of organisms and cell types according to the general theory of *hormesis*, that is, the beneficial effects of extremely low doses of agents that are otherwise toxic at higher doses.[17,18] In the usual environment, hormetic and detrimental effects of stress occur concomitantly in the same individual, and the outcome is the result of a balance between these two components. On the basis of these considerations, it is possible to hypothesize that a high individual capability of coping can systematically shift the effect of stress towards hormesis. On the contrary, a low individual capability of coping is a driving force towards the detrimental effects of stress. These general considerations

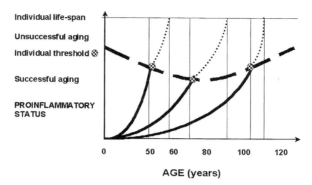

FIGURE 5. The individual thresholds of inflamm-aging. An individual threshold of the capability to cope with stress is hypothesized. If inflamm-aging trespasses on this level, the transition between successful (*continuous line*) and unsuccessful aging (*dotted line*) will occur. In accordance with epidemiologic data, the period of life during unsuccessful aging (disability) is maximal in the elderly (60–80 years) and minimal in young people and centenarians.

on stress and stress response can help in understanding a major paradox of aging and longevity. A comparison between data on inflammatory parameters in elderly affected by a variety of diseases and in healthy centenarians showed that an ever-increasing inflammatory status is shared by both groups.[19–22] Thus, the proinflammatory status is seemingly a characteristic of both successful and unsuccessful aging. To explain this apparent contradiction, we hypothesize that a threshold exists beyond which the adverse effects of stress become evident and drive the organism towards unsuccessful aging and death. This threshold varies among individuals and with age. FIGURE 5 suggests that to reach the pathological threshold, the proinflammatory process in young adults must be characterized by a high rate, but the period of disability before death will be short. In the elderly, this rate is slower and the period in which they will suffer disability before death will be quite long. In centenarians, we envisage a situation that is similar to that of young adults, but the time needed to reach the pathological threshold is extremely long. These hypotheses fit with recent data in the literature indicating that centenarians have lived in good shape and without disability until very old age,[24] despite the fact that in most of them the biochemical parameters related to inflammation can reach values that can be high.[19,25]

THE TWO HITS THEORY OF INFLAMM-AGING: AN INTERPLAY BETWEEN GENETIC AND ENVIRONMENTAL COMPONENTS

The individual capacity to cope with and adapt to stressors can be assumed as a complex trait having a genetic component. From this point of view, people can be subdivided into "robust" or "frail" according to the combination of gene variants that are involved in the variance of the trait.[26,27] We also assume that the rate of reaching the threshold of proinflammatory status over which disease/disability ensues has a genetic component, likely partially distinct from that responsible for frailty or ro-

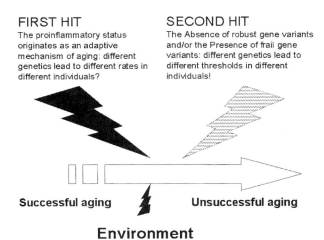

FIRST HIT
The proinflammatory status originates as an adaptive mechanism of aging: different genetics lead to different rates in different individuals?

SECOND HIT
The Absence of robust gene variants and/or the Presence of frail gene variants: different genetics lead to different thresholds in different individuals!

Successful aging   **Unsuccessful aging**

Environment

FIGURE 6. The two hits hypothesis of inflamm-aging. The persistence of inflamm-aging over time represents the *first hit* (inflammatory background) favoring the susceptibility to age-related diseases. A *second hit* is necessary in order to develop overt age-related diseases and disabilities, such as atherosclerosis, Alzheimer's dementia, osteoarthritis, and diabetes, currently considered organ-specific inflammatory diseases. The *second hit* can be identified in the absence of *robust* gene variants and/or the presence of *frail* gene variants.

bustness. In other words, we postulate that in each individual the rate of reaching the threshold of proinflammatory status can vary within a genetically constrained range, depending on the hormetic or detrimental intensity of the different lifelong (environmental) stressors to which he will be exposed. Environment, and particularly hormetic stressors, can therefore play a major role in extending longevity by increasing/inducing the efficiency of the anti-stress network. In any case, the interaction of these environmental factors with the genetic makeup of a given individual (or species) represents a biological constraint for the extension of longevity and for the capacity of stress to modulate the aging rate. The general situation that characterizes aging is illustrated in FIGURE 6, which tries to explain the progressive shift from successful to unsuccessful aging, assuming that the proinflammatory status is largely physiologic. Indeed, as first suggested by Metchnikoff,[10,11] inflammation is an evolutionary-conserved "positive" phenomenon, enabling the body to react to and neutralize foreign damaging agents. It can be speculated that the capability to mount a strong inflammatory process can contribute to fitness and survival and that people characterized by such a capacity have been positively selected for. In our two hits theory, we argue that inflammatory stimuli persist over time and inflammatory reactions add up, eventually representing a biological background (*first hit*) favoring the susceptibility to diseases. A *second hit* is necessary to develop overt age-related diseases and disabilities, which can be identified in the absence of *robust* gene variants and/or the presence of *frail* gene variants, both accounting for different disability/mortality thresholds in different individuals. The development of most age-related diseases can have two different types of genetic components besides environmental effects. The first component refers to the possible heritability of the mechanisms responsible for the first hit (inflammatory background), while the second concerns the genes conferring robustness or frailty at either a local or an organismal level. In particular, we predict that the concomitant presence of a high inflammatory status plus the presence of inherited variants of frailty genes or the absence of robust genes can explain why some individuals, but not others, are more susceptible to a certain type of age-related disease. Because in most cases such genes express their pathological effects in specific cells and organs, this hypothesis could help to explain the onset of a particular pathological condition, such as atherosclerosis, Alzheimer's dementia, and diabetes, currently considered organ-specific inflammatory diseases.[28,29]

INFLAMM-AGING AND THE CHALLENGE OF THE IL-6 PUZZLE AND OTHER PARADOXES IN CENTENARIANS

Most of the foregoing arguments apply to what has been defined as the "cytokine for gerontologists," that is, interleukin-6 (IL-6).[30] The plasma levels of IL-6 are low or undetectable in most young people and start to increase in healthy people at about 50–60 years of age. Accordingly, stimulated peripheral blood lymphocytes (PBL) from aged people produce higher levels of IL-6 than do PBL from young subjects.[31] The well established increase with age of IL-6 plasma levels appears to be unexpectedly present in both persons who enjoyed successful aging and those who suffered pathological aging. This increase continues with age, until the extreme limit of human life, and high levels of IL-6 are found in a high percentage of centenarians in good shape.[19] In these subjects other proteins, such as acute phase proteins, lipoprotein a (Lp(a)), fibrinogen and other coagulation factors, and proinflammatory cytokines are similarly increased (FIG. 7).[19,21,23,25] In sharp contrast, high levels of IL-6 have been referred to as the most powerful predictors of morbidity and mortality in the elderly.[32,33] This situation is either a challenging paradox or more simply what should be expected in the last phase of life in extremely old persons, whose probability of dying is very high. Indeed, we can assume that the increase in IL-6 with age is the consequence of the successful adaptation to a number of stresses, including infections, which unceasingly occur throughout life. This process occurs continuously, and in centenarians inflamm-aging eventually reaches levels very close to the threshold of morbidity and mortality, which indeed occur within months or a few years in

INCREASING OF:

- Coagulation factors
- Homocysteine
- IL6
- Proinflammatory cytokines
- Acute Phase Proteins
- Stress hormones
- ROS
- Lp(a)

A
PROINFLAMMATORY STATUS OF HEALTHY ELDERLY AND CENTENARIANS

FIGURE 7. The paradox of the proinflammatory status of healthy centenarians.

these subjects. Incidentally, despite the high sensitivity of IL-6 plasma levels to acute and chronic infections, as well as to other environmental conditions, strong genetic control of IL-6 plasma levels is emerging.[34] This latter finding, while confirming our hypothesis of a genetic component of the first hit, strongly indicates that future breakthroughs depend on genetic research concerning inflamm-aging. Thus, inflamm-aging could be the ultimate proof that the beneficial effect of the defense system network (innate immunity, stress, and inflammation), devoted to the neutralization of dangerous/harmful agents early in life and in adulthood, turns out to be detrimental late in life, in a period largely not foreseen by evolution. Thus, a tradeoff between early beneficial effects and late negative outcomes can occur at the genetic and molecular level. Similar phenomena have been predicted by evolutionary geneticists who proposed the theory of antagonistic pleiotropy.

ACKNOWLEDGMENTS

We acknowledge E.U. "GENAGE," M.U.R.S.T. 40%, M.U.R.S.T. Project "Genetic determinants of human longevity," and the Ministry of Health Projects project, "The prevention of chronic age-related diseases: the model of centenarians," for support.

REFERENCES

1. FRANCESCHI, C. 1989. Cell proliferation, cell death and aging. Aging. Clin. Exp. Res. **1:** 3–15.
2. KIRKWOOD, T.B.L. & C. FRANCESCHI. 1992. Is ageing as complex as it would appear? New perspectives in gerontological research. Ann. N.Y. Acad. Sci. **663:** 412–417.
3. JAZWINSKI, S.M. 1999. Molecular mechanisms of yeast longevity. Trends Microbiol. **7:** 247–252.
4. MURAKAMI, S. & T.E. JOHNSON. 1998. Life extension and stress resistance in *Caenorhabditis elegans* modulated by the tkr-1 gene. Curr. Biol. **8:** 1091–1094.
5. SOHAL, R.S. & R. WEINDRUCH. 1996. Oxidative stress, caloric restriction, and aging. Science **273:** 59–63.
6. BURKLE, A.1998. Poly(ADP-ribose) polymerase and aging. Exp. Gerontol. **33:** 519–523.
7. OTTAVIANI, E., E. CASELGRANDI, M. BONDI *et al.* 1991. The "immuno-mobile" brain: evolutionary evidence. Adv. Neuroimmunol. **1:** 27–39.
8. OTTAVIANI, E., A. COSSARIZZA, C. ORTOLANI *et al.* 1991. ACTH-like molecules in gastropod molluscs: a possible role in ancestral immune response and stress. Proc. Roy. Soc. Ser. B **245:** 215–218.
9. OTTAVIANI, E., S. VALENSIN & C. FRANCESCHI. 1998. The neuro-immunological interface in an evolutionary perspective: the dynamic relationship between effector and recognition systems. Front. Biosci. **15:** D431–D438.
10. OTTAVIANI, E. & C. FRANCESCHI. 1996. The neuroimmunology of stress from invertebrates to man. Progr. Neurobiol. **48:** 421–440.
11. OTTAVIANI, E. & C. FRANCESCHI. 1997. The invertebrate phagocytic immunocyte: clues to a common evolution of the immune and the endocrine systems. Immunol. Today. **18:** 169–174.
12. OTTAVIANI, E. & C. FRANCESCHI. 1998. A new theory on the common evolutionary origin of natural immunity, inflammation and stress response: the invertebrate phagocytic immunocyte as an eye-witness. Domest. Animal Endocrinol. **15:** 291–296.
13. OTTAVIANI, E., A. FRANCHINI & C. FRANCESCHI. 1997. Pro-opiomelanocortin-derived peptides, cytokines and nitric oxide in immune responses and stress: an evolutionary approach. Int. Rev. Cytol. **170:** 79–142.

14. FRANCESCHI, C., S. VALENSIN, F. FAGNONI et al. 1999. Biomarkers of immunosenescence: the challenge of heterogeneity and the role of antigenic load. Exp Gerontol. **34:** 911–921.
15. FRANCESCHI, C., M. BONAFÈ & S. VALENSIN. 2000. Human immunosenescence: the prevailing of innate immunity, the failing of clonotypic immunity, and the filling of immunological space. Vaccine **18:** 1717–1720.
16. FRANCESCHI, C., D. MONTI, P. SANSONI & A. COSSARIZZA. 1995. The immunology of exceptional individuals: the lesson of centenarians. Immunol. Today **16:** 12–16.
17. JOHNSON, T.E. & H. BRUUNSGAARD. 1998. Implications of hormesis for biomedical aging research. Hum. Exp. Toxicol. **17:** 263–265.
18. CALABRESE, E.J. & L.A. BALDWIN. 1999. The marginalization of hormesis. Toxicol. Pathol. **27:** 187–194.
19. BAGGIO, G., S. DONAZZAN, D. MONTI et al. 1998. Lipoprotein(a) and lipoprotein profile in healthy centenarians: a reappraisal of vascular risk factors. FASEB J. **12:** 433–437.
20. CHIRICOLO, M., G. BARTOLINI, M. ORLANDI et al. 1986. Prostaglandin and tromboxane biosynthesis in resting and activated platelet-free monocytes from aged subjects. Gerontology **32:** 69–73.
21. MARI, D., P.M. MANNUCCI, R. COPPOLA et al. 1995. Hypercoagulability in centenarians: the paradox of successful aging. Blood **85:** 3144–3149.
22. MARI, D., P.M. MANNUCCI, F. DUCA et al. 1996. Mutant factor V (Arg506Gln) in healthy centenarians. Lancet **347:** 1044.
23. MANNUCCI, P.M., D. MARI, G. MERATI et al. 1997. Gene polymorphisms predicting high plasma levels of coagulation and fibrinolysis proteins. A study in centenarians. Arterioscler. Thromb. Vasc. Biol. **17:** 755–759.
24. HITT, R., Y. YOUNG-XU, M. SILVER & T. PERLS. 1999. Centenarians: the older you get, the healthier you have been. Lancet **354:** 652.
25. BRUUNSGAARD, H., K. ANDERSEN-RANBERG, B. JEUNE et al. 1999. A high plasma concentration of TNF-alpha is associated with dementia in centenarians. J. Gerontol. A. Biol. Sci. Med. Sci. **54:** M357–M364.
26. YASHIN, A., J.W. VAUPEL, K.F. ANDREEV et al. 1998. Combining genetic and demographic information in population studies of aging and longevity. J. Epidemiol. Biostatist. **3:** 289–294.
27. YASHIN, A.I., G. DE BENEDICTIS, J.W. VAUPEL et al. 1999. Genes, demography and life span: the contribution of demographic data in genetic studies on aging and longevity. Am. J. Human Genet. **65:** 1178–1193.
28. WEISSBERG, P.L. & M.R. BENNETT. 1999. Atherosclerosis--an inflammatory disease. N. Engl. J. Med. **340:** 1928–1929.
29. EIKELENBOOM, P. & R. VEERHUIS. 1999. The importance of inflammatory mechanisms for the development of Alzheimer's disease. Exp. Gerontol. **34:** 453–461.
30. ERSHLER, W.B. 1993. Interleukin-6: a cytokine for gerontologists. J. Am. Geriatr. Soc. **41:** 176–181.
31. FAGIOLO, U., A. COSSARIZZA, E. SCALA et al. 1993. Increased cytokine production in mononuclear cells of healthy elderly people. Eur. J. Immunol. **23:** 2375–2378.
32. HARRIS, T.B., L. FERRUCCI, R.P. TRACY et al. 1999. Associations of elevated interleukin-6 and C-reactive protein levels with mortality in the elderly. Am. J. Med. **106:** 506–512.
33. FERRUCCI, L., T.B. HARRIS, J.M. GURALNIK et al. 1999. Serum IL-6 level and the development of disability in older persons. J. Am. Geriatr. Soc. **47:** 639–646.
34. FISHMAN, D., G. FAULDS, R. JEFFERY et al. 1998. The effect of novel polymorphisms in the interleukin-6 (IL-6) gene on IL-6 transcription and plasma IL-6 levels, and an association with systemic-onset juvenile chronic arthritis. J. Clin. Invest. **102:** 1369–1376.

The CA1 Region of the Human Hippocampus Is a Hot Spot in Alzheimer's Disease

M.J. WEST,[a,c] C.H. KAWAS,[b] L.J. MARTIN,[b] AND J.C. TRONCOSO[b]

[a]*Neurobiology, University of Aarhus, Aarhus, Denmark*

[b]*Neuropathology Laboratory, Johns Hopkins University, Baltimore, Maryland 21205, USA*

ABSTRACT: Data from an ongoing study of differences in the total number of neurons in the five major subdivisions of the hippocampal regions of the brains of patients with Alzheimer's disease (AD) and normal age-matched controls confirm an earlier finding from our laboratories of a pronounced loss of CA1 neurons associated with AD. In view of an earlier finding that the CA1 region does not suffer normal age-related neuronal loss, these data support the earlier conclusion that the neuropathologic mechanisms involved in the AD-related losses in CA1 are not related to normal aging and that the study of the cellular and molecular events involved in the AD-related loss of CA1 cells can aid in the identification of the unique pathologic processes associated with AD.

INTRODUCTION

Alzheimer's disease (AD) is a progressive neurodegenerative disease of the central nervous system characterized by disturbances in memory and higher cognitive functions.[1,2] Pathologically, it is characterized by abnormal numbers of senile plaques and neurofibrillary tangles and the abnormal loss of neurons in specific parts of the brain.[3] The hippocampal region is one of the first and most profoundly affected parts of the brain in AD patients.[4,5] The data presented here are from an ongoing series of studies designed to determine whether or not AD-related changes in neuron number are exaggerations of normal age-related changes, that is, whether AD is accelerated aging. The strategy used to make this determination is based on comparisons of the patterns of neuron loss in the hippocampal subdivisions. We have focused on neuronal loss because it is the pathologic feature that best expresses the cumulative affects of age and AD. Qualitative changes, defined in terms of differences in the subdivisions that show significant changes with normal aging and AD, are deemed evidence that the changes are not the same and that the neurodegenerative processes are not the same in normal aging and AD. Quantitative differences, expressed as differences in the degree of neuronal loss but not the subregions affected, are construed as evidence that AD is accelerated aging.

In earlier stereologic studies, we demonstrated that neurons in the various subdivisions of the hippocampal formation are lost to varying degrees during normal

[c]Address for correspondence: Mark J. West, Department of Neurobiology/Anatomy, University Park, University of Aarhus, DK 8000 Aarhus, Denmark. Phone: +45 8942 3011; fax: +45 8942 3060.

mjw@ana.aau.dk

aging[6] and AD.[7] In those studies, we found slight though significant normal age-related losses in the hilus of the fascia dentata and the subiculum and no evidence of age-related losses in the dentate granule cell layer, the CA3 region, and the CA1 region. We also found evidence of significant additional AD-related losses in the hilus and subiculum and a massive (68%) AD-related loss in the CA1 region, which did not show normal age-related loss. In the earlier study, we concluded that the regional patterns of hippocampal neuronal loss associated with normal aging and AD were different and therefore constituted evidence that the neurodegenerative processes associated with these two phenomena were different. The extent and uniqueness of the AD losses in CA1 led to the conclusion that this region would be the most appropriate place in the human hippocampal region in which to examine the cellular and molecular aspects of AD-related neurodegeneration.

MATERIALS AND METHODS

Brain Material

All brain material used in the ongoing study was obtained from the Alzheimer's Disease Research Center (ADRC) at Johns Hopkins University, Baltimore. Only the left hippocampal region of males was used. All individuals in the AD group ($n = 9$) had both clinical[8] and pathologic[3] diagnoses of AD. None of the subjects in the age-matched control group ($n = 6$) had a history of long-term illness, dementia, or neurologic disease.

Histology

All brains were fixed in formalin for a minimum of 3 months prior to being sliced in the frontal plane into 1-cm thick slabs. The cut surfaces of the slabs were photocopied for documentation, and the regions of the slabs containing the hippocampal formation were dissected free. The resulting blocks were embedded in paraffin and exhaustively sectioned at 70 mm, mounted on glass slides (thickness of section now 50 mm), and stained with cresylviolet.

Stereology

The optical fractionator[9] was used to estimate the total number of neurons in five of the major subdivisions of the hippocampus[10]: (1) granule cells of the dentate gyrus, (2) dentate hilar cells, (3) CA3-2 pyramidal cells, (4) CA1 pyramidal cells, and (5) principal neurons in the subiculum. A systematic random sample of 10–14 sections was taken at 30–40 section intervals from the entire set of sections obtained from the blocks. In accordance with fractionator sampling, these included partial sections from the edges of the blocks. The distance between optical disectors, on both x and y axes, was: (1) 475, (2) 823, (3) 770, (4) 2,000, and (5) 1,164 microns in the five respective subdivisions. The disectors were composed of counting frames with areas of (1) 414, (2) 10,208, (3) 3,710, (4) 3,710, and (5) 10,208 square microns in the respective subdivisions. All had heights of 20 mm.

TABLE 1. Estimates of the total number of neurons (N) x 10^6 in each subregion of each individual in the Alzheimer's disease group (AD, $n = 9$) and the age-matched control group (control, $n = 6$)

AD Gran N	Control Gran N	AD Hilus N	Control Hilus N	AD CA3-2 N	Control CA3-2 N	AD CA1 N	Control CA1 N	AD Sub N	Control Sub N
17.5	11.2	0.72	1.10	2.10	1.57	3.44	7.65	3.09	3.37
7.9	14.7	0.52	0.78	0.92	2.09	2.73	6.91	1.68	2.54
3.7	14.5	0.60	0.97	2.54	1.81	5.12	8.03	3.59	2.87
17.7	12.2	0.60	0.61	1.22	1.59	2.29	6.01	1.24	2.29
10.1	17.8	0.83	0.85	1.65	1.49	6.02	7.77	2.73	3.15
10.1	9.8	0.59	0.64	1.79	1.35	2.91	4.97	1.86	1.82
6.8		0.12		0.45		0.49		0.48	
7.7		0.80		1.34		2.05		1.09	
11.7		0.50		1.02		0.63		0.54	
Mean 11.5	13.4	0.59	0.83	1.45	1.65	2.85	6.89	1.81	2.67
CV 0.36	0.21	0.33	0.24	0.44	0.16	0.64	0.17	0.61	0.21
Difference	-1.9 (14%)		-0.24 (29%)		-20.1 (12%)		-4.03 (58%)		-0.86 (32%)
2p	0.35		0.042*		0.48		0.00039*		0.11

NOTE: Gran = granule cells of the dentate gyrus; Hilus = hilus of the dentate gyrus; CA3-2 = principal neurons of the CA3 and CA2 subregions combined; CA1 = principal neurons of the CA1 subregions; Sub = principal neurons of subiculum (for more detailed definitions of subregions, see West and Gundersen[10]); CV = coefficient of variation = SD/mean; Difference = difference between group means in absolute numbers and (% of controls).

Statistics

For each subdivision, differences between the mean total number of neurons (N) in the AD and control group were tested with Student's t test for significance. The $2p$ values of 0.05 or less were considered significant.

RESULTS

Estimates of the total number of neurons in each individual in the AD and control groups are shown in TABLE 1 along with the group means, the coefficient of variation of the group means, the difference in means, and the $2p$ values for the group comparisons.

The current data shows evidence of regionally specific AD-related neuronal losses similar to those observed in the earlier study. Focusing on the individual subdivisions, there is again no evidence of AD-related losses in the dentate granule cell layer and CA3 in the ongoing study. Although the current data still show a significant AD-related loss in the hilus (29%), there is now only a nonsignificant trend towards AD loss in the subiculum in the new data (29%, $2p = 0.11$). Notably, however, evidence still exists of a significant, large, AD-related loss of neurons (58%) in the CA1 subdivision.

DISCUSSION

Data from the ongoing study support the observation previously reported by our laboratory that the CA1 loses more neurons than does any other hippocampal subregion in AD patients and that immediately adjacent subregions do not have AD-related neuron loss. The evidence indicates no age-related loss of neurons in the CA1 region (see, however, Simic *et al.*[11]). The present data support our earlier conclusion that the CA1 subregion is particularly susceptible to whatever it is that is killing neurons in AD patients. Confirmation of this finding further supports the conclusion that the CA1 region is an appropriate part of the brain in which to study the pathology of AD at the cellular and molecular level. It is notable that: (1) the material used to obtain the current data came from one source, the ADRC at Johns Hopkins University, (2) the unbiased stereologic method used to obtain the data is different from that used in the previous study (i.e., the optical fractionator was used rather than the Vref² N_v method[12]), (3) the brain material analyzed in the two studies was embedded in two different media (glycolmethacrylate versus paraffin), and (4) counting in the two studies was performed by two different investigators blinded to the identity of the material.

One other finding of note in the ongoing study is the observation of a highly significant correlation ($r > 0.74$, $2p < 0.02$) between performance on the Mini Mental State Examination (MMSE) and the total number of neurons in CA1. This finding indicates that either the hippocampus is more involved in other cognitive processes than previously thought (the MMSE is not generally considered a hippocampal test) or the regionally specific loss of neurons in CA1, for some unknown reason, rigorously reflects critical AD-related changes in other parts of the brain.

REFERENCES

1. PRICE, D. 1986. New perspectives on Alzheimer's disease. Ann. Rev. Neurosci. **9:** 489–512.
2. JELLINGER, K. 1990. Morphology of Alzheimer's disease and related disorders. *In* Alzheimer's Disease, Epidemiology, Neuropathology, Neurochemistry, and Clinics. K. Maurer, P. Riederer & H. Beckman, Eds. : 61–77. Chapman and Hall. Vienna.
3. MIRRA, S.S., M.N. HART & R.D. TERRY. 1993. Making the diagnosis of Alzheimer's disease. Arch. Pathol. Lab. Med. **117:** 132–144.
4. BALL, M.J. 1988. Topographic distribution of neurofibrillary tangles and granulovacuolar degeneration in hippocampal cortex of ageing and demented patients. A quantitative study. Interdisc. Top. Gerontol. **25:** 16–37.
5. BRAAK, H. & E. BRAAK. 1991. Neuropathological stageing of Alzheimer-related changes. Acta Neuropathol. **82:** 239–259.
6. WEST, M.J. 1993. Regionally specific loss of neurons in the aging hippocampus. Neurobiol. Aging **14:** 287–293.
7. WEST, M.J., P.D. COLEMAN, D.G. FLOOD & J.C. TRONCOSO. 1994. Differences in the pattern of hippocampal neuronal loss in normal aging and Alzheimer's disease. Lancet **344:** 769–772.
8. MCKHANN, G., D. DRACHMAN, M. FOLSTEIN *et al.* 1984 Clinical diagnosis of Alzheimer's disease: report of the NINCDS-ADRDA Work Group. Neurology **42:** 939–944.
9. WEST, M.J., L. SLOMIANKA & H.J.G. GUNDERSEN. 1991. Unbiased stereological estimation of the total number of neurons in the subdivisions of the rat hippocampus using the Optical Fractionator. Anat. Rec. **231:** 482–497.
10. WEST, M.J. & H.J.G. GUNDERSEN. 1990. Unbiased stereological estimation of the number of neurons in the human hippocampus. J. Comp. Neurol. **296:** 1–22.
11. SIMIC, G., I. KOSTOVIC, B. WINBLAD & N. BOGDANOVIC. 1997. Volume and number of neurons of the human hippocampal formation in normal aging and Alzheimer's disease. J. Comp. Neurol. **379:** 482–494.
12. WEST, M.J. 1993. New stereological methods for counting neurons. Neurobiol. Aging **14:** 275–285.

Transgenic Mouse Models of Alzheimer's Disease

KLAUS D. BORNEMANN AND MATTHIAS STAUFENBIEL[a]

Novartis Pharma, Inc., Nervous System Research, CH-4002 Basel, Switzerland

ABSTRACT: Alzheimer's disease (AD) pathology is characterized by Aβ peptide-containing plaques, neurofibrillary tangles consisting of hyperphosphorylated tau, extensive neuritic degeneration, and distinct neuron loss. We generated several transgenic mouse lines expressing the human amyloid precursor protein (APP751) containing the AD-linked KM670/671NL double mutation (Swedish mutation) under the control of a neuron-specific Thy-1 promoter fragment. In the best APP-expressing line (APP23), compact Aβ deposits can be detected at 6 months of age. These plaques dramatically increase with age, are mostly Congo Red positive, and accumulate typical plaque-associated proteins such as heparansulfate proteoglycan and apolipoprotein E. Activated astrocytes and microglia indicative of inflammatory processes reminiscent of AD accumulate around the deposits. Furthermore, plaques are surrounded by enlarged dystrophic neurites as visualized by neurofilament or Holmes-Luxol staining. Strong staining for acetylcholinesterase activity is found throughout the plaques and is accompanied by local distortion of the cholinergic fiber network. All congophilic plaques contain hyperphosphorylated tau reminiscent of early tau pathology. Modern stereologic methods demonstrate a significant loss of neurons in the hippocampal CA1 region, correlating with an increasing Aβ plaque load. Interestingly, APP23 mice develop cerebral amyloid angiopathy in addition to amyloid plaques even though the APP transgene is only expressed in neurons. Crossbreeding of APP23 mice with transgenic mice carrying AD-linked presenilin mutations but not wild-type presenilin resulted in enhanced formation of pathology. In conclusion, our APP transgenic mice present many pathologic features, similar to those observed in AD and therefore offer excellent tools for studying the contribution of Aβ to AD pathogenesis.

INTRODUCTION

Alzheimer's disease (AD) is the most common form of progressive cognitive decline. Its clinical symptoms include well-known impairment of memory and learning but also deficiencies in many other intellectual functions such as orientation, abstract thinking, and speech. As originally described by Alois Alzheimer in 1907, this form of dementia is associated with amyloid plaque formation as well as neurofibrillary tangles in the brain of affected patients. Amyloid plaques are extracellular deposits of Aβ (or β-amyloid), a 39 to 43 amino acid peptide derived from a larger precursor protein, APP (β-amyloid precursor protein). In contrast to diffuse deposits that do

[a]Address for correspondence: Dr. Matthias Staufenbiel, Novartis Pharma, Inc., S-386.806, CH-4002 Basel, Switzerland. Phone: +41 61 324 9642; fax: +41 61 324 5524.
Matthias.Staufenbiel@pharma.novartis.com

not show any further pathology, compact plaques containing Congo Red-birefringent Aβ fibrils are generally associated with hypertrophic astrocytes, activated microglia cells, and various other typical features of inflammatory processes. In addition, dystrophic neurites are present in these amyloid structures. The intracellular neurofibrillary tangles consist of paired helical filaments formed by the cytoskeletal protein tau in an abnormal, hyperphosphorylated state. They are found both in association with fibrillar amyloid plaques as well as apart from them. A reduction in synapse number in the neocortex and neuron loss in distinct regions of AD brain has been observed. Several neurotransmitter systems such as the cholinergic system are impaired in this disease.

Although clinical phenotype and neuropathologic lesions are similar, AD is etiologically heterogeneous. Several genetic risk factors are known, most notably the apolipoprotein E ε4 allele. In addition, rare autosomal dominant mutations causing early familial forms of AD were first described for APP and later in the presenilin 1 and 2 genes. It can be assumed that the introduction of these dominant mutations may initiate pathogenic mechanisms in laboratory animals similar to those in human AD patients.

GENERATION OF APP23 TRANSGENIC MICE WITH Aβ DEPOSITS

Many of the initial attempts to generate transgenic mice using APP with the familial AD mutations did not lead to an AD-specific phenotype. The low expression levels of mutated APP in brain might have been insufficient to trigger AD-like pathogenesis during the short life span of mice. To optimize APP expression, we analyzed various promoters for high expression of the transgene. The Thy-1 promoter, which drives high expression in brain neurons, was found to be optimal.[1] Constructs were generated containing a human APP751 cDNA with the "Swedish" double mutation (KM670/671NL[2]) under the control of the mouse Thy-1 promoter. The best line, APP23, expresses human APP751 sevenfold higher than does endogenous mouse APP.[3] The highest levels of the human APP message are obtained in the deep cortical layers and the hippocampus, but expression at lower levels is also found in other brain regions such as the thalamus.

In APP23 mice the first Aβ deposits develop at 6 months of age in the cerebral cortex. They dramatically increase in size and number with age and occupy a substantial area of the cerebral cortex and hippocampus in 24-month-old mice. At this time, deposits are also found in the thalamus, caudate putamen, or olfactory bulb. As in AD, no deposits are found in the cerebellum, but a considerable amount of vascular amyloid is detected. This is particularly noteworthy, because APP expression is restricted to neurons. All deposits show immunoreactivity with a number of different Aβ antibodies, which is comparable to the staining of brain sections from AD patients. The Aβ deposits also react with end-specific antibodies recognizing the amino terminus of Aβ or its carboxy termini at amino acids 40, 42, and weakly at 43.[4]

Biochemical analyses of brain cortex lysates showed that the Aβ1-40 isoform is a little more prominent than the Aβ1-42 isoform. However, the amount of both Aβ1-

40 and 1-42 exponentially increases during aging in parallel to the increase in plaque volume, whereas the APP level remains constant.

PLAQUE CHARACTERISTICS

The Aβ deposits in APP23 mice are extracellular and mostly of the compact plaque type. Almost all plaques as well as the vascular amyloid deposits display Thioflavin S fluorescense. They also show Congo Red birefringence in polarized light, a characteristic feature of senile plaques in the brain of patients with Alzheimer's disease.[3] This as well as electron microscopic observations[5] indicate that Aβ is present in a fibrillar state in the deposits. Heparansulfate proteoglycan or apolipoprotein E typically associated with plaques in human AD brain are also components of the amyloid deposits in APP23 brain. There is no evidence of plaque maturation, because they are compact already at their first appearance. This result is difficult to reconcile with a widely presumed hypothesis[6] that deposits of the diffuse type are the precursors of congophilic neuritic plaques.

GLIOSIS AND INFLAMMATORY PROCESSES

Inflammatory reactions, which are thought to play a key role in the pathogenesis of AD, also occur in the APP23 transgenic mouse line. Massive astrogliosis can be demonstrated in plaque-containing brain areas such as neocortex and hippocampus.[3,7] Around the compact deposits an accumulation of astrocytes is found using glial fibrillary acidic protein (GFAP) staining.

Microgliosis can also be demonstrated in plaque-containing brain areas such as neocortex and hippocampus. In close vicinity to essentially all compact amyloid deposits an accumulation of microglia cells is found.[5,8] These cells are activated, as immunohistochemical analyses with MAC-1 antibodies, directed against the complement receptor 3 (CR3) or F4/80 that recognizes a macrophage/microglial-specific 160 kD protein, have shown.[8] The activated microglia cells have phagocytic capability, because they show increased expression of the lysosomal marker macrosialin. A subpopulation of the microglial cells further shows an elevation of MHC class II (IA) protein and is associated with dystrophic neurites. This was demonstrated by double-labeling for acetycholinesterase and IA. The accumulation of cholinesterase in swollen, dystrophic neurites in the vicinity of amyloid plaques is one of several typical pathologic features associated with Alzheimer's disease. The close association of MHC class II-positive microglia with dystrophic neurites could indicate that MHC class II molecules are involved in cellular interaction and adhesion of microglia with neurites. They may clear degenerating material or, alternatively, could be involved in the generation of these dystrophic structures. MHC class II reactivity is also found close to those capillaries in the neocortex, which contain vascular amyloid. On peripheral macrophages, MHC class II proteins are essential for antigen presentation to helper T lymphocytes. However, no evidence of the presence of T lymphocytes, neither in the neocortex nor in the hippocampus, could be obtained by immunohistochemical analyses with antibodies to several T-cell receptor (TCR)-associated protein components, such as CD3ε, CD4, and CD88. Additionally, no B lymphocytes were identified in the brain of APP23 mice, arguing against a contribution of cell-mediated immune responses to the inflammatory processes. Because of the very similar activation state of microglia in the brain of APP23 mice compared to those in AD, these mice serve as a very well suited model for further investigation into mechanisms, leading to chronic inflammatory processes.

DEGENERATION

Throughout the neocortex, deposition of Aβ plaques results in substantial disruption of the normal cytoarchitecture. In the hippocampal pyramidal cell layer, a reduction in neuron number adjacent to the Aβ deposits is apparent. This reduction does not appear to be caused by passive cell displacement owing to the deposited Aβ, because compression of cell bodies cannot be detected. Instead, observations suggest a local loss of neurons. Quantitative assessment of these changes by state-of-the-art stereologic methods has shown a significantly lower number of neuronal cell bodies in the CA1 field of the hippocampus of 14–18-month old APP 23 mice compared to age-matched controls.[9] The neuron number inversely correlates with plaque load. In contrast, global neuron loss does not occur in the neocortex of APP23 mice, although an apparently decreased number of neurons in distinct areas such as the entorhinal cortex was observed.[10] Similar assumption-free stereologic techniques could demonstrate selective neuronal loss in the hippocampal CA1 region of AD patients, whereas a general neuron loss could not be detected in the neocortex.[11,12] The effects found in the hippocampus of transgenic mice are smaller than those observed in postmortem AD brain. Among other factors, the shorter time of exposure to amyloid or increased APP levels suggested to be neuroprotective[13–16] may explain the difference. Nevertheless, these findings provide further evidence for a central role of Aβ in the pathogenesis of Alzheimer's disease and indicate that the APP23 model faithfully reflects this aspect of AD pathology.

Plaques in APP23 mouse brain are surrounded by enlarged dystrophic neurites visualized by neurofilament immunolabeling (NF200 antibody)[7] or Holmes-Luxol staining.[17] These methods also reveal distortion of neurites by the deposits, leading to disruption of the normal neuritic pattern, particularly in regions with a high plaque load. APP and synaptophysin antibodies react with dystrophic structures in the periphery of compact Aβ deposits very like the staining of plaques in human AD brain. Quantification using modern stereologic methods demonstrated a reduction in synaptophysin-positive buttons in the neocortex, which correlates with the plaque load.[10] However, we found an increased number of synapses in mice without or with only very few plaques, indicative of a synaptotrophic effect of the overexpressed APP.[15]

The cholinergic system is strongly implied in the cognitive decline of AD patients. Besides the loss of cholinergic neurons, an accumulation of cholinesterase in swollen neurites in the vicinity of plaques was described.[18,19] Specific staining for acetylcholinesterase in transgenic mouse brains visualizes strong labeling of plaque structures resembling swollen, dystrophic neurites. A local distortion of the cholinergic fiber network is also seen with this method. Stereologic data show a reduction in cholinergic fiber length in APP23 transgenic mice.[20] This degeneration of the cholinergic system replicates pathologic features associated with AD. As opposed to other parameters, this effect on the cholinergic system neither in AD nor in APP23 mice correlates with plaque load.[19]

TAU PATHOLOGY

The observation of dystrophic neurites prompted us to search for aspects of neurofibrillary pathology, particularly hyperphosphorylated microtubule-associated protein tau. Immunoreaction with the AT8 antibody directed against phosphorylated Ser 202/Thr 205 revealed hyperphosphorylated tau in the brain of APPP23 mice. The immunostaining is exclusively associated with Congo Red-positive plaques and highlights curly structures resembling distorted neurites which surround the core of the deposits. Comparable labeling patterns are obtained with a variety of antibodies specific for AD-typical tau phosphoepitopes[3]. Staining outside the plaque as observed in AD brain was not found with these antibodies. Analyses of mouse brain material by Western Blot using the AT8 antibody confirmed the histological data. An elevation of tau phosphorylation was detectable in APP23 mice at an age of 6 months and was further increased at 15 months.

We did not observe any neurofibrillay tangles or threads in the brain of APP23 mice using silver staining according to Gallyas.[21] Interestingly, however, similar structures as stained by the phosphorylation-dependent antibodies are labeled with Alz-50. This antibody recognizes a conformational change of tau associated with polymerization into filaments,[22] indicating some neurofibrillary pathology in addition to tau hyperphosphorylation. Despite the absence of neurofibrillary tangles or threads as detected by silver staining, these results indicate early neurofibrillary pathology.

FURTHER DEVELOPMENT OF THE MODEL

Homozygous mice expressing APP and Aβ at higher levels show pathology earlier and develop a higher plaque load. However, a qualitative advance in pathology such as the development of neurofibrillary tangles and threads has not been obtained. To test if human tau is required for tangle formation or detection, we crossed the APP23 mice with transgenic mice overexpressing human tau (Alz17). The resulting mice showed an addition of the pathologic features of both parent strains. Together with the characteristics of the APP23 mice, they displayed somatodendritic, hyperphosphorylated tau typical for the Alz17 line. Further AD-like tau pathology has not been developed.

We also intercrossed APP23 mice with animals expressing a mutated human presenilin cDNA (PS36). We observed earlier plaque formation and a more rapid increase in the plaque load compared to the findings in APP23 mice.

Ongoing studies with APP23 mice demonstrate that these mice are a very suitable model of cerebral amyloidosis, because they develop abundant neuritic plaques as well as amyloid angiopathy. Many of the degenerative mechanisms observed in AD, such as inflammatory processes or neuron degeneration and loss, are reproduced in the mice. A major missing feature is the tau-related pathology, which remains at an early stage. Attempts to enhance it up to now have not brought success and may hint at a general difficulty in obtaining neurofibrillary tangles in mice. Expression of the more recently detected mutations in tau, which are associated with frontal lobe dementia and parkinsonism,[23-25] hopefully leads to better insight into this issue. Aside from this aspect the APP23 mouse line is a very appropriate model for investigating various aspects of the pathogenic mechanisms active in AD. These studies eventually should allow the design of new approaches to interfere with processes essential for the formation or progression of AD.

REFERENCES

1. ANDRÄ, K. *et al.* 1996. Expression of APP in transgenic mice: a comparison of neuron-specific promoters. Neurobiol. Aging **17:** 183–190.
2. MULLAN, M. *et al.* 1992. A pathogenic mutation for probable Alzheimer's disease in the APP gene at the N-terminus of beta-amyloid. Nature Genet. **1:** 345–347.
3. STURCHLER-PIERRAT, C. *et al.* 1997. Two amyloid precursor protein transgenic mouse models with Alzheimer disease-like pathology. Proc. Natl. Acad. Sci. USA **94:** 13287–13292.
4. PAGANETTI, P.A. *et al.* 1996. Amyloid precursor protein truncated at any of the gamma-secretase sites is not cleaved to beta-amyloid. J. Neurosci. Res. **46:** 283–293.
5. STALDER, M. *et al.* 1999. Association of microglia with amyloid plaques in brains of APP23 transgenic mice. Am. J. Pathol. **154:** 1673–1684.
6. PROBST, A. *et al.* 1991. Alzheimer's disease: a description of the structural lesions. Brain Pathol. **1:** 229–239.
7. STAUFENBIEL, M. & B. SOMMER. 1998. Transgenic animal models in the development of therapeutic strategies for Alzheimer's disease. *In* The Molecular Biology of Alzheimer's Disease. Genes and Mechanisms Involved in Amyloid Generation. C. Haass, Ed. : 309–326. Harwood Academic Publishers. Amsterdam, Netherlands.
8. BORNEMANN, K.D. *et al.* Functional characteristics of activated microglia cells in APP23 transgenic mice. Submitted.
9. CALHOUN, M.E. *et al.* 1998. Neuron loss in APP transgenic mice. Nature **395:** 755–756.
10. CALHOUN, M.E. *et al.* 1997. Impact of amyloid plaques on neuron number, synaptic bouton number, and cholinergic fiber length in a transgenic mouse model of Alzheimer's disease. Soc. Neurosci. Abstr. **23:** 1637.
11. WEST, M.J. *et al.* 1994. Differences in the pattern of hippocampal neuron loss in normal ageing and Alzheimer's disease. Lancet **344:** 769–772.
12. REGEUR, L. *et al.* 1994. No global neocortical nerve cell loss in brains from patients with senile dementia of Alzheimer's type. Neurobiol. Aging **15:** 347–352.
13. MATTSON, M.P. *et al.* 1993. Evidence for excitoprotective and intraneuronal calcium-regulating roles for secreted forms of β-amyloid precursor protein. Neuron **10:** 243–254.
14. BOWES, M.P. *et al.* 1994. Reduction of neuronal damage by a peptide segment of the amyloid/A4 protein precursor in a rabbit spinal cord ischemia model. Exp. Neurol. **129:** 112–119.
15. MUCKE, L. *et al.* 1994. Synaptotrophic effects of human amyloid β protein precursors in the cortex of transgenic mice. Brain Res. **666:** 151–167.
16. ROCH, J.M. *et al.* 1994. Increase of synaptic density and memory retention by a peptide representing the trophic domain of the amyloid beta/A4 protein precursor. Proc. Natl. Acad. Sci. USA **91:** 7450–7454.
17. STAUFENBIEL, M. *et al.* 1997. Pathological features in APP transgenic mice resembling those of Alzheimer's disease. *In* Alzheimer's Disease II: Exploiting Mechanisms for Drug Development and Diagnosis. E. Friedman & L.M. Savage, Eds. : 4.2.1–4.2.12. IBC Library Series, International Business Communications. Southborough, MA, USA.

18. TAGO, H. *et al.* 1987. Acetycholinesterase fibers and the development of senile plaques. Brain Res. **406:** 363–369.
19. GEULA, C. & M. MESULAM. 1989. Cholinesterases and the pathology of Alzheimer's disease. Neuroscience **33:** 469–481.
20. CALHOUN, M.E. *et al.* Amyloid plaque formation in APP transgenic mice is associated with synaptic bouton loss and alterations in the cholinergic system. Submitted.
21. GALLYAS, F. 1971. Silver staining of Alzheimer's neurofibrillary changes by means of physical development. Acta Morphol. Acad. Sci. Hung. **19:** 1–8.
22. CARMEL, G. *et al.* 1996. The structural basis of monoclonal antibody Alz50's selectivity for Alzheimer's disease pathology. J. Biol. Chem. **271:** 32789–32795.
23. POORKAJ, P. *et al.* 1998. Tau is a candidate gene for chromosome 17 frontotemporal dementia. Ann. Neurol. **43:** 815–826.
24. HUTTON, M. *et al.* 1998. Coding and 5′ splice site mutations in tau associated with inherited dementia (FTDP-17). Nature **393:** 702–705.
25. SPILLANTINI, M.G. *et al.* 1998. Mutation in the tau gene in familial multiple system tau-opathy with presenile dementia. Proc. Natl. Acad. Sci. USA **95:** 7737–7741.

Molecular Misreading

A New Type of Transcript Mutation in Gerontology

FRED W. VAN LEEUWEN,[a,b] DAVID F. FISCHER,[b] ROB BENNE,[c] AND
ELLY M. HOL[b]

[b]Netherlands Institute for Brain Research, Meibergdreef 33, 1105 AZ Amsterdam,
The Netherlands

[c]Academic Medical Hospital, Department of Biochemistry, Meibergdreef 15, 1105 AZ
Amsterdam, The Netherlands

ABSTRACT: Molecular misreading is a novel process that causes mutations in
neuronal transcripts.[1] It is defined as the inaccurate conversion of genomic information from DNA into nonsense transcripts and the subsequent translation
into mutant proteins.[2] As a result of dinucleotide deletions (ΔGA, ΔGU, ΔCU)
in and around GAGAG motifs in mRNA the reading frame shifts to the +1
frame, and subsequently the so-called +1 proteins are synthetized. +1 Proteins
have a wild-type NH_2 terminus and from the site of the dinucleotide deletion
onwards an aberrant, nonfunctional COOH terminus. Molecular misreading
was found in the rat vasopressin gene associated with diabetes insipidus[3] and
in the human genes linked to Alzheimer's disease (AD), that is, β-amyloid precursor protein (βAPP) and ubiquitin-B (UBB).[1,2] Moreover, βAPP^{+1} and
UBB^{+1} proteins accumulate in the neuropathological hallmarks of AD. Inasmuch as these +1 proteins were also found in elderly, nondemented control patients, but not in younger ones (<72 years), molecular misreading may act as a
factor that becomes manifest in aged people. A hotspot for dinucleotide deletions is GAGAG motifs. Because statistically an average of 2.1 GAGAG motifs
per gene can be expected, other genes expressed in other tissues may undergo
molecular misreading as well. Indeed, we recently detected +1 proteins in proliferating cells present in tissues such as the liver, epididymis, parotid gland,
and neuroblastoma cell lines.[4] Therefore, molecular misreading can be regarded as a general biological source of transcript errors that may be involved in
cellular derangements in numerous age-related pathologic conditions apart
from Alzheimer's disease.

INTRODUCTION

Alzheimer's disease (AD) is the most common cause of cognitive decline in the
aged population. Worldwide, about 20 million people suffer from AD, and the incidence is expected to double over the next 30 years, thereby becoming one of the
leading causes of death (cf. www.alzforum.org). At present there is no effective therapy, because a clear elucidation of the initial cause of AD is lacking. During the past
decade, most attention in AD research has been focused on the familial types with
an autosomal dominant inheritance pattern, and this approach has met with consid-

[a]Corresponding author: Phone: 31-20-5665510; fax: 31-20-6961006.
f.van.leeuwen@nih.knaw.nl

erable success. In autosomal dominant forms, mutations in the genes for β-amyloid precursor protein (βAPP) on chromosome 21, presenilin 1 and 2 on chromosome 14 and 1, respectively, were identified. Nevertheless, together these autosomal dominant forms account only for less than 5% of all AD cases. Apolipoprotein E polymorphism on chromosome 19 and others are important risk factors.[5] For the majority of cases (about 95%), no mutation has yet been reported. For these nonautosomal dominant forms of AD, including the sporadic cases as the most frequent form, a different mechanism must exist that initiates neurodegeneration. Recently, we found evidence for the existence of such a mechanism that is not linked to mutated DNA, but is related to errors in transcripts. This process, that is, the inaccurate conversion of genomic information into mRNA and the subsequent translation into aberrant proteins, is called "molecular misreading."[1] Molecular misreading takes place during or shortly after transcription and is in an unknown manner correlated with transcriptional activity. The exact mechanism, however, is still not clear. Our recent discovery of dinucleotide deletions in βAPP and ubiquitin-B (UBB) mRNAs is a first example of molecular misreading in AD and Down's syndrome (DS) patients. Molecular misreading results in the partial loss of functional domains because in most cases the +1 proteins are truncated. These aberrant mRNAs (or so-called "nonsense transcripts") are translated in the +1 reading frame as "+1 proteins," that is, proteins with a wild-type NH_2 terminus and an aberrant, frameshifted COOH terminus. These +1 proteins coexist and accumulate as the hallmarks of AD.[1,2] Whether molecular misreading is a causal factor in the most frequent form of AD (i.e., the sporadic cases that account for at least 60% of all AD patients) is presently being investigated using a transgenic approach. Many factors may contribute to the increased frequency of AD during aging, for example, the activational state of neurons, oxidative damage, DNA damage, and synapse loss.[6–8] Molecular misreading is most probably another major factor, and it has been suggested that it could help explain why age is the greatest risk factor for the development of AD.[9]

DIFFERENT TYPES OF DEMENTIA AND ALZHEIMER'S DISEASE

A population-based, cross-sectional study of all dementias has shown that AD is the most frequently occurring neurodegenerative disorder (72%), followed by vascular dementia (16%), Parkinson's disease (6%), and other dementias (5%; e.g., Pick's disease).[10] Because not all studies supply data on the subdivisions of the different AD types, here the most recent insights are presented (TABLE 1; data composed in collaboration with Prof. C. van Broeckhoven and Dr. M. Cruts, both University of Antwerp, Belgium, and Dr. C.M. van Duijn, Erasmus University, Rotterdam, The Netherlands; see also a recently created website on AD mutations: http://molgen-www.uia.ac.be/ADmutations/). Apart from these, mutations in the *tau* gene located on chromosome 17 have recently been reported, but so far only in dementias different from AD (for a review see Wilhelmsen[11]). A *tau* mutations table is available at www.alzforum.org. Neurofibrillary tangles containing hyperphosphorylated *tau* correlate best with the onset of AD,[12,13] but βAPP, on the other hand, is thought to be central.[14] TABLE 1 clearly shows that the sporadic and non-autosomal forms of familial AD account for 95% of all AD cases. For these cases a different

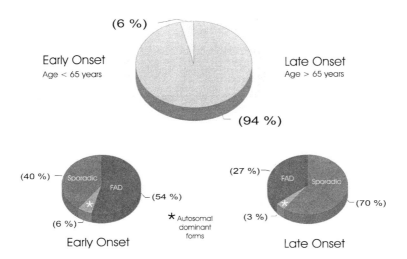

TABLE 1. Data on subdivision of different types of Alzheimer's disease[a]

Early (<65 years) and late (>65 years) onset (EOAD and LOAD) forms of Alzheimer's disease are distinguished. Familial means that AD was observed in relatives of the first degree. This study is based on references 10, 55, and 56. Data within *parentheses* represent a subpercentage.

In familial EOAD the majority (54%) is not yet linked to a chromosome, whereas 6% is inherited in an autosomal dominant way and linked to chromosome 1 (PS2, <1%), chromosome 14 (PS1, 33%), and 19 (APP, 5%); whereas 62% of the autosomal dominant forms is still not linked. PS1 and PS2 mutations can also be of the sporadic type.[55] In familial LOAD, the majority (90%) is not yet linked to a chromosome, whereas 10% is inherited in an autosomal dominant way. A subset has been linked to chromosome 12.[57]

Risk factors: 65% of all EOAD and 25% of all LOAD cases display an ApoE4 polymorphism (one or two E4 alleles), which is the best-known risk factor. ApoE4 data in early onset AD are based on a study by Van Duijn *et al.*[56] (n = 175 patients). Several other risk factors for late onset AD have been reported (for details, see Ref. 5).

Total picture (EOAD and LOAD): Familial Alzheimer's disease (FAD) represents about 40% of all AD cases. Most of these FAD cases (i.e., 35% of all AD patients) do not inherit AD as an autosomal dominant trait. Thus, the nonfamilial or sporadic form of AD comprises approximately 60% of all AD cases, and together with the nonautosomal dominant forms of FAD, it accounts for 95% of the cases.

For more information, please consult the website on AD mutations:
http://molgen-www.uia.ac.be/ADmutations/

[a]An important reason to present this table is that the terminology used regarding AD is sometimes inconsistent and thus confusing. For example, familial types and autosomal dominant forms of AD are often intermingled, although they represent different groups.

mechanism must exist. We have already reported that molecular misreading of correct genomic information is a major event in this respect.[1,2]

HOW MOLECULAR MISREADING WAS DISCOVERED

Some years ago, curious to understand the presence of vasopressin (VP) precursor products that theoretically could not exist,[15,16] we discovered a novel type of transcript variability in homozygous Brattleboro rats (di/di[3]). These rats suffer from hypothalamic diabetes insipidus as a result of a single-base germline mutation in the VP gene that encodes an aberrant VP precursor.[17] The mutant VP precursor is not admitted to the secretory pathway since it is arrested in the membranes of the endoplasmatic reticulum.[18] We surprisingly found that in a small number of solitary magnocellular neurons in the supraoptic and paraventricular nucleus an additional mutation (ΔGA) in their VP transcripts results in a restoration of the wild-type reading frame. The subsequently synthesized, still partially mutated VP protein (either 13 or 22 amino acids are still out of frame) is able to enter the secretory pathway and undergoes axonal transport, during which enzymatic processing into functional VP takes place.[3,19,20] Two sites of the dinucleotide deletion (ΔGA) were found that occurred preferentially in GAGAG motifs of the VP RNA. Moreover, the mutation rate may depend on transcriptional activity.[3] It was subsequently demonstrated that a similar mutation also occurs in wild-type rat and human VP gene transcripts, meaning that these mutations are not restricted to the VP-producing cells of di/di rats.[3,21] In wild-type VP transcripts, the reading frame is not restored as in the di/di rat, but instead the outcome is a frameshift and the synthesis of a mutated protein. As the number of cells with a +1 reading frame in rats increases with age, these findings suggest the existence of a novel mechanism that may very well be related to aging and age-related diseases in general. These observations prompted us to initiate a study on the presence of dinucleotide deletions in other neuronal genes associated with the age-related neuropathology of AD (TABLE 2). Dinucleotide deletions have since been shown to occur much more widely both in other neuronal transcripts of postmitotic neurons and in proliferating tissues and neuroblastoma cells.[1,2,4]

THE FREQUENCY OF GAGAG MOTIFS IN TRANSCRIPTS

In the VP transcripts of the di/di rat, the GAGAG motif is a hotspot for the occurrence of a dinucleotide deletion.[3,22] In the human transcript population, GAGAG is also a hotspot for this type of deletion. The chance that a GAGAG motif occurs in a sequence is one in every 1024 (i.e. $1:4^5$) bases. The human haploid genome consists of 3×10^9 bases, 9×10^7 (3%) of which code for transcripts. This implies that there are 9×10^4 GAGAG motifs within coding sequences. In addition, the human genome consist of 65,000–80,000 genes with a mean length of 2.2 kb, so that an average of 2.1 GAGAG motifs per gene can be expected.[23] We selected a panel of AD-associated genes, calculated the expected number of GAGAG motifs (1:1024), and determined the actual number of these motifs (TABLE 2). Both coding sequences of βAPP and UBB showed a higher number of actual GAGAG motifs compared to the calculated number of GAGAG motifs: βAPP contains seven motifs instead of the calcu-

TABLE 2. Genes associated with Alzheimer's disease

Gene	Number of base pairs Coding sequence, longest form	GAGAG motifs Expected number	GAGAG motifs Actual number	Exon from which +1 peptides are derived	+1 Peptide sequence
βAPP	2234	2.2	7	9/10	RGRTSSKELA
Ubiquitin-B	687	0.7	2	2	DHHPGSGAQ
Tau	1056	1.1	—	13	HGRLAPARHAS[a]
Presenilin I	1392	1.4	3	9	SIQKFQV
Presenilin II	1346	1.3	3	3	VEKPGERGGR
Apolipoprotein E4	951	0.9	—	4	GAPRLPPAQAA[a]
MAP2b[b]	5595	5.6	13	18	KTRFQRKGPS
Neurofilament-light	1625	1.6	3	1	PGNRSMPGHE
Neurofilament-medium	2748	2.8	3	3	EAEGEGSPS
Neurofilament-heavy	3063	3.1	2	1	VGAARDSRAA
GFAP[c]	1299	1.3	6	6	EDRGDAGWRGH
Huntington	12432	12.4	9	12	RGFHSRVSRS

[a]Based on a GAGA motif.
[b]Microtubule-associated protein 2B.
[c]Glial fibrillary acidic protein.

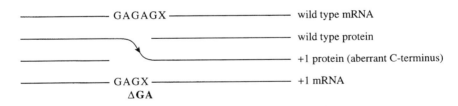

FIGURE 1. A schematic representation of a frameshift mutation. A +1 protein is formed as a result of a dinucleotide deletion (for example ΔGA) and consists of a wild-type NH$_2$ terminus. Downstream of the dinucleotide deletion, the transcript is translated in the +1 reading frame and gives rise to the aberrant COOH terminus of the +1 protein. (From van Leeuwen *et al.*[2] Reproduced by permission.)

lated 2.2, and UBB contains two motifs instead of 0.7. A dinucleotide deletion in or adjacent to the GAGAG motif results in a translation of the protein in the +1 reading frame (i.e., a "+1 protein"; FIG. 1). The presence of βAPP and UBB frameshift mutations and +1 proteins was studied in the brains of AD, DS, and Parkinson patients and age-matched, nondemented controls (see http://www.knaw.nl/nih/index.htm[1])

β-AMYLOID PRECURSOR PROTEIN

The βAPP gene is located on chromosome 21 and has 18 exons; by alternative splicing three different proteins are formed: APP 695 (695 amino acids (aa) lacking exon 7 and 8), APP 751 (751 aa, lacking exon 8), and APP 770 (770 aa.)[24] APP 695 is abundantly expressed in neurons and lacks the Kunitz-type serine protease domain. Mutations found in the FAD cases in the APP gene are all located in exons 16 and 17, which is the region in the mRNA coding for $A\beta_{42-43}$.[5] Based on calculations, the coding sequence of βAPP (2234 bases) should contain 2.2 GAGAG-motifs. However, in the sequence seven GAGAG-motifs are present, indicating that βAPP has a relatively high risk of dinucleotide deletions around GAGAG motifs, especially in exons 9 and 10, where three GAGAG motifs are clustered (see TABLE 2).

Dinucleotide deletions in exon 9 or 10 give rise to two very similar +1 proteins of 38 kDa (running at MW 60 kDa on DAA gels); antibodies were raised against a part of the identical COOH terminus of these +1 proteins: RGRTSSKELA (see TABLE 2). Paraffin sections of brains (temporal and frontal cortex and hippocampus) of AD and DS patients and age, sex, and postmortem delay-matched controls were stained with this βAPP[+1] antibody.[1] The antibody stained the dystrophic neurites of the neuritic plaques, tangles, and neuropil threads in brain sections of early (<65 years) and late onset (>65 years) AD and DS patients and one Parkinson patient with initial AD neuropathology (FIG. 2). No βAPP[+1] immunoreactivity was found in paraffin sections

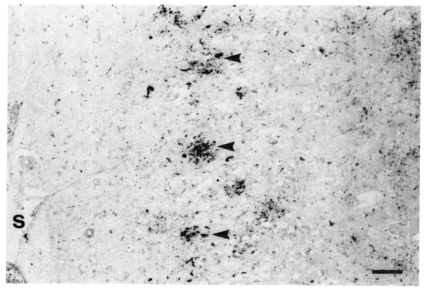

FIGURE 2. APP[+1] immunoreactivity in a paraffin section of the frontal cortex (BA11) of an Alzheimer patient. Note the intense immunoreactivity present in dystrophic neurites forming a neuritic plaque in layers 2 and 3 (*arrowheads*). s = sulcus. In layer 5 scattered APP[+1] immunoreactivity is also present and in between the neuritic plaques, neuropil threads are visible. Scale bar = 50 μm.

of cerebral cortex and hippocampus of nondemented controls and in the nigrostriatal system of 10 Parkinson patients[1] (see http://www.knaw.nl/nih/index.htm). The combined data of the frontal and temporal cortex and the hippocampus show that indeed a frameshift in the translation of βAPP occurred that led to the production of the aberrant form in βAPP in 71% of the AD patients and in all of the demented DS patients studied.

A deletion in exon 9 results in the loss of the growth-promoting (neurotrophic) domain in βAPP,[25] and because of the premature termination codon in principle no $A\beta_{42-43}$ is produced. However, the production of $A\beta_{42-43}$ may not depend completely on that of βAPP. There is an additional translation initiation site in βAPP downstream of the mutation[26] that might be activated by means of the dinucleotide deletion, leading to misprocessing of βAPP and the deposition of $A\beta_{42-43}$. Furthermore, wild-type βAPP is still expressed and accounts for the $A\beta_{42-43}$ production as well.

ENHANCED GENE TRANSCRIPTION IN DOWN'S SYNDROME PATIENTS AND MOLECULAR MISREADING

Down's syndrome (DS) patients are of particular interest as almost all of these patients exhibit a similar neuropathology to that of AD patients.[27] The AD neuropathology in DS patients usually becomes apparent 40 years earlier than the pathology in sporadic AD. DS patients highly express βAPP, because of an extra copy of chromosome 21. However, the expression level of the βAPP gene in DS patients is fivefold higher than that of controls and thus exceeds the expression level expected from the gene dosage (trisomy 21) alone.[28,29] In line with the high βAPP mRNA level, we detected a transcript mutation and the expression of APP^{+1} protein in almost all DS patients[1] (see http://www.knaw.nl/nih/index.htm).

UBIQUITIN-B

Ubiquitin is an evolutionarily highly conserved protein of 76 amino acids (aa) that plays an essential role in a large number of processes including apoptosis and ATP-dependent proteasomal breakdown of at least 60% of the proteins.[30,31] Before protein degradation, ubiquitin is tagged to the target protein through an isopeptide bond between a lysine moiety in the target protein and the COOH-terminal glycine of ubiquitin.[30] The COOH-terminal glycine of one ubiquitin molecule can be connected to the amino group of a lysine in a ubiquitin tag leading to multi-ubiquitylation. The presence of a multi-ubiquitin chain on an aberrant protein triggers the proteasomal breakdown of that particular protein. Ubiquitin is a component of neurofibrillary tangles,[32–34] and ubiquitin is conjugated to Tau protein in paired helical filaments. Most ubiquitin in paired helical filaments appears to represent mono-ubiquitylated Tau,[35] indicating that multi-ubiquitylation does not occur and consequently the proteasomal breakdown of these mono-ubiquitylated proteins is inefficient.

Ubiquitin is encoded by a multigene family (i.e., ubiquitin-A, -B, and -C; *cf.* Hol *et al.*[36]), with some genes containing long repeats of ubiquitin-coding sequences.[37] UBB and UBC are expressed in the human brain.[38] UBB consists of three repeats[39]

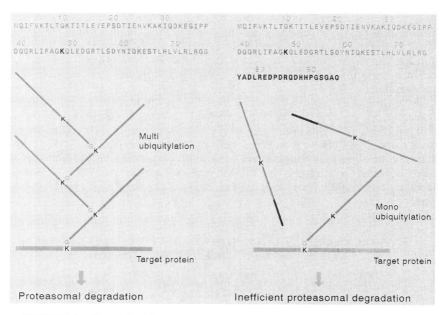

FIGURE 3. The ubiquitin–proteosomal pathway to degrade aberrant proteins. Under normal circumstances (*left side*), a lysine (K) residue of the target protein (*thick line*, for example, hyperphosphorylated Tau or any other protein that is degraded via the proteosomal pathway) is recognized by the COOH-terminal glycine (G) residue of ubiquitin-B (UBB). In turn, a lysine residue (K) of UBB at position 48 is recognized by another UBB molecule. If this process is repeated, multiubiquitylation occurs, the proteasomal pathway is triggered and the protein degraded. The right panel shows that a GU deletion in the UBB transcript results in the loss of the G lysine residue and the formation of an aberrant COOH terminus (*bold sequences*, Y–Q), which blocks the process of multiubiquitylation. Only the wild-type transcripts are still able to ubiquitylate. In Alzheimer's disease, however, the Tau protein is mainly mono-ubiquitylated.[35] It is hypothesized that in time (Alzheimer's disease lasts for approximately eight years) the proteasomal degradation becomes increasingly inefficient, resulting in a piling up of aberrant proteins in the neuron whose function is increasingly disturbed. (From van Leeuwen *et al.*[2] Reproduced by permission.)

and comprises two GAGAG motifs, one in the first repeat and one in the second at amino acid position 75. A dinucleotide deletion in either of the repeats will lead to a similar aberrant protein of 93 aa instead of 76 aa (FIG. 3). This UBB^{+1} protein misses the essential COOH-terminal glycine and is unable to multi-ubiquitylate target proteins. Consequently, the presence of UBB^{+1} may result in an inefficient proteasomal breakdown and thus could account for the lack of poly-ubiquitylation of Tau protein in paired helical filaments.[35] In fact, UBB^{+1} can be ubiquitylated, thereby weakening the proteasomal degradation.

An antibody was raised against the COOH terminus of the UBB^{+1} sequence (RQDHHPGSGAQ[1]; TABLE 2). In early- and late-onset sporadic AD cases, we found UBB^{+1} in neuritic plaques, neuropil threads, and tangles (FIG. 4). Brain sections of DS patients revealed a similar result.[1] In addition, in paraffin sections of hippocampi

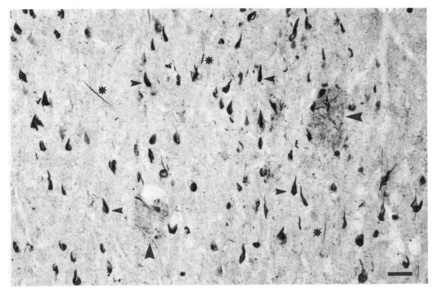

FIGURE 4. Ubiquitin-B^{+1} immunoreactivity in a 50-μm-thick vibratome section of the temporal cortex of a 92-year-old Alzheimer patient. Intense immunoreactivity was present in the neurofibrillary tangles (*small arrowheads*), neuropil threads, and the dystrophic neurites (*asterisks*) surrounding the neuritic plaques (*large arrowheads*). Scale bar = 50 μm.

(CA1 and subiculum) of elderly control patients (>72 years) and of one Parkinson patient with initial AD neuropathology UBB^{+1} immunoreactivity was found. In paraffin sections of young nondemented controls (<72 years), no UBB^{+1} immunoreactivity was detected.[1] Recent studies on vibratome sections, however, which yield better results than paraffin sections, showed that UBB^{+1} immunoreactivity was already present at an earlier age (>51 years; #94119; see website cited below) in the hippocampus and temporal cortex of young, nondemented controls. For a detailed overview of the neuropathological status of the patients studied and the presence of UBB^{+1} immunostaining in frontal cortex, temporal cortex, and hippocampus, see http://www.knaw.nl/nih/index.htm. Taking all these data together, UBB^{+1} was expressed in 100% of the AD and demented DS patients studied. UBB^{+1} is probably an early marker for neurodegeneration as immunoreactivity is also present in nondemented elderly controls in brain structures that are known to be an early target for AD neuropathology, such as the entorhinal cortex, CA1, and subiculum.

GENOMIC MUTATION OR TRANSCRIPT MODIFICATION

In the di/di rat, an age-dependent increase in the number of VP cells with a revertant phenotype was found, which we originally interpreted as somatic mutations in the genes occurring at an exceptionally high frequency.[16,22] However, we were unable to show a mutation at the genomic level, whereas in transcripts the mutation

(ΔGA) could be readily determined.[20] The same results were obtained in transcripts and genomic DNA of βAPP and UBB using even more sensitive approaches.[1] Using the most sensitive approach, we were able to amplify as a positive control 10 copies of mutant plasmid DNA out of a background of 500 ng genomic DNA corresponding to about 160,000 copies of DNA or 80,000 cells. Using this very sensitive method, a total of 5 µg DNA for each control, AD, and DS patient was investigated (i.e., a total of 1.6×10^6 genome equivalents, in DS even $2.4.10^6$ βAPP equivalents), but no mutations were found.[1] Consequently, it is very unlikely that these dinucleotide deletions take place in the genome. We therefore favor the opinion that transcript mutations occur, the more so since different +1 proteins coexist in the same neurons.[1,2]

THE MECHANISM OF RNA MUTATIONS AND RNA QUALITY CONTROL

In principle, RNA mutations could be generated either co- or post-transcriptionally. A number of different RNA editing processes have been described in different organisms that also result in RNA sequences that are different from those encoded by the genome.[40] A base substitution form of RNA editing that converts specific adenosines of certain mRNAs into inosines by hydrolytic deamination has indeed been found in the nervous system.[41] However, the mechanism of the posttranscriptional A to I conversions appears to be completely different from the processes that generate RNA deletions. The only type of RNA editing that seems to possess similar mechanistic features is the cotranscriptional insertion of Gs by slippage or stuttering of certain viral RNA-dependent RNA polymerases.[42,43] So far, however, it is unclear whether mammalian DNA-dependent RNA polymerases can indeed slip and skip dinucleotides while transcribing repeats such as GAGAG or CTCT, resulting in deletions in the transcript.

The finding that two different +1 proteins, arising from two different transcripts, coexist in neurons of AD patients points to a defective general controlling mechanism (the "common denominator"), resulting in a failure to detect mutated mRNA in AD patients.[1] It is well known that such a proofreading mechanism for DNA acts during DNA replication.[44,45] In the case of transcripts, mRNA surveillance has been described, which is a mechanism that checks for premature stop codons in RNA.[46,47] This system increases the fidelity of gene expression by eliminating incompletely translated RNAs and could act as the above-mentioned common denominator. mRNA surveillance genes seem to be conserved during evolution since a human homologue of a yeast gene with potential "mRNA surveillance" activity (i.e., human up frameshift protein, HUPF1) has recently been cloned.[48,49]

A declining accuracy of such an mRNA surveillance system later in life could therefore be an important aging factor, in addition to those already known (e.g., oxidative stress and DNA damage; cf. Refs. 8, 50). Recently, a similar mechanism has been suggested to play a role in sporadic amyotrophic lateral sclerosis, which is associated with the appearance of aberrant RNA of the astrocyte glutamate transporter ($EEAT_2$), as a result of RNA processing errors (e.g., exon skipping). The resulting protein may exert a dominant negative effect and seems to escape from mRNA surveillance and proteasomal degradation.[51]

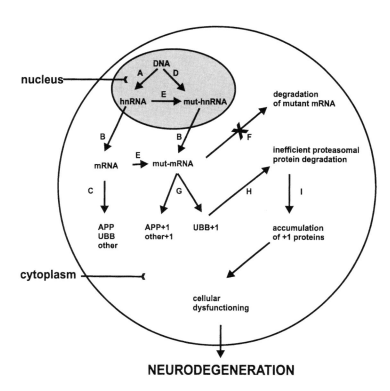

FIGURE 5. Working hypothesis. This model explains how frameshift mutations in RNA and the subsequent +1 proteins could accumulate in the neurons of DS and AD patients. In the cell nucleus, the genomic information is intact and is transcribed into heteronuclear RNA (hnRNA) (**A**). This hnRNA is spliced and transported to the cytoplasm (**B**), where it is translated into wild-type βAPP, UBB, and other proteins (**B**). In the event that the coding sequence contains GAGAG motifs, dinucleotide deletions may occur (**D,E**). Molecular misreading can be caused by stuttering or slippage of the RNA polymerase during transcription (**D**) or by editing of the mRNA or hnRNA after transcription (**E**). In DS and AD patients, the mRNA decay pathway, termed mRNA surveillance, may be impaired (**F**) and therefore the +1 mRNA is translated into +1 protein (**G**). In particular, UBB[+1] may have a prominent role, as it will probably directly interfere with the ubiquitin/proteasome system and lead to an inefficient protein breakdown through the proteasomal pathway (**H**). Together, these changes lead to a gradual accumulation of aberrant proteins in neurons, cellular dysfunction, and ultimately neurodegeneration. (From Hol *et al.*[54] Reproduced by permission).

OTHER NEURODEGENERATIVE AND AGE-RELATED DISEASES

Other neurodegenerative and age-related diseases may also be associated with molecular misreading. Candidates are diseases with inclusion bodies such as frontal lobe dementia or Pick's disease, argyrophilic grain disease, amyotrophic lateral sclerosis, diffuse Lewy body disease, progressive supranuclear palsy, multisystem atrophy, and Huntington's disease, all of which contain UBB inclusions. Indeed, in some of these diseases UBB^{+1} and APP^{+1} proteins are also present, which implies that molecular misreading of the genome is much more widespread than assumed previously.[52] In addition, for the diseases not associated with the appearance of APP^{+1} and UBB^{+1} proteins, it cannot be excluded that molecular misreading of other genes occurs, since several genes thought to be involved in these diseases indeed contain GAGAG motifs (e.g., neurofilament genes in Parkinson's disease[2]). The resulting clinical and pathological phenotypes of these diseases may be determined by various other factors (such as other +1 proteins, cell type, and risk factors[6]).

Molecular misreading might also occur in genomes transcribed outside the nervous system, namely in proliferating tissues.[4] Recently, we have indeed obtained evidence that molecular misreading occurs in the male reproductive tract.[4] These data suggest that molecular misreading may contribute to a variety of age-related diseases and nonneural pathologic conditions, such as alcoholic liver disease[53] and cancer (Van Leeuwen, unpublished data).

CONCLUDING REMARKS

The novel mechanism of molecular misreading has already raised a large number of key questions, four of which will be briefly addressed below. Our major challenge in the coming years will be to show that molecular misreading of genomic information is an early event in AD. Transgenic mice overexpressing +1 proteins or viral vectors expressing +1 proteins in cell lines and *in vivo* will be instrumental to finding out whether they are causal in the onset of AD.

A second key question is the mechanism of molecular misreading in simple sequence repeats such as GAGAG or CUCU.[1] Presently we are studying the mechanism in neuroblastoma and peripheral neuroepithelioma cell lines that have been observed to display molecular misreading.[4]

A third important question is whether molecular misreading (e.g., of the UBB gene) is a process restricted to Alzheimer's pathology or occurs more generally also in other neurodegenerative disorders and other age-related diseases (*cf.* van Leeuwen *et al.*[2]). This issue is presently being investigated with the use of the antibodies mentioned in TABLE 2 (*cf.* Sonnemans *et al.*[52]).

A fourth essential question is what the ratio is between mutated +1 proteins and wild-type protein (e.g., UBB^{+1}/UBB wild-type). This issue is presently addressed in single neurons isolated from vibratome sections of Alzheimer patients.

We anticipate that molecular misreading of genes will result in a loss of cellular function. The presence of various +1 proteins in one neuron points to loss of control over mRNA quality and protein degradation by different common denominators (*cf.* Hol *et al.*[54]). mRNA and protein surveillance systems are likely candidates for this role: if their activity decreases during aging, this might explain why age is the greatest risk factor for developing neural and nonneural pathologic conditions.

ACKNOWLEDGMENTS

The authors are greatly indebted to the support of the Internationale Stichting Alzheimer Onderzoek (ISAO), NWO-GPD, NWO Program Support and the Hersenstichting.

REFERENCES

1. VAN LEEUWEN, F.W., D.P. V. DE KLEIJN, H.H. VAN DEN HURK, *et al.* 1998. Frameshift mutants of β amyloid precursor protein and ubiquitin-B in Alzheimer's and Down patients. Science **279:** 242–247.
2. VAN LEEUWEN, F.W., J.P.H. BURBACH & E.M. HOL. 1998. Mutations in RNA: a first example of molecular misreading in Alzheimer's disease. TINS **21:** 331–335.
3. EVANS, D.A.P., A.A.M. VAN DER KLEIJ, M.A.F. SONNEMANS, *et al.* 1994. Frameshift mutations at two hotspots in vasopressin transcripts in post-mitotic neurons. Proc. Natl. Acad. Sci. USA **91:** 6059–6063.
4. VAN LEEUWEN, F.W., E.M. HOL, R.W.H. HERMANUSSEN, *et al.* 2000. Molecular misreading in non-neuronal cells. FASEB J. In press.
5. VAN BROECKHOVEN, C. 1998. Alzheimer's disease: identification of genes and genetic risk factor. *In* Neuronal Degeneration and Regeneration: From Basic Mechanisms to Prospects for Therapy. F.W. van Leeuwen *et al.*, Eds. Prog. Brain Res. **117:** 315–326.
6. MANN, D.M.A. 1997. Molecular biology's impact on our understanding of aging. Br. Med. J. **315:** 1078–1081.
7. SWAAB, D.F., P.J. LUCASSEN, A. SALEHI, *et al.* 1998. Reduced neuronal activity and reactivation in Alzheimer's disease. Progr. Brain Res. **177:** 343–379.
8. NUNOMURA, A., G. PERRY, M.A. PAPOLLA, *et al.* 1999. RNA oxidation is a prominent feature of vulnerable neurons in Alzheimer's disease. J. Neurosci. **19:** 1959–1964.
9. VOGEL, G. 1998. Possible new cause of Alzheimer's disease found. Science **279:** 174.
10. OTT, A., M.M.B. BRETLER, F. VAN HARSKAMP, *et al.* 1995. Prevalence of Alzheimer's disease and vascular dementia: association with education. The Rotterdam study. Br. Med. J. **310:** 970–973.
11. WILHELMSEN, K. C. 1999. The tangled biology of tau. Proc. Natl. Acad. Sci. USA **96:** 7120–7121.
12. ARRIAGADA, P.V., J.H. GROWDON, E.T. HEDLEY-WHITE & B.T. HYMAN. 1992. Neurofibrillary tangles but not senile plaques parallel duration and severity of Alzheimer's disease. Neurology **42:** 631–638.
13. BRAAK, H., E. BRAAK, J. BOHL & R. REINTJES. 1996. Age, neurofibrillary changes, Aβ-amyloid and the onset of Alzheimer's disease. Neurosci. Lett. **210:** 87–90.
14. SELKOE, D.J. 1999. Translating cell biology into therapeutic advances in Alzheimer's disease. Nature **399:** 23–31.
15. VAN LEEUWEN, F.W., E.M. VAN DER BEEK, M. SEGER, *et al.* 1989. Age-related development of a heterozygous phenotype in solitary neurons of the homozygous Brattleboro rat. Proc. Natl. Acad. Sci. USA **86:** 6417–6420.
16. FINCH, C.E. & M.F. GOODMAN. 1997. Relevance of "adapative" mutations arising in non-dividing cells of microorganisms to age-related changes in mutant phenotypes of neurons. TINS **20:** 501–507.
17. VALTIN, H. 1982. The discovery of the Brattleboro rat, recommended nomenclature, and the question of proper controls. Ann. N.Y. Acad. Sci. **394:** 1–9.
18. SCHMALE, H., B. BOROWIAK, H. HOLT-GREZ & D. RICHTER. 1989. Impact of altered protein structures on the intracellular traffic of a mutated vasopressin precursor from Brattleboro rats. Eur. J. Biochem. **182:** 621–627.

19. SONNEMANS, M.A.F., D.A.P. EVANS, J.P.H. BURBACH & F.W. VAN LEEUWEN. 1996. Immunocytochemical evidence for the presence of vasopressin in intermediate sized neurosecretory granules of solitary neurohypophyseal terminals in the homozygous Brattleboro rat. Neuroscience **72:** 225–231.

20. VAN LEEUWEN, F.W., D.A.P. EVANS, R.W.H. VERWER & J.P.H. BURBACH. 1998. The magnocellular neurons of the hypothalamo-neurohypophyseal system display remarkable neuropeptidegic phenotypes leading to novel insights in neuronal biology. *In* Brain Vasopressin and Related Peptides, Life Science Advances on Neuropeptides. J.P.H. Burbach, I.J.A. Urban & D. De Wied, Eds. Progr. Brain Res. **119:** 115–126. Elsevier Biomedical Press. Amsterdam.

21. EVANS, D.A.P., J.P.H. BURBACH, D.F. SWAAB & F.W. VAN LEEUWEN. 1996. Mutant vasopressin precursors in the human hypothalamus: evidence for neuronal somatic mutations in man. Neuroscience **71:** 1025–1030.

22. EVANS, D.A.P., J.P.H. BURBACH & F.W. VAN LEEUWEN. 1995. Somatic mutations in the brain: relationship to aging? Mutat. Res. **338:** 173–182.

23. STRACHAN, T. & A.P. READ. 1996. Human Molecular Genetics. Bios Scientific Publishers Ltd. Oxford.

24. YOSHIKAI, S.-I., H. SASAKI, K. DOH-URA, *et al.* 1990. Genomic organization of the human amyloid beta-protein precursor gene. Gene **87:** 257–263.

25. JIN, L.-W., H. NINOMIYA, J.-M. ROCH, *et al.* 1994. Peptides containing the RERMS sequence of amyloid β/A4 protein precursor bind cell surface and promote neurite extension. J. Neurosci. **14:** 5461–5470.

26. CITRON, M., C. HAASS & D.J. SELKOE. 1993. Production of amyloid-β-peptide by cultured cells: no evidence for internal initiation of translation at Met_{596}. Neurobiol. Aging **14:** 571–573.

27. WISNIEWSKI, K.E., H.M. WISNIEWSKI & W. GY. 1985. Occurrence of neuropathological changes and dementia of Alzheimer's disease in Down's syndrome. Ann. Neurol. **17:** 278–282.

28. NEVE, R.L., E.A. FINCH & L.R. DAWES. 1988. Expression of the Alzheimer amyloid precursor gene transcript in the human brain. Neuron **1:** 669–677.

29. RUMBLE, B., R. RETALLACK, C. HILBICH, *et al.* 1989. Amyloid A4 protein and its precursor in Down's syndrome and Alzheimer's disease. N. Engl. J. Med. **320:** 1446–1452.

30. VARSHAVSKY, A. 1997. The ubiquitin system. Trends Biochem. Sci. **22:** 383–387.

31. SCHWARTZ, A.L. & A. CIECHANOVER. 1999. The ubiquitin–proteasome pathway and pathogenesis of human diseases. Annu. Rev. Med. **50:** 57–74.

32. MORI, H., J. KONDO & Y. IHARA. 1987. Ubiquitin is a component of paired helical filaments in Alzheimer's disease. Science **235:** 1641–1644.

33. MAYER, R.J., J. ARNOLD, L. LÁSLÓ, *et al.* 1991. Ubiquitin in health and disease. Biochim. Biophys. Acta **1089:** 141–157.

34. MAYER, R.J., C. TIPLER, J. ARNOLD, *et al.* 1996. Endosome-lysosomes, ubiquitin and neurodegeneration. Adv. Exp. Med. Biol. **389:** 261–269.

35. MORISHIMA-KAWASHIMA, M., M. HASEGAWA, K. TAKIO, *et al.* 1993. Ubiquitin is conjugated with amino-terminally processed tau in paired helical filaments. Neuron **10:** 1151–1160.

36. HOL, E.M., A. NEUBAUER, D.P.V. DE KLEIJN, *et al.* 1998. Dinucleotide deletions in neuronal transcripts: a novel type of mutation in non-familial Alzheimer's disease and Down syndrome patients. Progr. Brain Res. **117:** 379–394.

37. WIBORG, O., M.S. PEDERSON, A. WIND, *et al.* 1985. The human ubiquitin multigene family: some genes contain multiple directly repeated ubiquitin coding sequences. EMBO J. **4:** 755–759.

38. ADAMS, M.D., M. DUBNICK, A.R. KERLAVAGE, *et al.* 1992. Sequence identification of 2,375 human brain genes. Nature **355:** 632–634.

39. BAKER, R.T. & P.G. BOARD. 1987. The human ubiquitin gene family: structure of a gene and pseudogenes from the UbB subfamily. Nucleic Acids Res. **15:** 443–463.

40. BENNE, R. 1996. RNA editing: how a message is changed. Curr. Opin. Genet. Dev. **6:** 221–231.

41. BASS, B.L. 1997. RNA editing and hypermutation by adenosine deamination. Trends Biochem. Sci. **22:** 157–162.

42. JACQUES, J.P. & D. KOLAKOFSKY. 1991. Pseudo-templated transcription in prokaryotic and eukaryotic organisms. Genes Dev. **5:** 707–713.
43. KOLAKOFSKY, D. & S. HAUSMANN. 1998. Co-transcriptional paramyxovirus mRNA editing; a contradiction in terms? *In* Modification and Editing of RNA. H. Grosjean & R. Benne, Eds.: 413–420. ASM Press. Washington, DC.
44. RADMAN, M. & R. WAGNER. 1988. The high fidelity of DNA duplication. Sci. Am. **259(2):** 24–30.
45. KUNKEL, T.A. 1995. The intricacies of eukaryotic spell-checking. Curr. Biol. **5:** 1091–1094.
46. PULAK, R. & P. ANDERSON. 1993. mRNA surveillance by the *Caenorhabditis elegans smg* genes. Genes Dev. **7:** 1885–1887.
47. RUIZ-ECHEVARRÍA, M.J., K. CZAPLINSKI & S.W. PELTZ. 1996. Making sense of nonsense in yeast. Trends Biochem. Sci. **21:** 433–438.
48. APPLEQUIST, S.E., M. SELG, C. RAMAN & H.-M. JÄCK. 1997. Cloning and characterization of HUPF1, a human homolog of the *Saccharomyces cerevisiae* nonsense mRNA-reducing UPF1 protein. Nucleic Acids Res. **25:** 814–821.
49. PERLICK, H.A., S.M. MEDGHALCHI, F.A. SPENCER, *et al.* 1996. Mammalian orthologues of a yeast regulator of nonsense transcript stability. Proc. Natl. Acad. Sci. USA **93:** 10928–10932.
50. FINCH, C.E. 1990. Longevity, Senescence, and the Genome. University of Chicago Press. Chicago.
51. LIN, C.L.G., L.A. BRISTOL, L. JIN, *et al.* 1998. Aberrant RNA processing in a neurodegenerative disease: the cause for absent EAAT2, a glutamate transporter, in amyotrophic lateral sclerosis. Neuron **20:** 589–602.
52. SONNEMANS, M.A.F., R.A.I. DE VOS, E.H.N. JANSEN, *et al.* 1999. Molecular misreading of β amyloid precursor protein and ubiquitin B genes in various tauo- and ubiquitopathies. Society of Neurosciences Meeting, Miami, Florida, USA, October 23–28. **25:** 836.
53. MCPHAUL, L., J. WANG, S. YUAN, *et al.* 1999. Frameshift mutants of ubiquitin-B in Mallory body formation in human liver. FASEB J. **13:** A736.
54. HOL, E.M., J.A. SLUIJS, M.A.F. SONNEMANS, *et al.* 1999. Chapter 9. Molecular misreading and accumulation of +1 protein in Alzheimer and Down syndrome patients. *In* Proceedings of the 6th International Conference on Alzheimer's Disease and Related Disorders. K. Iqbal *et al.*, Eds.: 59–67. Wiley. New York.
55. CRUTS, M., C.M. VAN DUIJN, H. BACKHOVENS, *et al.* 1998. Estimation of the genetic contribution of presenilin-1 and -2 mutations in a population-based study of presenile Alzheimer disease. Hum. Mol. Genet. **7:** 43–51.
56. VAN DUIJN, C.M., P. DE KNIJFF, M. CRUTS, *et al.* 1994. Apolipoprotein E$_4$ allele in a population-based study of early onset Alzheimer's disease. Nat. Genet. **7:** 74–78.
57. PERICAK-VANCE, M.A., M.P. BASS, L.H. YA,AOKA, *et al.* 1997. Complete genomic screen in late-onset familial Alzheimer disease. Guidance for a new locus on chromosome 12. JAMA **278:** 1237–1241.

Biogerontology: The Next Step

SURESH I.S. RATTAN[a]

Laboratory of Cellular Ageing, Danish Centre for Molecular Gerontology, Department of Molecular and Structural Biology, University of Aarhus, Gustav Wieds Vej, DK-8000 Aarhus - C, Denmark

ABSTRACT: After a long period of collecting empirical data describing the changes in organisms, organs, tissues, cells, and macromolecules, biogerontological research is now able to develop various possibilities for intervention. Because aging is a stochastic and nondeterministic process characterized by a progressive failure of maintenance and repair, it is reasoned that genes involved in homeodynamic repair pathways are the most likely candidate gerontogenes. A promising approach for the identification of critical gerontogenic processes is through the hormesis-like positive effects of mild stress. Stimulation of various repair pathways by mild stress has significant effects on delaying the onset of various age-associated alterations in cells, tissues, and organisms.

The study of the biological basis of aging, biogerontology, is now ready to take the next step. After decades of systematic collection of data describing the changes in organisms, organs, tissues, cells and macromolecules, it is clear that although the basic biochemistry of all living systems is very much alike, there are no universal patterns of aging and age-related alterations (TABLE 1). Collecting more descriptive data by using more powerful techniques will only strengthen this fact. Most importantly, like any other mature field of intellectual enquiry, biogerontology is now able to accept such differences in the progression of aging, which is indicative of the fact that biogerontology is now a mature science. Hence, the practitioners of biogerontology (biogerontologists) do not need to anymore hide their identities behind any other "-logy"!

ESSENTIAL LIFESPAN AND AGING

Aging is a nondeterministic stochastic phenomenon occurring mainly as a result of the failure of homeodynamics and an organism's inability to prevent it.[1–3] The evolutionary theories strongly argue against the existence of genes that may have evolved specifically to cause aging and to determine maximum lifespan of an organism. The genetic regulation of lifespan is primarily in terms of determining what can be called as the essential lifespan (ELS) of a species, which is the time required to fulfill the Darwinian purpose of life, that is, successful reproduction and continua-

[a]Address for correspondence: Phone: +45 8942 5034; fax: +45 8612 3178. rattan@imsb.au.dk

TABLE 1. General observations on the aging phenotype

The phenotype and rate of progression of aging are:
- different in different species;
- different in different individuals;
- different in different organs, systems, and tissues;
- different in different cells;
- different in different organelles; and
- different in different macromolecules.

tion of generations. For example, species undergoing fast maturation and early onset of reproduction with large reproductive potential generally have a short ELS.[4–6] In contrast, slow maturation, late onset of reproduction, and small reproductive potential of a species is concurrent with its long ELS.[4–6] Therefore, from an evolutionary point of view, the term "maximum lifespan" is meaningless because of the fact that the extended survival of a small number of individuals to very late ages, as compared with the majority of the population in the wild, has no significance in the context of the declining force of natural selection with age.

The genes that do influence longevity are those that have evolved in accordance with the life history of a species for assuring the ELS. Such genes are termed longevity assurance genes[7] or vitagenes[8] and are considered to constitute various maintenance and repair pathways, including antioxidative defenses, DNA repair, fidelity of genetic information transfer, and stress response pathways. There are several examples of genes, particularly in DNA repair and antioxidant pathways, whose activities have been reported to correlate directly with species lifespan.[9–11] Further evidence that the maintenance and repair pathways are the main determinants of longevity comes from experiments performed to retard aging and to increase the lifespan of organisms.[11,12]

It is the longevity-assuring vitagenes in the maintenance and repair pathways that set limits to the ELS of a species and that are integral to an evolutionarily determined developmental period. Aging, however, is the postdevelopmental and often postreproductive period, observed most commonly in protected environments that permit the survival of an individual much beyond its ELS. A progressive impairment of functional ability and increased chances of death generally characterize this extended period of life. Furthermore, the diversity of the forms and variations in which age-related alterations are manifested underline the fact that the progression of aging is not programmed or deterministic but is stochastic in nature (TABLE 1). A large body of descriptive data clearly shows that individually no tissue, organ, or system becomes functionally exhausted, even in very old organisms. Instead, it is their combined interaction and interdependence that determines the survival of the whole. It has been suggested that age-related alterations observed at all levels of organization are the sign of *continuous remodeling* of the body,[13] which is a kind of a *survival instinct response* for counteracting the ill effects of progressively failing repair and maintenance processes.

GERONTOGENES

In an evolutionary sense, it is meaningless to use the term gerontogenes. For practical purposes, however, it may be appropriate to invoke this term in order to focus any discussion about the gene-based biochemical processes involved in aging and to describe genes whose altered activity influences aging and longevity.[12,14] Two kinds of gerontogene action are postulated to be responsible for the emergence of the aging phenotype. The first considers the role of late-acting mutations that are already present at the time of fertilization and birth and show their deleterious effects after the period of growth, development, and maturation. The second category of gene action is referred to as the antagonistic pleiotropy, which involves genes selected for beneficial effects during early development, but have harmful effects in postreproductive life when they escape the force of natural selection.[15–17] In both cases, these genes were not selected as the real genes to cause aging, but manifest themselves only as virtual gerontogenes because of their involvement in the progression of aging.[8,12]

Until now, several putative gerontogenes have been reported for various aging systems, including yeast, nematodes, insects, and mammals. The molecular identities of some of these genes have been established. In the case of the budding yeast, the nematode, and the fruitfly, these genes are longevity-determining genes, but the molecular pathways affected by them have little or no similarity among different organisms. For example, in *S. cerevisiae*, the functions of the LAG, RAS, uth, and Sir complex range from being transmembrane proteins to transcriptional silencing of telomeres.[7,15,18–22] In *C. elgans*, the normal functions of various gerontogenes include PI3-kinase activity, tyrosine kinase receptor activity, transcription factor activity, and insulin receptor-like activity, and it is only when mutated that a loss or alteration in the activity of their gene products is associated with increased longevity.[23–26] There are several other candidate gerontogenes, such as four clock genes *clk-1, clk-2, clk-3,* and *gro-1*27, and *age-2*28 in *C. elegans*, discovered for their effects on longevity, but whose molecular identities are yet to be established. In *Drosophila*, the *methuselah (mth)* gene, whose predicted protein sequence has homology to several GTP-binding protein–coupled receptors, is also associated with increased lifespan and enhanced resistance to various forms of stress.[29] However, in the case of the mouse *klotho* gene, which is a membrane protein β-glucosidase,[30] and the human Werner's gene, which is a DNA helicase,[31] the phenotype of premature aging is manifested along with a plethora of diseases.

Additionally, genetic linkage studies for longevity in mice have identified major histocompatibility complex (MHC) regions,[32] and quantitative trait loci (QTL) on chromosomes 7, 10, 11, 12, 16, 18, and 19[33,34] as putative gerontogenes. In human centenarians, certain alleles of the HLA locus on chromosome 6, different alleles of ApoE and ApoB, and the DD genotype of angiotensin–converting enzyme (ACE) have been linked to their long lifespan.[32,35,36] The diversity of the genes associated with aging and longevity of different organisms indicates that whereas the genes involved in repair and maintenance pathways may be important from an evolutionary point of view (the so-called "public" genes), each species may also have additional "private"[37] gerontogenic pathways that influence its aging phenotype.

TABLE 2. Major age-related alterations during cellular aging *in vitro*

Structural changes

Increase in cell size; change of shape from thin, long, and spindle-like to flattened and irregular; loss of fingerprint-like arrangement in parallel arrays; increased number of vacuoles and dense lysosomal residual bodies containing UV-fluorescent pigments; rodlike polymerization of the cytoskeletal actin filaments and disorganized microtubules; and increased level of chromosomal aberrations and multinucleation.

Physiological changes

Reduced response to growth factors and other mitogens; increased sensitivity to toxins, drugs, irradiation, and other stress; altered calcium flux, pH, viscosity, and membrane potential; reduced respiration and energy metabolism; and increased duration of G1 phase of the cell cycle.

Biochemical and molecular changes

Decreased activity, specificity, and fidelity of various enzymes; accumulation of post-translationally modified and inactivated proteins; reduced rates of protein synthesis and degradation; increased levels of oxidative damage in nuclear and mitochondrial DNA; reduced levels of methylated cytosines; reduced length of telomeres; and altered (increased or decreased) expression of several genes.

CELLULAR AGING *IN VITRO*

An experimental system that has been used extensively for the identification of genes involved in aging is the so-called Hayflick system of cellular aging *in vitro*. It is now well established that normal diploid cells of various kinds and from various species have a limited proliferative potential *in vitro*. The ultimate phenotype of the serially passaged diploid cell is the permanent arrest of late-passage or high population doubling level (PDL) cells in late-G_1/S phase boundary of the cell cycle. Most of the studies on the identification of gerontogenes in this model system have focussed on this endpoint phenotype, also known as replicative senescence. Several genes have been identified whose products either act as active regulators of cell cycle-arrest or correlate with replicative senescence.[38–44] These genes are generally involved in either the activation or the inhibition of the protein phosphorylation and dephosphorylation cascade involving various transcription factors and cyclin complexes in association with the p53 and Rb genes.[45–49] Experimentally induced or spontaneous immortalization of such cells *in vitro* and cancer cells *in vivo* is almost always accompanied by the deregulation of these genes.

It is important to realize that the irreversible cell cycle–arrest is the ultimate phenotype reached after a long period of active cell proliferation. Furthermore, the replicative senescent phenotype may never be achieved *in vivo* for many cell types, such as fibroblasts, epithelial cells, endothelial cells and osteoblasts. Numerous studies have been performed that show that cellular aging *in vitro* is accompanied by a whole range of physiological, metabolic, biochemical, and molecular changes that occur progressively throughout the replicative lifespan of cells *in vitro*, culminating in the cessation of cell proliferation. Some of the major characteristics of cellular aging are summarized in TABLE 2.

Progressive accumulation of stochastically occurring damage is the hallmark of cellular aging *in vitro* and is the signal for the activation and upregulation of various genes whose products then act towards arresting and maintaining the senescent cell in a permanent state of growth arrest.[50–52] Experimental induction of damage by free radicals and excessive loss of telomeres by severe growth arrest results in the premature onset of the senescent phenotype (both in normal and transformed cells) by the resultant rapid activation of the so-called senescence-specific genes.[53,54] Thus the significance of the cellular aging system *in vitro* can be best realized as a model to study progressive accumulation of damage in cells resulting in their physiological impairment.

CELLULAR RESPONSIVENESS AND HORMESIS

Using the model system of cellular aging, it has been shown that, whereas the cellular response to various growth factors and mitogens is significantly reduced during aging, their sensitivity to toxins, antibiotics, irradiation, oxidants, and heat shock is increased.[55,56] Furthermore, induction of high levels of stress, particularly oxidative stress, has been used as a tool to induce replicative senescence or irreversible growth arrest by increasing the intracellular damage resulting in the upregulation of the so-called senescence-specific genes described above.[52] Although such an experimental approach can facilitate the understanding of the regulation and interactions of various cell cycle-checkpoint genes, this approach has only a limited value in identifying genes involved in maintaining various pathways of maintenance and repair. Therefore, other experimental strategies are required for identifying gerontogenes that influence the basic process of aging.

It has been suggested that if cells and organisms are exposed to brief periods of stress so that their stress response–induced gene expression is upregulated and the related pathways of maintenance and repair are stimulated, one should observe anti-aging and longevity-promoting effects. Such a phenomenon, in which stimulatory responses to low doses of otherwise harmful conditions improve health and enhance lifespan, is known as hormesis.[57–60]

Anti-aging and life-prolonging hormesis-like effects of thermal stress have been reported for *Drosophila*,[61] nematodes,[62,63] and yeast.[64] In *Drosophila*, the mild stress of hypergravity exposure has been shown to delay aging, to increase longevity, and to increase resistance to heat shock.[65] Similarly, low doses of ionizing radiation have been shown to increase the lifespan of mice[66] and may have several beneficial effects on human health, especially in terms of prevention and treatment of neurodegenerative diseases and cancers.[67–69] The beneficial effects of exercise, which stimulates the production of reactive oxygen species,[70,71] and the life-extending effects of calorie restriction have been suggested to work through hormesis.[72] In the case of cellular aging *in vitro*, repeated low doses of γ- and X-rays have been shown to prolong the cellular lifespan.[73–75] Anti-aging effects of repeated mild heat shock on human skin fibroblasts have also been reported.[76]

Therefore, it may be possible to use the approach of hormesis in order to identify genes that are important for aging and longevity. For example, if repeated mild heat-shock treatment has life-prolonging and anti-aging effects in cells and organisms, it is likely that the genetic pathways of the heat-shock response are also associated with longevity determination. Similarly, other chemical, physical, and biological treatments can be used to unravel various pathways of maintenance and repair whose sustained activities improve the physiological performance and survival of cells and organisms. This will be helpful not only for having a complete understanding of the mechanistic aspects of the aging process, but also for preventing the onset of various age-related diseases by maintaining the efficiency of repair processes.

Some of the main targets for prevention of age-related pathology include the following biochemical processes which may be accessible to modulation through hormesis: (1) the appearance and accumulation of abnormal proteins and proteolytic products leading to, for example, Alzheimer's disease; (2) posttranslational modifications and crosslinks between macromolecules, leading to, for example, cataracts and atherosclerosis; (3) reactive oxygen species-induced mitochondrial defects leading to, for example, Parkinson's disease, Huntington's disease, and amyotropic lateral sclerosis; and (4) genomic instability leading to, for example, cancers.

The clinical implications of the hormesis-like stress response in the diagnosis and treatment of several diseases including arthritis, Duchenne muscular dystrophy, multiple sclerosis, myocardial ischemia, mitochondrial encephalomyopathy, some cancers, and autoimmune diseases such as systemic lupus erythematosis are being increasingly realized.[77] In the case of aging research, although at present there are only a few studies performed that utilize mild stress as a modulator of aging and longevity, hormesis appears to be the next step and represents a promising experimental approach in biogerontology.

ACKNOWLEDGMENT

This project is a part of the shared cost action program GENAGE under the EU Biomed and Health Programme Projects.

REFERENCES

1. ROSE, M.R. 1991. Evolutionary Biology of Aging. Oxford University Press. New York.
2. KIRKWOOD, T.B.L. 1977. Evolution of ageing. Nature **270:** 301–304.
3. PARTRIDGE, L. & N.H. BARTON. 1993. Optimality, mutation and the evolution of ageing. Nature **362:** 305–311.
4. FINCH, C.E. 1990. Longevity, Senescence, and the Genome. The University of Chicago Press. Chicago.
5. FINCH, C.E. 1998. Variations in senescence and longevity include the possibility of negligible senescence. J. Gerontol. Biol. Sci. **53A:** B235–B239.
6. HOLLIDAY, R. 1994. Longevity and fecundity in eutherian mammals. *In* Genetics and Evolution of Aging. M.R. Rose & C.E. Finch, Eds. Kluwer Academic Publishers. Amsterdam, The Netherlands.
7. D'MELLO, N.P., A.M. CHILDRESS, D.S. FRANKLIN, et al. 1994. Cloning and characterization of *LAG1*, a longevity-assurance gene in yeast. J. Biol. Chem. **269:** 15451–15459.
8. RATTAN, S.I.S. 1998. The nature of gerontogenes and vitagenes. Antiaging effects of repeated heat shock on human fibroblasts. Ann. N.Y. Acad. Sci. **854:** 54–60.
9. HOLLIDAY, R. 1995. Understanding Ageing. Cambridge University Press. Cambridge, U.K.
10. RATTAN, S.I.S. 1989. DNA damage and repair during cellular aging. Int. Rev. Cytol. **116:** 47–88.
11. RATTAN, S.I.S. 1995. Ageing—a biological perspective. Mol. Aspects Med. **16:** 43–508.

12. RATTAN, S.I.S. 1995. Gerontogenes: real or virtual? FASEB J. **9:** 28–286.
13. FRANCESCHI, C., D. MONTI, P. SANSONI & A. COSSARIZZA. 1995. The immunology of exceptional individuals: the lesson of centenarians. Immunol. Today **16:** 12–16.
14. RATTAN, S.I.S. 1985. Beyond the present crisis in gerontology. BioEssays **2:** 226–228.
15. JAZWINSKI, S.M. 1996. Longevity, genes, and aging. Science **273:** 54–59.
16. FINCH, C.E. & R.E. TANZI. 1997. Genetics of aging. Science **278:** 407–411.
17. MILLER, R.A. 1999. Kleemeir award lecture: Are there genes for aging? J. Gerontol. Biol. Sci. **54A:** B297–B307.
18. JAZWINSKI, S.M., S. KIM, C.-Y. LAI & A. BENGURIA. 1998. Epigenetic stratification: the role of individual change in the biological aging process. Exp. Gerontol. **33:** 571–580.
19. JAZWINSKI, S.M. 1998. Genetics of longevity. Exp. Gerontol. **33:** 773–783.
20. JAZWINSKI, S.M. 1999. Longevity, genes, and aging: a view provided by a genetic model system. Exp. Gerontol. **34:** 1–6.
21. GUARENTE, L. 1997. Link between aging and the nucleolus. Gene. Dev. **11:** 2449–2455.
22. SINCLAIR, D.A., K. MILLS & L. GUARENTE. 1997. Accelerated aging and nucleolar fragmentation in yeast *sgs1* mutants. Science **277:** 1313–1316.
23. MORRIS, J.Z., H.A. TISSENBAUM & G. RUVKUN. 1996. A phosphatidylinositol-3-OH kinase family member regaulating longevity and diapause in *Caenorhabditis elegans.* Nature **382:** 536–539.
24. KIMURA, K.D., H.A. TISSENBAUM, Y. LIU & G. RUVKUN. 1997. *daf-2,* an insulin receptor-like gene that regulates longevity and diapause in *Caenorhabditis elegans.* Science **277:** 942–946.
25. LIN, K., J.B. DORMAN, A. RODAN & C. KENYON. 1997. *daf-16:* an HNF-3/forkhead family member that can function to double the life-span of *Caenorhabditis elegans.* Science **278:** 1319–1322.
26. OGG, S., S. PARADIS, S. GOTTLIEB, *et al.* 1997. The fork head transcription factor DAF-16 transduces insulin-like metabolic and longevity signals in *C. elegans.* Nature **389:** 994–999.
27. LAKOWSKI, B. & S. HEKIMI. 1996. Determination of life-span in *Caenorhabditis elegans* by four clock genes. Science **272:** 1010–1013.
28. YANG, Y. & D.L. WILSON. 1999. Characterization of life-extending mutation in *age-2,* a new aging gene in *Caenorhabditis elegans.* J. Gerontol. Biol. Sci. **54A:** B13–B142.
29. LIN, Y.-J., L. SEROUDE & S. BENZER. 1998. Extended life-span and stress resistance in the *Drosophila* mutant *methuselah.* Science **282:** 943–946.
30. KURO-O, M., *et al.* 1997. Mutation of the mouse *klotho* gene leads to a syndrome resembling ageing. Nature **390:** 45–51.
31. YU, C.-E., *et al.* 1996. Positional cloning of the Werner's syndrome gene. Science **272:** 258–P262.
32. GELMAN, R., A. WATSON, R. BRONSON & E. YUNIS. 1988. Murine chromosomal regions correlated with longevity. Genetics **118:** 693–704.
33. MILLER, R.A., C. CHRISP, A.U. JACKSON & D. BURKE. 1998. Marker loci associated with life span in genetically heterogeneous mice. J. Gerontol. Med. Sci. **53A:** M257–M263.
34. DE HAAN, G., R. GELMAN, A. WATSON, *et al.* 1998. A putative gene causes variability in lifespan among gentoypically identiacal mice. Nat. Genet. **19:** 114–116.
35. SCHÄCHTER, F., L. FAURE-DELANEF, F. GUÉNOT, *et al.* 1994. Genetic associations with human longevity at the APOE and ACE loci. Nature Genet. **6:** 29–32.
36. JIAN-GANG, Z., M. YONG-XING, W. CHUAN-FU, *et al.* 1998. Apolipoprotein E and longevity among Han Chinese population. Mech. Ageing Dev. **104:** 159–167.
37. MARTIN, G.M. 1997. The Werner mutation: does it lead to a "public" or "private" mechanism of aging? Mol. Med. **3:** 35–358.
38. NODA, A., Y. NING, S.F. VENABLE, *et al.* 1994. Cloning of senescent cell-derived inhibitors of DNA synthesis using an expression screen. Exp. Cell. Res. **211:** 90–98.
39. WONG, H. & K. RIABOWOL. 1996. Differential CDK-inhibitor gene expression in aging human diploid fibroblasts. Exp. Gerontol. **31:** 311–325.

40. WHITAKER, N.J., T.M. BRYAN, P. BONNEFIN, et al. 1995. Involvement of RB-1, p53, p16^{INK4} and telomerase in immortalisation of human cells. Oncogene **11:** 971–976.
41. STEIN, G.H., L.F. DRULLINGER, A. SOULARD & V. DULIC. 1999. Differential roles for cyclin-dependent kinase inhibitors p21 and p16 in the mechanisms of senescence and differentiation in human fibroblasts. Mol. Cell. Biol. **19:** 2109–2117.
42. ALCORTA, D.A., Y. XIONG, D. PHELPS, et al. 1996. Involvement of the cyclin-dependent kinase inhibitor p16 (INK4a) in replicative senescence of normal human fibroblasts. Proc. Natl. Acad. Sci. USA **93:** 13742–13747.
43. KAMB, A. et al. 1994. A cell cycle regulator potentially involved in genesis of many tumor types. Science **264:** 436–440.
44. JACOBS, J.J.L., K. KIEBOOM, S. MARINO, et al. 1999. The oncogene and polycomgroup gene bmi-1 regulates cell proliferation and senescence through ink4a locus. Nature **397:** 164–168.
45. DERVENTZI, A., E.S. GONOS & S.I.S. RATTAN. 1996. Ageing and cancer: a struggle of tendencies. In Molecular Gerontology—Research Status and Stratgies. S.I.S. Rattan & O. Toussaint, Eds.: 15–23. Plenum Press. New York.
46. DERVENTZI, A., S.I.S. RATTAN & E.S. GONOS. 1996. Molecular links between cellular mortality and immortality. Anticancer Res. **16:** 2901–2910.
47. GAO, C.Y. & P.S. ZELENKA. 1997. Cyclins, cyclin-dependent kinases and differentiaion. BioEssays **19:** 307–315.
48. DULIC´, V., L.F. DRULLINGER, E. LEES, et al. 1993. Altered regulation of G1 cyclins in senescent human diploid fibroblasts: accumulation of inactive cyclin E—Cdk2 and cyclin D1—Cdk2 complexes. Proc. Natl. Acad. Sci. USA **90:** 11034–11038.
49. HENGST, L., V. DULIC, J.M. SLINGERLAND, et al. 1994. A cell cycle-regulated inhibitor of cyclin-dependent kinases. Proc. Natl. Acad. Sci. USA **91:** 5291–5295.
50. REDDEL, R.R. 1998. A reassessment of the telomere hypothesis of senescence. BioEssays **20:** 977–984.
51. FARAGHER, R.G.A., C.J. JONES & D. KIPLING. 1998. Telomerase and cellular lifespan: ending the debate? Nat. Biotechnol. **16:** 701–702.
52. TOUSSAINT, O., et al. 1998. Reciprocal relationship between the resistance to stresses and cellular aging. Ann. N. Y. Acad Sci. **851:** 450–465.
53. VON ZGLINICKI, T., G. SARETZKI, W. DÖCKE & C. LOTZE. 1995. Mild hyperoxia shortens telomeres and inhibits proliferation of fibroblasts: a model for senescence? Exp. Cell Res. **220:** 186–193.
54. SARTEZKI, G., J. FENG, T. VON ZAGLINICKI & B. VILLEPONTEAU. 1998. Similar gene expression pattern in senescent and hyperoxic-treated fibroblasts. J. Gerontol. Biol. Sci. **53A:** B438–B442.
55. DERVENTZI, A. & S.I.S. RATTAN. 1991. Homeostatic imbalance during cellular ageing: altered responsiveness. Mutat. Res. **256:** 191–202.
56. RATTAN, S.I.S. & A. DERVENTZI. 1991. Altered cellular responsiveness during ageing. BioEssays **13:** 601–606.
57. NEAFSEY, P.J. 1990. Longevity hormesis: a review. Mech. Ageing Dev. **51:** 1–31.
58. POLLYCOVE, M. 1995. The issue of the decade: hormesis. Eur. J. Nucl. Med. **22:** 399–401.
59. CALABRESE, E.J. 1997. Hormesis revisited: new insights concerning the biological effects of low-dose exposure to toxins. Environ. Law Rep. **27:** 10526–10532.
60. CALABRESE, E.J. & L.A. BALDWIN. 1999. Chemical hormesis: its historical foundations as a biological hypothesis. Toxicol. Pathol. **27:** 195–216.
61. KHAZAELI, A.A., M. TATAR, S.D. PLETCHER & J.W. CURTSINGER. 1997. Heat-induced longevity extension in Drosophila. I. Heat treatment, mortality, and thermotolerance. J. Gerontol. Biol Sci. **52A:** B48–B52.
62. LITHGOW, G.J., T.M. WHITE, S. MELOV & T.E. JOHNSON. 1995. Thermotolerance and extended life-span conferred by single-gene mutations and induced by thermal stress. Proc. Natl. Acad. Sci. USA **92:** 7540–7544.
63. LITHGOW, G.J. 1996. Invertebrate gerontology: the age mutations of Caenorhabditis elegans. BioEssays **18:** 809–815.
64. SHAMA, S., C.-Y. LAI, J.M. ANTONIAZZI, et al. 1998. Heat stress-induced life span extension in yeast. Exp. Cell Res. **245:** 379–388.

65. LE BOURG, E. & N. MINOIS. 1999. A mild stress, hypergravity exposure, postpones behavioral aging in *Drosophila melanogster*. Exp. Gerontol. **34:** 157–172.
66. CARATERO, A., M. COURTADE, L. BONNET, *et al.* 1998. Effect of continuos gamma irradiation at a very low dose on the life span of mice. Gerontology **44:** 272–276.
67. BUSCIGLIO, J., J.K. ANDERSEN, H.M. SCHIPPER, *et al.* 1998. Stress, aging, and neurodegenerative disorders. Ann. N.Y. Acad. Sci. **851:** 429–443.
68. WYNGAARDEN, K.E.V. & E.K.J. PAUWELS. 1995. Hormesis: are low doses of ionizing radiation harmful or beneficial? Eur. J. Nucl. Med. **22:** 481–486.
69. JOHNSON, T.E., G.J. LITHGOW & S. MURAKAMI. 1996. Interventions that increase the response to stress offer the potential for effective life prolongation and increased health. J. Gerontol. Biol. Sci. **51A:** B392–B395.
70. RADÁK, Z., K. ASANO, A. NAKAMURA, *et al.* 1998. The effect of high altitude and calorie restriction on reactive carbonyl derivatives and activity of glutamine synthetase in rat brain. Life Sci. **62:** 1317–1322.
71. RADÁK, Z., A. NAKAMURA, H. NAKAMOTO, *et al.* 1998. A period of anaerobic exercise increases the accumulation of reactive carbonyl derivatives in the lungs of rats. Eur. J. Physiol. **435:** 439–441.
72. MASORO, E.J. 1998. Hormesis and the antiaging action of dietary restriction. Exp. Gerontol. **33:** 61–66.
73. HOLLIDAY, R. 1991. A re-examination of the effects of ionizing radiation on lifespan and transformation of human diploid fibroblasts. Mutat. Res. **256:** 295–302.
74. ICARD, C., R. BEUPAIN, C. DIALTOFF & A. MACIEIRA-COELHO. 1979. Effect of low dose rate irradiation on the division potential of cells in vitro. VI. changes in DNA and in radiosensitivity during aging of human fibroblasts. Mech. Ageing Develop. **11:** 269–278.
75. YANG, Z., S. KODAMA, K. SUZUKI & M. WATANABE. 1998. Telomerase activity, telomere length, and chromosome aberrations in the extension of life span of human embryo cells induced by low-dose X-rays. J. Radiat. Res. **39:** 35–51.
76. RATTAN, S.I.S. 1998. Repeated mild heat shock delays ageing in cultured human skin fibroblasts. Biochem. Mol. Biol. Int. **45:** 753–759.
77. VAN EDEN, W. & D.B. YOUNG, Eds. 1996. Stress Proteins in Medicine. Marcel Dekker. New York.

Nuclear-Mitochondrial Interactions Involved in Aging in *Podospora anserina*

CORINA BORGHOUTS AND HEINZ D. OSIEWACZ[a]

Molekulare Entwicklungsbiologie und Biotechnologie, Johann Wolfgang Goethe-Universität, Botanisches Institut, Marie-Curie-Strasse 9, D-60439 Frankfurt am Main, Germany

A GENETIC BASIS OF AGING

All wild-type strains of the filamentous ascomycete *Podospora anserina* display a limited life span. In the past, the mitochondrial DNA (mtDNA) was found to play an important role in the aging process of this fungus. In aged cultures, large parts of the mtDNA are deleted, leading to energy deficits and finally to cell death. In addition, several regions of the mtDNA become amplified as circular molecules. The most prominent element of this type is the so-called plDNA, which is derived from the first intron of the mitochondrial cytochrome-c-oxidase I gene (*CoI*).[1,2]

Recently, several nuclear genes (*Grisea, Su12, As6,* and *As1*) were found to play a role in mtDNA reorganizations as well as in life span. Whereas a mutation in *As1* leads to severe site-specific mtDNA deletions, resulting in a reduced life span, mutations in *Grisea, Su12,* and *As6* prevent plDNA amplification and simultaneously lead to an increased life span.[3–5]

GRISEA, ENCODING A TRANSCRIPTION FACTOR INVOLVED IN CELLULAR COPPER HOMEOSTASIS, AFFECTS mtDNA STABILITY

The detailed mechanisms involved in nuclear-mitochondrial interactions leading to stabilization of the mtDNA and to the observed life span extension are currently speculative. To unravel these mechanisms, we set out to characterize the long-lived mutant grisea in some detail.

Analysis revealed that a point mutation in the single intron of the *Grisea* gene leads to a splicing defect (loss-of-function mutation).[5] In the wild strain, *Grisea* was found to code for a protein (GRISEA) that shows significant homology to three yeast transcription factors (MAC1, ACE1 from *Saccharomyces cerevisiae*, and AMT1 from *Candida glabrata*).[5,6] Subsequent reporter gene assays performed in yeast confirmed that GRISEA also encodes a transcription factor that is an ortholog of MAC1.[7] The DNA-binding domain was localized within the first 168 amino acids at the N-terminus, allowing binding of GRISEA to specific promoter sequences of *Ctr1*, a target gene of MAC1 (TABLE 1, line 7–9; FIG. 1). In yeast, *Ctr1* encodes a high affinity copper transporter involved in copper uptake.[8] Thus, a *Ctr1* homolog

[a]Corresponding author. Phone: +49 69 79829264; fax: +49 69 79829363.
Osiewacz@em.uni-frankfurt.de

TABLE 1. Heterologous reporter gene assays in a MAC1-deficient strain of *S. cerevisiae* (mac1-1)

	Reporter vector[a]	Expression vector[b]	Amino acids present		β-Galactosidase activity[d]	
			GRISEA	GALA4-AD[c]	+Cu[d]	+BCS[f]
1	pCtr1-Lacz	—	—	—	0.6 ± 0.1	0.8 ± 0.1
2	—	pAdh-gr597	1-597	—	0.1 ± 0.1	0.6 ± 0.1
3	pCtr1-Lacz	pAdh-gr597	1-597	—	0.6 ± 0.1	12.6 ± 3.0
4	pCtr1-Lacz	pAdh-gr588	1-588	—	n.d.[g]	11.4 ± 1.5
5	pCtr1-Lacz	pAdh-gr574	1-574	—	n.d.	4.7 ± 0.5
6	pCtr1-Lacz	pAdh-gr379	1-379	—	n.d.	2.1 ± 0.4
7	pCtr1-Lacz	pAdh-gr75	1-75	761-881	3.0 ± 0.6	4.2 ± 0.5
8	pCtr1-Lacz	pAdh-gr168	1-168	761-881	31.8 ± 5.8	27.5 ± 5.7
9	pCtr1-Lacz	pAdh-gr260	1-260	761-881	22.4 ± 1.3	25.2 ± 2.3
10	pCtr1-lacz	pAdh-gr260Gly[h]	1-260 Gly^{38}→Cys^{38}	—	1.4 ± 0.4	1.6 ± 0.3
11	pCtr1ΔBS-Lacz[i]	pAdh-gr	1-597	—	0.3 ± 0.2	0.4 ± 0.2

[a]Construct containing the *Ctr1* promoter of *S. cerevisiae* cloned upstream of the *LacZ* reporter gene.
[b]Constructs expressing (parts of) the *Grisea* gene, regulated by the *Adh1* promotor.
[c]Some expresion vectors contain parts of the *Grisea* gene fused to the activation domain of GAL4 to allow reporter gene expression.
[d]β-galactosidase activity in Miller units ± SD.
[e]Addition of 100 μM CuSO$_4$.
[f]Addition of 33 μM bathocuproinedisulfonic acid (BCS) and 1 mM ascorbic acid to reduce copper availability in normal medium.
[g]Not determined.
[h]The glycine at position 38 in the DNA binding domain of the GRISEA protein was substituted for cysteine. As previously shown for ACE1, this specific mutation prevents binding of the protein to the promotor.
[i]A small part of the *Ctr1* promotor (−346 → −296), shown to be necessary for the binding of MAC1, was deleted in this construct.

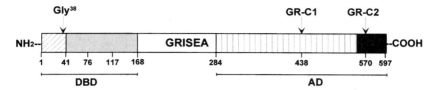

FIGURE 1. Schematic presentation of transcription factor GRISEA. The N-terminal region contains the DNA-binding domain (a.a. 1-168). The first 41 amino acids show high homology to the three yeast copper regulated transcription factors MAC1, ACE1, and AMT1.[6] The position of an essential glycine (Gly[38]) is indicated (TABLE 1, line 10). The acidic activation domain is located at the C-terminus (a.a. 493-597), whereas highest activity is achieved by using the complete C-terminal half of the protein (a.a. 284-597).[7] The position of two cysteine-rich motifs (GR-C1 and GR-C2) with high homology to two cysteine-rich motifs of MAC1 are indicated. For MAC1 these motifs were shown to bind copper at a high cellular copper concentration causing a conformational change, repressing transactivation activity of the protein.[10]

of *P. anserina* represents a good candidate target gene of GRISEA. Additionally, an activation domain was identified at the C-terminal end of the protein (TABLE 1, line 3–6; FIG. 1). As for MAC1, the activity of this domain is repressed by high cellular copper levels. In contrast, copper deficiency leads to activation of GRISEA and to strong expression of the target genes (TABLE 1, line 1–3; FIG. 1). These data unraveled the molecular basis of the phenotype of mutant grisea which can be rescued by the addition of high amounts of copper to the growth medium.[9] A mutation in *Grisea* affects the expression of a copper transporter, leading to a cellular copper deficiency.

To find out whether cellular copper affects the stability of the mtDNA, we analyzed mtDNA preparations from senescent cultures from both the wild-type strain and the long-lived mutant grisea. Growth of grisea cultures in copper-supplemented medium led to a reversion of the mutant phenotype, to the wild-type specific mtDNA deletions, and particularly to amplified plDNA molecules.[11]

AGING, MOLECULAR DAMAGE, AND REMODELING CAPACITY

These and additional data led us to speculate that the increase in life span in mutant grisea is the result of a reduced generation of reactive oxygen species (ROS). First, ROS generation via Fenton chemistry should be reduced because of lower cellular copper levels. Second, and most importantly, stabilization of mtDNA as it is found in mutant grisea guarantees stable expression of mitochondrial genes over time. As a consequence, defective proteins of the respiratory chain can be efficiently replaced, and functional respiratory chains can be maintained longer. In the wild type, due to extensive mtDNA reorganizations such an efficient exchange of damaged, mtDNA-encoded proteins is not possible, and defective mitochondria accumulate much faster than in mutant grisea.

Interestingly, the role of mitochondria and mtDNA instabilities in aging is not restricted to *P. anserina*; it was also demonstrated in higher systems including mammals. It is clear today that the involved mechanisms necessarily need to differ in their details from system to system. In particular, instabilities due to the activity of mobile introns cannot be expected to occur in animal systems, because introns are absent in the mtDNA of these organisms. However, general clues about the role of mitochondria in the aging process of biological systems can be expected to emerge from investigations with lower organisms. In particular, more details about nuclear-mitochondrial interactions, which are essential for mitochondrial biogenesis and function, need to be elucidated in the future. Species such as *P. anserina* with an accentuated mitochondrial etiology of aging appear to be the best systems to address this important issue.

ACKNOWLEDGMENTS

This work was supported by a grant of the Deutsche Forschungsgemeinschaft (Bonn Bad-Godesberg) to H.D.O.

REFERENCES

1. STAHL, U., P.A. LEMKE, P. TUDZYNSKI *et al.* 1987. Evidence for plasmid like DNA in a filamentous fungus, the ascomycete *Podospora anserina*. Mol. Gen. Genet. **162:** 341–343.
2. OSIEWACZ, H.D. & K. ESSER. 1984. The mitochondrial plasmid of *Podospora anserina*: a mobile intron of a mitochondrial gene. Curr. Genet. **8:** 299–305.
3. BELCOUR, L., O. BEGEL & M. PICARD. 1991. A site-specific deletion in mitochondrial DNA of *Podospora anserina* is under the control of nuclear genes. Proc. Natl. Acad. Sci. USA **88:** 3579–3583.
4. SILAR, P., F. KOLL & M. ROSSIGNOL. 1997. Cytosolic ribosomal mutations that abolish accumulation of circular intron in the mitochondria without preventing senescence of *Podospora anserina*. Genetics **145:** 697–705.
5. BORGHOUTS, C., E. KIMPEL & H.D. OSIEWACZ. 1997. Mitochondrial DNA rearrangements of *Podospora anserina* are under the control of the nuclear gene *Grisea*. Proc. Natl. Acad. Sci. USA **94:** 10768–10773.
6. OSIEWACZ, H.D. & U. NUBER. 1996. *Grisea*, a putative copper-activated transcription factor from *Podospora anserina* involved in differentiation and senescence. Mol. Gen. Genet. **152:** 115–124.
7. BORGHOUTS, C. & H.D. OSIEWACZ. 1998. Grisea, a copper-modulated transcription factor from *Podospora anserina* involved in senescence and morphogenesis, is an ortholog of mac1 in *Saccharomyces cerevisiae*. Mol. Gen. Genet. **260:** 492–502.
8. DANCIS, A., D. HAILE, D.S. YUAN & R.D. KLAUSNER. 1994. The *Saccharomyces cerevisiae* copper transport protein (ctr1p). biochemical characterization, regulation by copper, and physiologic role in copper uptake. J. Biol. Chem. **269:** 25660–25667.
9. MARBACH, K., J. FERNANDEZ-LARREA & U. STAHL. 1994. Reversion of a long-living, undifferentiated mutant of *Podospora anserina* by copper. Curr. Genet. **26:** 184–186.
10. JENSEN, L.T. & D.R. WINGE. 1998. Identification of a copper-induced intramolecular interaction in the transcription factor Mac1 from *Saccharomyces cerevisiae*. EMBO J. **17:** 5400–5408.
11. BORGHOUTS, C., S. KERSCHNER & H.D. OSIEWACZ. 2000. Copper-dependence of mitochondrial DNA rearrangements in *Podospora anserina*. Curr. Genet. In press.

Inherited Frailty

ApoE Alleles Determine Survival after a Diagnosis of Heart Disease or Stroke at Ages 85+

E.H. CORDER,[a,b,c] H. BASUN,[d,e] L. FRATIGLIONI,[d,e] Z. GUO,[d,e] L. LANNFELT,[d] M. VIITANEN,[d,e] L.S. CORDER,[c] K.G. MANTON,[c] AND B. WINBLAD[d,e]

[b]The University of Southern Denmark, Odense University, DK-5000 Odense C, Denmark

[c]Center for Demographic Studies, Duke University, Durham, North Carolina 27708, USA

[d]Division of Geriatric Medicine, Neurotec, Karolinska Institutet, S-141 86, Huddinge, Stockholm, Sweden

[e]Gerontology Research Center, S-113 82, Stockholm, Sweden

Apolipoprotein E (ApoE) polymorphism was found to predict life expectancy within the Kungsholmen Project cohort.[1] It was the oldest-old subjects aged 85+ who demonstrated mortality differentials, not the relatively younger subjects, aged 75 to 84, when first examined as part of the cohort. For the oldest-old, there was a 2-year difference in life expectancy: It was shorter for those who carried the ApoE ε4 allele (i.e., the ε3/4 genotype) versus ε3/3, and shorter for ε3/3 versus ε2/3.

The spectrum of vascular disorders was prospectively investigated within the Kungsholmen cohort to explain the ApoE variation in life expectancy. Both disease occurrence and subsequent mortality in affected subjects were investigated.

STUDY SUBJECTS

The Kungsholmen cohort consists of persons aged 75+ living in a neighborhood of Stockholm on October 1, 1987.[2,3] Eleven hundred twenty-four of 1810 subjects had ApoE information. ApoE genotype information was used for 1077 subjects who carried ε2/3, ε3/3, or ε3/4; that is, the sample investigated for mortality from the date of entry into the cohort until December 31, 1994.[1]

DIAGNOSES

ICD-8 codes 410 to 438 denote ischemic heart disease, nonischemic heart disease, and cerebrovascular disease.[4] The first date of diagnosis for each code in this spectrum was obtained from computerized hospital discharge records for the Stock-

[a]Address for correspondence: Professor Elizabeth H. Corder, Ph.D., Duke University, Center for Demographic Studies, 2117 Campus Drive, Durham, NC 27709-0408, USA. Phone: 919-684-6126; fax: 919-684-3861.

Beth@cds.duke.edu

holm region. Cohort members were identified by national person number. The diagnosis information covered 1969 through 1994, that is, prospective with respect to entry date into the cohort between 1987 and 1989.

STATISTICAL METHODS

Kaplan-Meier plots were constructed to describe (1) disease occurrence as disease-free survival and (2) prognosis for each genotype. Log-rank tests were used to compare the genotype-specific plots (two-sided, alpha = 0.05).

DISEASE OCCURRENCE

Incidence for heart disease and stroke was unrelated to ApoE genotype for the oldest-old and relatively younger subjects. There were 72 (33%) affected subjects among those aged 85+ at entry into the cohort ($\chi^2 = 0.56$ with 2 df, $p = 0.75$). There were 229 (26%) affected among those aged 75 to 84 at entry ($\chi^2 = 1.09$ with 2 d.f., $p = 0.13$). The first date of diagnosis in the interval from the time of entry into the cohort until December 31, 1994 was used in this comparison.

PROGNOSIS

ApoE genotype determined survival from disease onset for subjects aged 85+ ($\chi^2 = 7.38$ with 2 d.f., $p = 0.02$). There were 131 subjects diagnosed at ages 85+ during the follow-up interval: Three-month mortality was 8% for subjects carrying the ε2/3 genotype, 29% for ε3/3, and 40% for ε3/4—a fivefold variation. Prognosis was unrelated to ApoE polymorphism for occurrences at ages 75 to 84 (170 subjects, $\chi^2 = 1.09$, d.f. = 2, $p = 0.58$).

HEART DISEASE AND STROKE

ApoE genotypic predicted prognosis for both cardio- and cerebrovascular disease. FIGURE 1a displays prognosis for cardiovascular disease at ages 85+ in terms of survival from the date of diagnosis. There were 97 subjects: Three-month mortality was 5% for ε2/3, 31% for ε3/3, and 39% for ε3/4—a more than fivefold variation. Nonischemic heart disease diagnoses predominated and were found for 91 of 97 subjects. Only 28 of the 97 carried ischemic heart disease diagnoses.

FIGURE 1b displays the prognosis for cerebrovascular disease at ages 85+. Three-month mortality was 14%, 22%, and 39% for subjects having the respectively ApoE genotypes ε2/3, ε3/3, and ε3/4. Twelve subjects had both cardio- and cerebrovascular diagnoses.

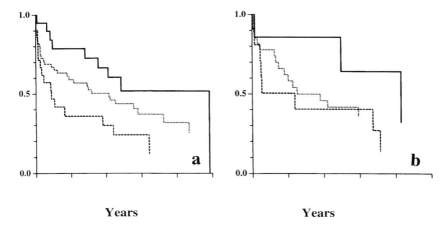

FIGURE 1. Survival after a diagnosis of cardiovascular (**a**) or cerebrovascular (**b**) disease at ages 85+. The three lines represent subjects with the ApoE ε2/3 genotype (*top*), ε3/3 (*middle*), and the ε3/4 genotype (*lower*).

GENDER AND COGNITION

ApoE genotype predicted prognosis at ages 85+ for women ($n = 97$) and for men ($n = 34$). Three-month mortality was 11%, 31%, and 39%, for the respective genotypes among women and 0%, 21%, and 43% for men. There were 35 subjects who were cognitively impaired when first examined as part of the cohort. Their three-month mortality was 17%, 45%, and 57%, depending on ApoE genotype. The 96 who were not cognitively impaired (i.e., MMSE score 24 to 30) had better prognosis: Mortality was 5%, 21%, and 40%.

DISCUSSION

Several population samples demonstrate that ApoE polymorphism predicts, in part, risk of death in late ages.[5,6] The Kungsholmen cohort demonstrated a two-year variation in life expectancy among the oldest old, age 85+, predicted by ApoE genotype: ε3/4 (worst) < ε3/3 < ε2/3 (best).[1,7] Closer investigation of that sample indicated that poor prognosis for vascular disorders, both heart disease and stroke, likely explains a large fraction of the difference in life expectancy: A third of the sample carried a diagnosis. There was fivefold variation in three-month mortality depending on ApoE genotype: ε3/4 (worst) < ε3/3 < ε2/3 (best). The prognostic variation was clinically meaningful ranging from 40% for ε3/4 to 8% for ε2/3.

REFERENCES

1. CORDER, E.H. *et al.* 1996. Apolipoprotein E genotype determines survival in the oldest old (85 years or older) who have good cognition. Arch. Neurol. **53:** 418–422.
2. FRATIGLIONI, L. *et al.* 1991. Prevalence of Alzheimer's disease in an elderly urban population: relationship with age, sex, and education. Neurology **41:** 1886–1892.
3. FRATIGLIONI, L. *et al.* 1992. Occurrence of dementia in advanced age: the study design of the Kungsholmen project. Neuroepidemiology **1:** 229–236.
4. WORLD HEALTH ORGANIZATION. 1977. International Classification of Diseases 1975. Revision. Vol 1. World Health Organization. Geneva.
5. SCHACHTER, F. *et al.* 1994. Genetic associations with human longevity at the APOE and ACE loci. Nat. Genet. **6:** 29–32.
6. LOUHIJA, J. *et al.* 1994. Ageing and genetic variation of plasma apolipoproteins: relative loss of the apolipoprotein ε4 phenotype in centenarians. Arterioscler. Thromb. **14:** 1084–1089.
7. CORDER, E.H. *et al.* 1997. Does apolipoprotein E polymorphism determine the risk or survivability of cardio- and cerebrovascular diseases? Implications for mortality differentials in late ages [abstract]. Med. Genet. **9:** P2.090.

MtDNA Deletions in Aging and in Nonmitochondrial Pathologies

A. CORMIO,[a,d] A.M.S. LEZZA,[a] J. VECCHIET,[b] G. FELZANI,[b] L. MARANGI,[c] F.W. GUGLIELMI,[c] A. FRANCAVILLA,[c] P. CANTATORE,[a] AND M.N. GADALETA[a]

[a]Department of Biochemistry and Molecular Biology, University of Bari, Via Orabona 4, 70125 Bari, Italy

[b]Department of Internal Medicine and Aging, "G. D'Annunzio" University, Chieti, Italy

[c]Department of Gastroenterology, University of Bari, 70100 Bari, Italy

INTRODUCTION

According to the "mitochondrial theory of aging," the age-related increase in re-active oxygen species (ROS) is responsible for the bioenergetic decay of mitochondria in aging through a mutagenic effect on mitochondrial DNA (mtDNA).[1] Such an effect should be more relevant in nervous tissue and skeletal muscle, which are post-mitotic tissues that are highly dependent on oxidative metabolism.

RESULTS AND DISCUSSION

To establish a reliable set of reference data for use as the healthy age-matched counterpart of potential or assessed pathologic cases with mitochondrial involvement, we performed a wide search for mtDNA deletions (mtDNA[4977], mtDNA[7436], mtDNA[10422], and others) in the skeletal muscle in five age-classes of healthy individuals (135 subjects). A similar search was carried out in different nonmitochondrial pathologic situations such as oculopharyngeal muscular dystrophy (OPMD) and cirrhosis with severe asthenia (AC).

A summary of the results in FIGURE 1 indicates that healthy individuals have an age-related increase in the number of mtDNA deleted species and in the average level of mtDNA[4977] (FIG. 1, bottom line). Comparison with a healthy age-matched class shows in both OPMD cases a higher number of deleted species and a higher level of mtDNA[4977] than in the counterpart.[2,3] This strongly suggests that even in nonmitochondrial pathologies of nuclear origin such as OPMD, in which no defect of the mitochondrial respiratory chain is detected, mitochondrial suffering may be present, causing an increase in mtDNA deletions. Conversely, comparison with age-matched healthy classes of the four AC cases shows similar numbers of mtDNA deletions and lower levels of mtDNA[4977] in patients as in their counterparts, suggesting that asthenia is probably related to a loss of muscle fibers rich in mitochondria, as also suggested by a reduction in the mtDNA/nuclearDNA ratio found in the same specimens

[d]Phone: +39-080-5443310; fax: +39-080-5443317-3403.
a.cormio@biologia.uniba.it

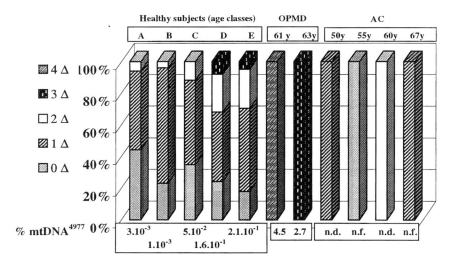

FIGURE 1. Frequency of subjects harboring different PCR-detected mtDNA deletions in the skeletal muscle. Data refer to three different groups: (1) 135 healthy subjects subdivided into five age-classes (A = <40 yr, B = 40–50 yr, C = 51–60 yr, D = 61–70 yr, E = >70 yr); (2) 2 OPMD patients; and (3) 4 cirrhotic patients with severe asthenia. *Bottom line* shows the mean value of the $mtDNA^{4977}$ expressed as $(mtDNA^{4977}/total\ mtDNA) \times 100$ in five age-classes of healthy subjects and in single patients. n.d. = not determined; n.f. = not found.

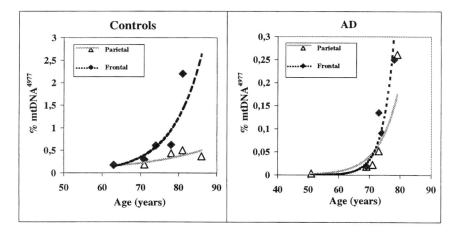

FIGURE 2. Increase with age of the $mtDNA^{4977}$ level in the frontal and parietal cortices of controls and AD patients.[4]

(V. Pesce, personal communication). In the brain, we measured the level of mtDNA^{4977} in autoptic samples from the parietal and frontal cortices of five aged healthy subjects and five patients with Alzheimer's disease (AD). Results (FIG. 2) show a similar exponential increase with age in the mtDNA^{4977} level in both lobes of controls and AD patients. However, values in the patients are always lower than those in age-matched controls, suggesting that the formation of mtDNA deleted molecules and, perhaps, the overall replication of mtDNA, eventually due to a high level of 8-hydroxydeoxyguanosine in the patient's mtDNA, may be hampered in the disease.[4]

In conclusion, our data support the quantitative and qualitative increase of mtDNA deletions in aging. Furthermore, mtDNA deletions may be a good marker of mitochondrial suffering in nonmitochondrial pathologies.

ACKNOWLEDGMENTS

This work was accomplished partially with funds from ESF, P.O.P. '94–'99.

REFERENCES

1. LINNANE, A.W. *et al.* 1989. Lancet **i:** 642–645.
2. LEZZA, A.M.S. *et al.* 1997. FEBS Lett. **418:** 167–170.
3. LEZZA, A.M.S. *et al.* 1994. Biochem. Biophys. Res. Commun. **205:** 772–779.
4. LEZZA, A.M.S. *et al.* 1999. FASEB J. **13:** 1083–1088.

Transcriptome and Proteome Analysis in Human Senescent Fibroblasts and Fibroblasts Undergoing Premature Senescence Induced by Repeated Sublethal Stresses

JEAN-FRANÇOIS DIERICK,[a,b,c,d,e] THIERRY PASCAL,[b,e]
FLORENCE CHAINIAUX,[b] FRANÇOIS ELIAERS,[b] JOSÉ REMACLE,[b]
PETER MOSE LARSEN,[c] PETER ROEPSTORFF,[d] AND OLIVIER TOUSSAINT[a,b]

[b]Department of Biology, Unit of Cellular Biochemistry and Biology, University of Namur (FUNDP), 5000 Namur, Belgium

[c]The Centre for Proteome Analysis in Life Sciences, International Science Park Odense, 5230 Odense M, Denmark

[d]Department of Molecular Biology, University of Southern Denmark, 5230 Odense M, Denmark

INTRODUCTION

Several studies have shown that a few days to several weeks after sublethal stresses, *in vitro* human diploid fibroblasts (HDFs) keep marks of their exposure to stresses of nature as different as exposure to free radicals, ethanol, and UV radiation. Human diploid fibroblasts exposed to sublethal stresses under reactive oxygen species (H_2O_2, *tert*-butylhydroperoxide (*t*-BHP), hyperoxia), or ethanol irreversibly display various biomarkers of replicative senescence.[1] From these findings, it has been proposed that sublethal stresses could induce a premature replicative senescence.[2]

In this work, using both transcriptome and proteome analysis, we first compared the gene and protein expression of HDFs at early cumulative population doublings, termed young, and HDFs at the end of their replicative potential, termed senescent. Second, we compared the gene and protein expression of young cells two days after exposure to five successive sublethal stresses under 30 μM *t*-BHP or 5% ethanol (vol/vol) versus control cells. Transcriptome analysis was performed by RT-PCR differential display (DD RT-PCTR) and proteome analysis by high-resolution, two-dimensional gel electrophoresis (2DGE).

[a]Address for correspondence: Olivier Toussaint, Department of Biology, Unit of Cellular Biochemistry and Biology, University of Namur (FUNDP), Rue de Bruxelles, 61 B-5000 Namur, Belgium. Phone: +32 (0) 81 72 41 32; fax : +32 (0) 81 72 41 35.
olivier.toussaint@fundp.ac.be
[e]These authors contributed equally to this work.

MATERIALS AND METHODS

Cell Culture and Stress Conditions Inducing Premature Senescence

Confluent WI-38 normal human diploid lung fibroblasts at early population doublings were exposed five times to noncytotoxic concentrations of 30 µM *t*-BHP (1 hour) or 5% ethanol (vol/vol) (2 hours) diluted in culture medium (BME) with 10% (vol/vol) fetal calf serum (FCS). After each stress, the cells were rinsed with BME and then replaced in BME + 10% FCS before the next stress. Noncytotoxic stress conditions were selected by estimating the cell survival by assaying cellular protein using the Lowry method. The proportion of cells positive for the senescence-associated β-galactosidase activity[3] was determined. After the last stress, the cells were allowed to recover either for 48 hours before protein labeling or for 72 hours before RNA extraction.

Differential Display RT-PCR (DD RT-PCR)

The RNA samples were submitted to DNase treatment before RT-PCR amplification and labeling with [33]P dATP. The differential display was performed as described in Liang and Pardee.[4]

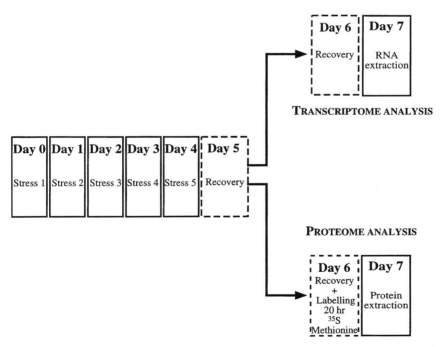

FIGURE 1. **Presentation of the stress model used.** Confluent HDFs were exposed to daily sublethal stresses consisting of *t*-BHP, 30 µM, for 1 hour or 5% vol/vol ethanol for 2 hours with one stress per day for 5 days. Transcriptome and proteome analyses were performed at 72 hours after the end of the last stress.

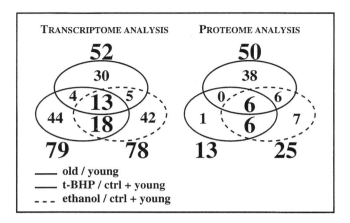

FIGURE 2. After transcriptome analysis, among the approximately 2000 bands located, 52 showed a significant differential expression between young and senescent cells, 79 between *t*-BHP-treated and control HDFs and 78 between ethanol-treated and control cells. During the proteome analysis, among the 1819 spots located, 50 showed a significantdifferential expression between young and senescent cells, 13 between *t*-BHP-treated and control HDFs and 25 between ethanol-treated and control cells.

Two-Dimensional Gel Electrophoresis

At 48 hours after the end of the last stress or trypsination, the cells were labeled for 20 hours with 150 µCi [^{35}S]methionine/well and lysed with a 9.5 mol/l urea, 2% (wt/vol) Nonidet P-40, 5% (vol/vol) β-mercaptoethanol and 2% (vol/vol) ampholyte pH 7–9 buffer. For each situation independent triplicates were prepared. The whole experiment was repeated twice with different batches of reagents and cells in order to discriminate among the reproducible changes, ending up with 6 samples per situation. The two-dimensional gel electrophoresis (2DGE) was performed as described in Byrjalsen *et al.*[5] Image acquisition was performed with a phosphorimager (Agfa Diagnostic Centre, Agfa-Gevaert AG, Germany), and the analysis was done with the Bio-Image system (Millipore, USA).

For each spot the integrated intensity was calculated as the ratio in percent between its intensity and the sum of the intensities given by all the spots located on this gel (%I.I.). A statistical analysis based on a *t*-test ($n = 6$, $p > 0.05$) allowed us to discriminate between the spots with a significant difference in %I.I. between two situations. Then the spots that present a ratio of %I.I. > 30% between those two situations were selected. We also checked that the %I.I. of the selected spots were not significantly different ($n = 3$, $p > 0.05$) between the two sets of three samples obtained in each situation.

RESULTS

DD RT-PCR allowed us to make comparisons of the intensities of approximately 2000 bands, and at the proteome level 1819 spots were followed in each experimental condition.

Among the bands/spots presenting a variation of intensity in at least one situation (156 in DD RT-PCR and 64 in 2DGE), only a small number showed a similar variation during senescence and after the two types of stress performed (13 in DD RT-PCR and 6 in 2DGE) (FIG. 2). On the other hand, 18 bands and 6 spots displayed a variation of intensity specific to the long-term effects of the two kinds of sublethal stresses studied.

CONCLUSION

From the patterns of DD RT-PCR bands and 2DGE, these preliminary results suggest that HDFs treated under sublethal stresses and senescent HDFs are not identical cellular states. The next aim of this study will be to identify the genes/proteins that share common expressional changes in senescent HDFs when compared to HDFs undergoing stress-induced premature senescence. Indeed these genes/proteins may be responsible for the establishment of the observed phenotype. Then, we shall identify the genes/proteins that display common expressional changes in cells undergoing stress-induced premature senescence because they may represent long-term molecular scars of sublethal stresses independently of the nature of the applied stress.

ACKNOWLEDGMENTS

J-F. Dierick has a fellowship from the FRIA, Belgium and O. Toussaint has a fellowship from the FNRS, Belgium.

REFERENCES

1. TOUSSAINT, O. *et al.* 1998. Reciprocal relationships between the resistance to stresses and cellular aging. Ann. N.Y. Acad. Sci. **851:** 450–465.
2. CHEN, Q. & B.N. AMES. 1994. Senescence-like growth arrest induced by hydrogen peroxide in human diploid fibroblast F65 cells. Proc. Natl. Acad. Sci. USA **91:** 4130–4134.
3. DIMRI, G.P. *et al.* 1995. A biomarker that identifies senescent human cells in culture and in aging skin in vivo. Proc. Natl. Acad. Sci. USA **92:** 9363–9367.
4. LIANG, P. & A.B. PARDEE. 1992. Differential display of eukaryotic messenger RNA by means of the polymerase chain reaction. Science **257:** 967–971.
5. BYRJALSEN, I. *et al.* 1995. Human endometrial proteins with cyclic changes in the expression during the normal menstrual cycle: characterization by protein sequence analysis. Hum. Reprod. **10:** 13–18.

Human Diploid Fibroblasts Display a Decreased Level of *c-fos* mRNA at 72 Hours after Exposure to Sublethal H₂O₂ Stress

PATRICK DUMONT,[a,b] MAGGI BURTON,[b] QIN M. CHEN,[c]
CHRISTOPHE FRIPPIAT,[b] THIERRY PASCAL,[b] JEAN-FRANÇOIS DIERICK,[b]
FRANÇOIS ELIAERS,[b] FLORENCE CHAINIAUX,[b] JOSÉ REMACLE,[b]
AND OLIVIER TOUSSAINT[a,b]

[b]*The University of Namur (FUNDP), Department of Biology, Laboratory of Cellular Biochemistry and Biology, Rue de Bruxelles, 61, B-5000 Namur, Belgium*

[c]*University of Arizona, Pharmacology & Toxicology, 1703 E. Mabel Street, Tucson, Arizona 85721, USA*

INTRODUCTION

Human diploid fibroblasts (HDFs) exposed to sublethal stresses under *tert*-butyl-hydroperoxide display a very low proliferative rate and morphological changes[1,2] typical of senescent cells at 72 hours after the stresses. They also display the senescence-associated β-galactosidase activity,[2] overexpress senescence-associated genes,[2] and fail to phosphorylate the retinoblastoma protein after serum stimulation.[2] HDFs stressed under 450 μM H₂O₂ also undergo premature replicative senescence.[3] The increase in the mRNA level of the cyclin-dependent kinase inhibitors (CDKIs) p21^{waf-1}, p16^{INK-4a}, and p14/15^{INK-4b} partly explains the irreversible growth arrest occurring in HDFs at 72 hours after sublethal H₂O₂ stress.[4] In addition to the overexpression of CDKIs, senescent HDFs also display a decrease in the mRNA of *c-fos*, a component of the heterodimeric AP-1 (activator protein 1) transcription factors, positively regulating cell growth.[5] In this study, we tested whether H₂O₂-treated HDFs also downregulate their level of *c-fos* mRNA at 72 hours after stress, using semiquantitative RT-PCR.

MATERIALS AND METHODS

Cell Culture

IMR-90 HDFs (Cornell Institute for Medical Research, USA) were classically cultivated in 100-mm culture dishes (Corning, Corning, NY) containing 10 ml of Dulbeco's modified Eagle medium (DMEM) + 10% fetal calf serum (FCS) (Flow Laboratories, U.K.).

[a]Address for correspondence: Fax: +32 81 724135.
patrick.dumont, olivier.toussaint@fundp.ac.be

TABLE 1. Primers (A) and RT-PCR conditions (B) used in this study

A.	Gene	Sequence	Positions
	c-fos	5'-ACGCAGACTACGAGGCGTCA-3'	908–927 (exon 1)
		5'-TTCACAACGCCAGCCCTGGA-3'	1960–1979 (exon 2)
	GAPDH	5'-CGTCTTCACCACCATGGAGA-3'	360–379
		5'-CGGCCATCACGCCACAGTTT-3'	640–659

B.	Gene	
	c-fos	*GAPDH*
RT	48°C/60 min	48°C/60 min
PCR		
Denaturation	94°C/2 min	94°C/2 min
Cycle		
Denaturation	94°C/30 sec	94°C/30 sec
Annealing	62°C/30 sec	60°C/30 sec
Elongation	68°C/45 sec	68°C/45 sec
Termination	68°C/7 min	68°C/7 min

Semiquantitative RT-PCR

Total RNA was extracted with Trizol (Gibco, USA). Fifty nanograms total RNA were used in each RT-PCR performed with the Access RT-PCR kit (Promega, USA) in the presence of 0.025 µCi [α-^{32}P] dCTP and 50 pmoles of each primer. Primers and RT-PCR conditions are described in TABLE 1. Glyceraldehyde-3-phosphate dehydrogenase (*GAPDH*) was used as reference gene as previously described in this experimental model.[4] The common range of exponential PCR amplification was determined as 27 cycles for *c-fos* and *GAPDH*. RT-PCR products were electrophoresed on 5% polyacrylamide gels. The dried gels were analyzed on an Instant Imager (Packard Instruments, USA) and exposed on β-max autoradiography films (Amersham, U.K.). The bands were quantified, and the ratio between the quantifications obtained for *c-fos* and *GAPDH* was calculated. The results were normalized, considering *c-fos* mRNA level in HDFs at early cumulative population doublings (CPDs) as 100%.

RESULTS AND DISCUSSION

As expected, *c-fos* mRNA was poorly induced in IMR-90 HDFs at late CPDs (CPDs > 50; between 95% and 100% of proliferative life span) stimulated for growth with fresh medium + 10% FCS (FIG.1). The amount of the *c-fos* mRNA was found to be about threefold higher in HDFs at early CPDs when compared to their aged counterparts. Subconfluent IMR-90 HDFs at early CPDs (CPDs < 25; between 40% and 50% of proliferative life span) were exposed for 2 hours to 450 µM H_2O_2 diluted in culture medium + 10% FCS and then allowed to recover for 72 hours. Interestingly, these cells also expressed a low *c-fos* mRNA level when compared to the untreated control cells at similar early CPDs (FIGS.1 and 2).

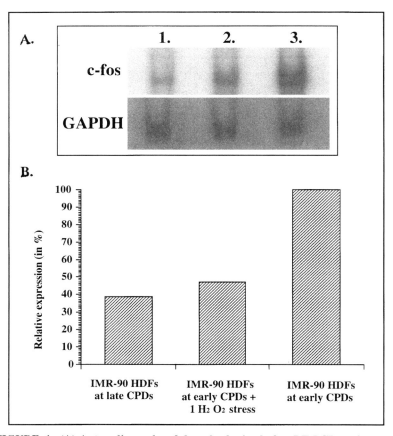

FIGURE 1. (A) Autoradiography of the gels obtained after RT-PCR performed on c-fos and GAPDH mRNA. For both genes, bands of the expected lengths were obtained (*c-fos*: 317 bp; *GAPDH*: 299 bp) given the sets of primers. Negative controls (RT-PCR without template) were performed (not shown). *Lane 1*: IMR-90 HDFs at late CPDs. *Lane 2*: IMR-90 HDFs at early CPDs exposed for 2 hours to 450 μM H_2O_2 diluted in DMEM + 10% FCS. *Lane 3*: control HDFs at early CPDs, for which culture medium was changed for 2 hours. **(B) Quantifications of the bands.** The bands were quantified, and the ratio between the quantifications obtained for *c-fos* and *GAPDH* was calculated. The signal obtained in non-stressed HDFs at early CPDs was considered to be 100%.

After activation, AP-1 (c-fos, c-Jun) induces the transcription of genes required for entering the S phase of the cell cycle. Growth factors induce the transcription of the *c-fos* mRNA in quiescent cells at early CPDs, whereas this induction disappears in senescent cells. From our results, the decrease in *c-fos* mRNA level at 72 hours after sublethal H_2O_2 stress is suggested to participate in the H_2O_2-induced irreversible growth arrest, together with the overexpression of several CDKIs, as observed in replicative senescence.

ACKNOWLEDGMENTS

P. Dumont, C. Frippiat, F. Chainiaux, and J.-F. Dierick have a fellowship from the FRIA, Belgium. O. Toussaint is a Scientific Collaborator of the FNRS, Belgium. We thank the European Union Biomed and Health Research Programme, Concerted Action Programme Molgeron (BMH1 CT94) and Shared-Cost Action Programme "Genage" (BMH2 CT98), the NATO scientific commissions, and the Fulbright programme.

REFERENCES

1. TOUSSAINT, O., A. HOUBION, *et al.* 1992. Aging as a multi-step process characterized by a lowering of entropy production leading the cell to a sequence of defined stages. II. Testing some predictions on aging human fibroblasts in culture. Mech. Ageing Dev. **65:** 65–83.
2. DUMONT, P., M. BURTON, *et al.* 2000. Induction of replicative senescence biomarkers by sublethal oxidative stresses in normal human fibroblast. Free Rad. Biol. Med. **28:** 361–373.
3. CHEN, Q.M., J.C. BARTHOLOMEW, *et al.* 1998. Molecular analysis of H_2O_2-induced senescent-like growth arrest in normal human fibroblasts: p53 and Rb control G1 arrest but not cell replication. Biochem. J. **332:** 43–50.
4. FRIPPIAT, C., Q.M. CHEN, *et al.* 2000. Cell cycle regulation in H_2O_2-induced premature senescence of human diploid fibroblasts. Exp. Gerontol. In press.
5. SESHADRI, T. & J. CAMPISI. 1990. Repression of *c-fos* transcription and an altered genetic program in senescent human fibroblasts. Science **247:** 205–209.

Replicative Senescence of Interleukin-2-Dependent Human T Lymphocytes

Flow Cytometric Characteristics of Phenotype Changes

EWA JARUGA,[a] JANUSZ SKIERSKI,[b] EWA RADZIESZEWSKA,[a] AND EWA SIKORA[a,c]

[a]Molecular Bases of Aging Laboratory, Nencki Institute of Experimental Biology, Polish Academy of Sciences, Warsaw, Poland

[b]Flow Cytometry Laboratory, Drug Institute, Warsaw, Poland

The dramatic decline in immune function with age in T-cell proliferative activity has been documented in animal models and in numerous studies of the elderly. A similar proliferative decline is also seen in human interleukin (IL-2)-dependent T-lymphocyte cultures. Results from a number of groups demonstrate that normal human T lymphocytes have a finite lifespan even if the number of doublings varies depending on culture conditions.[1] Previously, we showed that under our conditions polyclonal T cells cease proliferation after about three weeks.[2] Now, we present results of a more detailed analysis of growth *in vitro* as well as phenotypic changes of T cells.

FIGURE 1 shows rates of growth and cell cycle analysis of T-cell cultures. T lymphocytes were isolated from the blood of young (20–30 years old), healthy donors. After T-cell stimulation with phytohemagglutinin (PHA 10 µg/ml) for 3 days, followed by continuous cultivation in the presence of IL-2 (10 IU/ml) at a density that provided the conditions for logarithmic growth, cells incorporated [^{3}H]thymidine at the same high level until the 16th–18th day of culture (young). Afterwards, a progressive decline of [^{3}H]thymidine incorporation (presenescent) to a level almost equal for nonstimulated cells (about 30th day, senescent) was observed (FIG.1A). Population doublings per 24 hours was highest at the beginning of the culture (0.8 in 5th day) and then progressively slowed down to zero after the 25th day of culture (FIG.1B). Cell cycle analysis showed that about 20% of cells were in the S phase until the 17th day of culture (FIG. 1C), which corresponds to the results of [^{3}H]thymidine incorporation. The highest number of mitotic cells (phase G2/M; 10%) was obtained during the first week of culture. All nondividing senescent cells were stopped in the G_1 phase (FIG. 1D). Altogether, these results prove that cessation of T-cell growth is gradual and that the growth retardation proceeds through an S phase of longer duration.

It can be supposed that after cessation of proliferation cells undergo death by apoptosis. Assuming that the T cells in the culture have an S phase of fixed duration, one could expect that massive apoptosis would occur at the end of culture. Nonethe-

[c]Corresponding author: esik@nencki.gov.pl

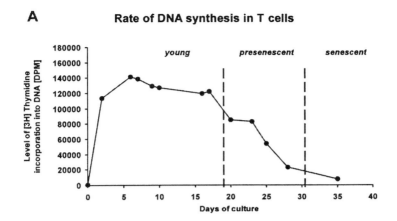

A   **Rate of DNA synthesis in T cells**

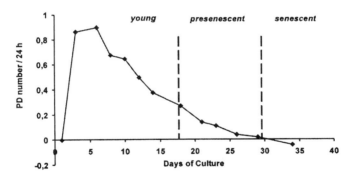

B   **Number of T cell population doublings per 24 h**

FIGURE 1A and B. *In vitro* **growth characteristics of T lymphocytes. (A)** [³H] Thymidine incorporation into DNA. Cells (1.4×10^5) at a density of 0.7×10^6 cells/ml were incubated with [³H]thymidine (2.5 mCi/ml, specific activity 27 Ci/mmol) for 20 hours. The
radioactivity level [DPM] of the cells harvested with a SKATRONAS COMBI cell harvester
was measured with a Beckman LS 6500 counter. The rate of DNA synthesis was monitored
at every culture passage. **(B) Density of T lymphocytes.** To establish the number of cell population doublings per 24 hours, the cell density was monitored each day of the culture. At
the same time, the percentage of the sub-G_1 fraction was measured by flow cytometry in
cells stained with DAPI (1 mg/ml, Molecular Probes) and sulforhodamine 101 (20 mg/ml,
Molecular Probes) (not shown).

less, we observed that the sub-G_1 fraction did not exceed 8% during the whole period
until the 30th day of culture. The only exception was third day of culture in which
cells were PHA-treated (15% of sub-G_1) (not shown). This observation provides additional proof that replicative senescence of T cells is due to a gradual prolonging of
cell cycle phase duration.

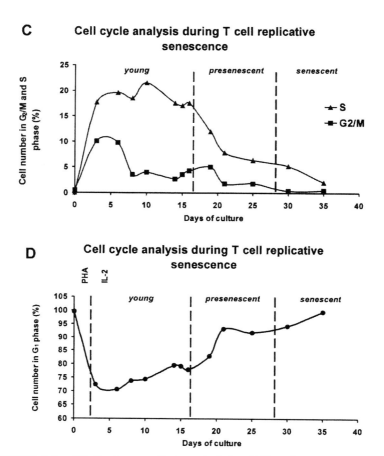

FIGURE 1C and D. *In vitro* growth characteristics of T lymphocytes. T-cell distributions in the cell cycle. Cell distribution in the cell cycle was analyzed by DAPI and sulforhodamine staining, using a Vantage Flow Cytometer (Becton Dickinson). The percentage of cells representing the G_1, S, and G_2/M phases was established using MacCycle Programme; and the mean numbers of cells in the given phase were plotted against the culture duration.

During *in vitro* T-lymphocyte growth, a quite rapid selection to CD3CD8 cells occurs. In the pool of young cells, we observed 30% CD4, but in the presenescent and senescent populations of cells the majority of them (above 90%) were CD8. Although senescent cells, despite having IL-2 in the medium, are no longer dividing, we checked the alpha-chain IL-2 receptor (CD25) expression. In young and presenescent cells, about one-third of the cell population was CD25 positive, but only 15% CD25 was present in the pool of senescent cells. The same distribution of CD25-positive cells as in the whole population was observed in the CD8-positive subpopulation. Although we did not observe dramatic differences in the percentage of CD25-positive cells between the young and presenescent and senescent subpop-

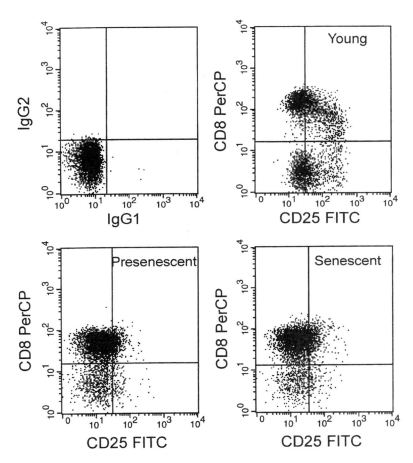

FIGURE 2. Distribution of CD25-positive cells in the population of CD8-positive lymphocytes. Expression of CD25 in the CD8-positive subpopulation of T lymphocytes was analyzed by an immunofluorescence assay that was performed by incubating the T cells with anti-CD25 (FITC, FL-1) and anti-CD8 (PerCP, FL-3) according to Becton Dickinson procedures for Simultest™. Double negative control of cells stained with IgG1 and IgG2 is also shown.

ulations, the analysis of cell brightness reflecting the amount of anti-CD25 antibodies bound to the surface of a given cell revealed that the number of CD25 receptors on a single young cell is higher than that observed on presenescent and senescent cells (FIG. 2). As expected, CD25 receptors were present on the surfaces of dividing cells, and these are distinguished on the FL-1 (FITC-CD25) versus FL-2 (PI) plot as a subpopulation of CD25-positive cells (FL-1, FITC) shifted to higher values of FL-2 corresponding to S and G2/M phases, shown by staining the cells with propidium iodide and anti-CD25 (not shown).

Our results show that the selection to CD8 T cells takes place during *in vitro* T-cell growth. These cells can be characterized as having a progressively longer S phase, a gradual loss of alpha-chain IL-2 receptor (CD25), and the same propensity as young cells to undergo apoptosisis.

REFERENCES

1. EFFROS, R.B. 1996. Insights on immunological aging derived from the T lymphocyte cellular senescence model. Exp. Gerontol. **31:** 21–27.
2. RADZISZEWSKA, E., K. PIWOCKA, J. SKIERSKI & E. SIKORA. 1999. UVC-induced cell death of IL-2-dependent human lymphocytes. Cell Biol. Int. **23:** 97–103.

Age-Related Changes in Irradiation-Induced Apoptosis and Expression of p21 and p53 in Crypt Stem Cells of Murine Intestine

KAREEN MARTIN,[b,c] CHRISTOPHER S. POTTEN,[a,c] AND
THOMAS B.L. KIRKWOOD[b]

[b]Biological Gerontology Group, University of Manchester, 3.239 Stopford Building,
Oxford Road, Manchester M13 9PT, United Kingdom

[c]CRC Epithelial Biology Group, Paterson Institute for Cancer Research, Christie
Hospital NHS Trust, Wilmslow Road, Manchester M20 4BX, United Kingdom

Aging is associated with a progressive deterioration of many organs within the body. In tissues with high cell turnover, the maintenance of the stem cells is of particular importance. Any accumulation of damage in stem cells may affect their progeny and function and hence threaten the homeostasis and regenerative capacity of the tissue. Unfortunately, stem cells do not have any specific markers, but the intestine is a useful model for studying stem cells and their function because of its hierarchical and spatial organization.[1,2] The stem cells produce cells that divide rapidly and migrate outwards, differentiating into mature cells on the surface of the villi, from where they are eventually shed into the lumen of the gut. Stem cells are particularly sensitive to damage, and previous studies have shown that low doses of irradiation induced an acute apoptotic response shortly after exposure to the radiation, which was found to be *p53* dependent.[3,4] We recently showed that in old mice (29 months) the level of apoptosis following low-dose irradiation was increased twofold compared with young (5 months) and middle-aged (15–18 months) mice.[5]

To further study the molecular basis of such events, the levels of *p53* and *p21* expression were determined by immunohistochemistry in crypts from the proximal and distal small intestines of young (6–7 months) and old (28–30 months) ICRFa mice irradiated with 1 Gy of γ irradiation. The animals were then culled at different times after exposure, that is, at 2, 4.67, 9, 12, 18, and 24 hours. Some sections were stained with hematoxylin and eosin to enable the assessment of the cell positional incidence of apoptosis. Apoptotic and *p53*- and *p21*-positive cells were scored in good longitudinal crypt sections, and the frequency of immunohistochemistry positive cells was plotted against cell position.

The apoptotic index rose quickly, reaching a peak at 4 to 9 hours (FIG. 1 and TABLE 1). The apoptotic response was enhanced in old mice, with significant differences at 4.7 hours. The level of apoptosis declined between 9 and 12 hours. Contrary to young mice, old mice exhibited a second wave of apoptosis at 24 hours. Expression of *p53* was found to occur rapidly and preferentially at the stem cell position

[a]Address for correspondence: phone: +44 161 446 3179; fax: +44 161 446 3181.
kmartin@picr.man.ac.uk

Apoptosis

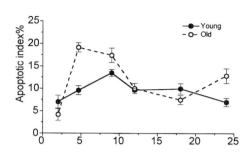

p53

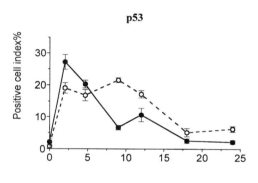

p21

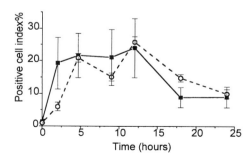

FIGURE 1. Curves showing the average number of apoptotic, *p53-*, and *p21-*positive cells as a function of time after 1 Gy of γ irradiation in the small intestine of young (*solid line*) and old (*dashed line*) ICRFa mice.

TABLE 1. Apoptosis, p53 and p21 expressions in young and old mice after exposure to 1Gy of γ-irradiation[a]

Time (hr)	Age (mo)	Mice (n)	Apoptosis				p53 expression				p21 expression			
			AI%	Max AI%	Peak AI%	AI% at position 3–7	PI%	Max PI%	Peak PI%	PI% at position 3–7	PI%	Max PI%	Peak PI%	PI% at position 3–7
0	Young	4					2.1	6.5	2	2.6	1.9	2.9	9	1.0
	Old	4					0.8	1.5	2	1.1	1.4	2.1	11	0.7
2	Young	3	7.0	8.3	8	8.1	27.2	53.3	4	49.3	21.1	29.7	7	25.1
	Old	3	4.2	7.1	6	6.1	19.1	29.5	5	29.0	7.0	9.9	7	8.0
4.67	Young	5	9.6	29.9	5	23.4	20.3	39.2	3	35.8	22.6	35.7	8	32.6
	Old	5	19.1	40.9	5	39.3	16.9	31.7	5	29.0	21.4	34.5	5	32.1
9	Young	4	13.5	22.1	4	21.7	6.6	12.6	6	12.4	21.8	34.2	5	32.4
	Old	3	17.4	29.6	5	29.8	21.5	39.1	4	36.1	16.4	27.8	6	24.8
12	Young	4	9.7	19.1	4	18.0	10.6	22.1	2	20.1	24.4	37.3	9	29.9
	Old	3	10.0	17.0	4	16.5	ND	ND	ND	ND	ND	ND	ND	ND
18	Young	4	10.0	17.3	5	17.1	2.5	4.42	3	3.8	15.7	15.0	8	12.4
	Old	4	7.5	14.3	6	13.0	ND	ND	ND	ND	16.0	25.3	8	20.0
24	Young	4	7.0	11.4	7	11.0	2.1	4.0	8	3.0	10.4	13.2	8	10.8
	Old	4	12.9	20.4	6	19.5	6.3	9.8	9	8.2	10.8	19.9	8	9.7

[a]Mean values for the average number of apoptotic cells (AI%), p53 and p21 positive cells per crypt (PI%), maximum apoptotic (max AI%) and positive indices (max PI%), cell position of the maximum apoptotic or positive indices (Peak), and the average of positive indices at cell position 3 to 7 from the base of the crypt in ICRFa male mice from different age groups in the proximal and distal small intestine and in both regions. ND = not done. Numbers in **bold** represent significant differences between values for young and old animals ($p < 0.05$).

after irradiation, with peak values at 2 hours (FIG. 1 and TABLE 1). In older mice, the level of *p53* expression also increased, but the peak was lower than in younger mice at 2 hours and higher at later times than in young mice. *p21* expression increased rapidly and reached a plateau, decreasing only at the later time (FIG. 1 and TABLE 1). *p21* expression was found to be slightly delayed in old mice. In both ages, *p21* expression was mainly found at about cell position 7.

DISCUSSION

These results showed that a clear alteration in the apoptotic response to ionizing radiation occurs with age in the mouse intestinal crypt cells. Crypt stem cells from older mice were more sensitive to irradiation, suggesting an alteration with age in the regulation of the pathway responsible for detection of damage and the decision between repair and in induction of cell death. *p53* expression has been described as playing a key role in the induction of apoptosis following exposure to ionizing radiation.[3,4,6] *p53* expression correlated with the first wave of apoptosis, whereas the second wave was *p53*-independent. *p53* inhibits cell-cycle progression (via *p21*) while damaged DNA is repaired. If the damage is too extensive, p53 may induce apoptosis.[7] The delayed *p53* expression in older animals implies a delay in DNA damage screening and repair, which is in accordance with the delayed expression of *p21*. Unrepaired cells that escaped from the first apoptosis peak may continue to progress into S phase, and they would reach the G_2/M checkpoint with DNA damage. The elevated first apoptotic peak and the presence of a second peak at 24 hours, together with the delayed expression of both *p53* and *p21*, clearly reveal an alteration in the cellular response to injury in the murine intestinal crypts in older mice.

REFERENCES

1. POTTEN, C.S. & M. LOEFFLER. 1990. Stem cells: attributes, cycles, spirals, pitfalls and uncertainties. Lessons for and from the crypt. Development **110:** 1001–1020.
2. POTTEN, C.S. *et al.* 1997. The intestinal epithelial stem cell: the mucosal governor. Int. J. Exp. Pathol. **78:** 219–243.
3. CLARKE, A.R. *et al.* 1997. *p53*, mutation frequency and apoptosis in the murine small intestine. Oncogene **14:** 2015–2018.
4. MERRITT, A.J. *et al.* 1997. Apoptosis in small intestinal epithelial from *p53*-null mice: evidence for a delayed, *p53*-independent G2/M-associated cell death after gamma-irradiation. Oncogene **14:** 2759–2766.
5. MARTIN, K. *et al.* 1998. Age changes in stem cells of murine small intestinal crypts. Exp. Cell. Res. **241:** 316–323.
6. ARAI, T. *et al.* 1996. Comparative alterations in *p53* expression and apoptosis in the irradiated rat small and large intestine. Br. J. Cancer **74:** 406–412.
7. WILSON, J.W. *et al.* 1998. Radiation-induced *p53* and *p21*WAF-1/CIP1 expression in the murine intestinal epithelium: apoptosis and cell cycle arrest. Am. J. Pathol. **153:** 899–909.

Testing Evolutionary Theories of Aging

GAWAIN McCOLL,[a] NICOLE L. JENKINS, DAVID W. WALKER, AND
GORDON J. LITHGOW

School of Biological Sciences, University of Manchester, Oxford Road, United Kingdom

The physiological and molecular processes that specify life span await complete elucidation, despite numerous theories of the evolutionary origins of aging. Aging is considered to be nonadaptive, a product of decreasing natural selection with age.[1] The pleiotropy theory proposes that aging results from the late-life detrimental effects of alleles selected via beneficial effects in early life.[1] Furthermore, life history theory predicts that a trade-off may occur between extensions in longevity and reductions in early life traits such as fecundity.[2] Evidence for this theory has been provided by the correlated responses observed in laboratory selection experiments in *Drosophila*. For example, populations selected for increased longevity have decreased fecundity,[3] and populations selected for increased stress resistance have increased longevity and decreased fecundity.[4]

These previous studies, however, are associative. Pleiotropy, which implicates the multiple effects of a single gene, has yet to be rigorously tested. Although many genes contribute to life span, the recent identification of single genes of major effect in *Caenorhabditis elegans* allows this theory of aging to be re-examined.

As a model organism, *C. elegans* provides an experimental system well suited to testing evolutionary theories of aging. Reproducing as facultative, self-fertilizing hermaphrodites, isogenic populations of *C. elegans* can be maintained without inbreeding depression of fitness traits.[5] Multiple genetic pathways determine the life span of *C. elegans*.[6] In particular, an insulin-like signaling pathway has been identified that has several components with demonstrable effects on life span.[7] This pathway provides a defined molecular and mechanistic basis to examine potential trade-offs between longevity and life history traits.

We have used a mutation of the *C. elegans age-1* gene (encoding a phosphatidylinositol 3-kinase in the insulin-like signal pathway), which exhibits markedly increased mean and maximum life span with little or no apparent cost under standard (20°C) culturing conditions.[8] Direct comparisons of competitive fitness were made in mixed populations of wild-type and mutant genotypes.[9] Changes in allele frequency across generations were assessed by utilizing the altered development at 27°C of mutants to a morphologically distinct (dauer) larval stage. Under conditions of environmental stress (cyclical starvation) a life history trade-off was apparent.[9] These findings support the pleiotropy theory of aging.

[a]Address for correspondence: Gawain McColl, School of Biological Sciences, University of Manchester, 3.239 Stopford Building, Oxford Road, Manchester M13 9PT, United Kingdom.

REFERENCES

1. MEDAWAR, P.B. 1946. Old age and natural death. Mod. Quart. **1:** 30–56.
2. WILLIAMS, G.C. 1957. Pleiotropy, natural selection, and the evolution of senescence. Evolution **11:** 398–411.
3. ZWAAN, B., R. BIJLSMA & R.F. HOEKSTRA. 1995. Direct selection of life span in *Drosophila melanogaster*. Evolution **49:** 649–659.
4. PARSONS, P.A. 1995. Stress resistance and longevity: a stress theory of ageing. Heredity **75:** 216–221.
5. JOHNSON, T.E. & E.W. HUTCHINSON. 1993. Absence of strong heterosis for life span and other life history traits in *Caenorhabditis elegans*. Genetics **134:** 465–474.
6. HEKIMI, S., B. LAKOWSKI, T.M. BARNES & J.J. EWBANK. 1998. Molecular genetics of life-span in *C. elegans*: how much does it teach us? TIG **14:** 14–20.
7. THOMAS, J.H. & T. INOUE. 1998. Methuselah meets diabetes. Bioessays **20:** 113–115.
8. JOHNSON, T.E., P.M. TEDESCO & G.J. LITHGOW. 1993. Comparing mutants, selective breeding, and transgenics in the dissection of aging processes of *Caenorhabditis elegans*. Genetica **91:** 65–77.
9. WALKER, D.W., G. McCOLL, N.L. JENKINS, *et al.* 2000. Evolution of lifespan in *C. elegans*. Nature. In press.

A Novel *in Vitro* Model of Conditionally Immortalized Human Vascular Smooth Muscle Cells

A Tool for Aging Studies

HARRIS PRATSINIS,[a] CATHERINE DEMOLIOU-MASON,[b] ALUN HUGHES,[b] AND
DIMITRIS KLETSAS[a,c]

[a]*Institute of Biology, National Centre for Scientific Research "Demokritos,"
153 10, Athens, Greece*

[b]*Department of Clinical Pharmacology, National Heart & Lung Institute, Imperial
College of Science Technology and Medicine, St.Mary's Hospital, London W2 1NY, UK*

INTRODUCTION

It is well established that serial *in vitro* propagation of normal human cells eventually leads to replicative senescence.[1] However, certain viral and cellular oncogenes have the ability to immortalize normal cells. Conditionally immortalized rodent[2] or human[3] fibroblast lines have proved very useful for the study of cellular senescence.

We have developed a conditionally immortalized human vascular smooth muscle cell (HVSMC) line, designated SM1, after transfection of primary vein-derived SMCs with a nonreplicative retroviral vector containing a temperature-sensitive (tsA58) mutant of simian virus 40 (SV40) large T-antigen (LT-antigen).[4] In contrast to normal HVSMCs, which stop dividing after 10–15 passages, SM1 cells have been in culture at the permissive temperature (36°C) for over 200 population doublings. Their growth rate at the permissive temperature is much higher than that of normal HVSMCs; furthermore, they achieve a higher cell density. At the nonpermissive temperature (39°C), however, LT-antigen expression is downregulated, a phenomenon accompanied by growth arrest and acquisition of a senescent cell-phenotype: diminished BrdU incorporation, increase in cell size, appearance of perinuclear vacuoles, increased SA-β-Gal staining, and expression of the cell cycle inhibitor p21[WAF1/SDI1] at high levels.[4]

AUTOCRINE REGULATION OF SM1 PROLIFERATION

We previously showed that normal human fibroblasts control their proliferation by secreting autocrine growth factors and that this mechanism persists during *in vitro* aging.[5] Because serum withdrawal under permissive conditions does not lead SM1 cells to growth arrest, we tested whether their transformation was accompanied by

[c]Corresponding author. Phone: + 30 1 6503565; fax: + 30 1 6511767.
dkletsas@mail.demokritos.gr

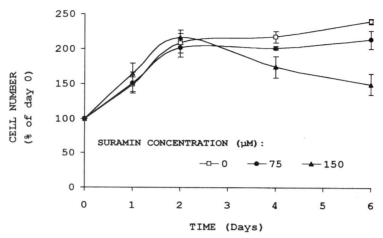

FIGURE 1. SM1 proliferation in the presence or absence of suramin. SM1 cells were plated in 24-well clusters in NCTC-109 medium containing 10% fetal calf serum. Twenty-four hours after plating, the medium was changed to serum-free NCTC-109, containing the indicated concentrations of suramin. Cell number was monitored at the indicated time-points after trypsinization of the cultures by using a Coulter counter. Medium was renewed every second day.

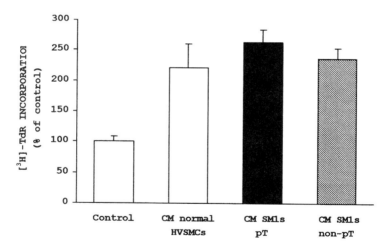

FIGURE 2. Stimulation of DNA synthesis in normal HVSMC cultures by conditioned medium (CM). Primary human saphenous vein smooth muscle cells[4] were cultured in DMEM containing 10% fetal calf serum (FCS) in 24-well clusters. After they reached confluency, the medium was changed to DMEM 0.1% FCS, and 48 hours later they were stimulated with media conditioned by normal HVSMCs or by SM1 cells at either the permissive temperature (pT) or the non-pT (conditioned media were produced by a 48-hour incubation of subconfluent cell cultures with serum-free DMEM) along with [^{3}H]thymidine (0.2 μCi/ml). Twenty-four hours later the cells were fixed and the incorporated radioactivity was measured as described.[5] Values represent mean (error bar = standard deviation) of three individual experiments performed in triplicate wells.

enhanced secretion of autocrine growth factors. As shown in FIGURE 1, in the presence of suramin, which blocks growth factor-receptor interactions,[5] the proliferation of SM1 cells in serum-free medium is inhibited, indicating that the cells secrete autocrine growth factors. On the other hand, comparison of the stimulatory activity of SM1-conditioned medium (CM) to that of the CM of normal HVSMCs (FIG. 2) does not reveal significant differences, suggesting that the increased growth rate of SM1 cells is not due to subversion of HVSMC-autocrine machinery. Inasmuch as SM1 cells at the permissive temperature accumulate p53 protein in the nucleus due to formation of a stabilization complex with LT-antigen,[4] the increased growth rate of SM1 cells probably reflects deregulation of the cell-cycle machinery in general. Moreover, CM from SM1s at the permissive and nonpermissive temperatures possess comparable mitogenic activiy (FIG. 2), indicating that the autocrine growth regulatory mechanism persists even at the senescent-like state of these cells.

ACKNOWLEDGMENT

This work was supported by the "Britain-Greece: Joint Research and Technology Programmes."

REFERENCES

1. HAYFLICK, L. 1965. The limited *in vitro* lifetime of human diploid cell strains. Exp. Cell. Res. **37:** 614–636.
2. JAT, P.S. & P.A. SHARP. 1989. Cell lines established by a temperature-sensitive simian virus 40 large-T-antigen gene are growth restricted at the nonpermissive temperature. Mol. Cell. Biol. **9:** 1672–1681.
3. WRIGHT, W.E., O.M. PEREIRA-SMITH & J.W. SHAY. 1989. Reversible cellular senescence: implications for immortalization of normal human fibroblasts. Mol. Cell. Biol. **9:** 3088–3092.
4. HSIEH, J.-K., D. KLETSAS, G. CLUNN *et al.* 2000. p53, p21$^{WAF1/CIP1}$ and MDM2 involvement in the proliferation and apoptosis in an *in vitro* model of conditionally immortalized human vascular smooth muscle cells. Arterioscler. Thromb. Vasc. Biol. **20:** 973–981.
5. PRATSINIS, H., D. KLETSAS & D. STATHAKOS. 1997. Autocrine growth regulation in fetal and adult human fibroblasts. Biochem. Biophys. Res. Commun. **237:** 348–353.

Using Stress Resistance to Isolate Novel Longevity Mutations in *Caenorhabditis elegans*

JAMES N. SAMPAYO, NICOLE L. JENKINS, AND GORDON J. LITHGOW

School of Biological Sciences, University of Manchester, Manchester, UK

INTRODUCTION

Up to 30 single gene mutations have been identified in the free-living nematode worm *Caenorhabditis elegans* that extend both mean and maximum life span (Age mutations).[1] All Age mutations tested also confer increased thermotolerance (Itt).[2] The Itt phenotype allows for the rapid and economic isolation of mutant lines with extended life spans.

We are currently using 4,5′, 8-trimethylpsoralen (TMP)-induced mutagenesis to facilitate cloning of novel Age genes. This mutagen requires activation by ultraviolet light[3] and cross-links DNA. TMP was successfully used in *C. elegans*[4] and more recently in zebrafish *Danio rerio*[5] and was shown to produce small deletions in DNA, ranging in size from 0.1–15 kb in length.

Presented here is preliminary analysis of genetic variants of *C. elegans* produced by mutagenesis using TMP. A mutagenesis protocol was modified from that of Yandell *et al.*[4] and successfully produced mutants with visible mutant phenotypes (e.g., uncoordinated, dumpy, slow development, embryonic lethality, and increased male frequency). To identify mutants with extended life span (Age mutants), the associated phenotype of Itt was used as the selection method.

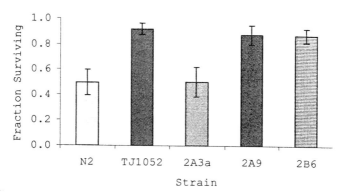

FIGURE 1. Lock-in assay of thermotolerance. Mutant strains were compared to wild-type strains (N2) for survival of a 7-hour heat shock at 35°C. TJ1052 is [*age-1(hx546)*].

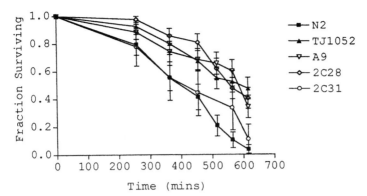

FIGURE 2. Longitudinal thermotolerance assay at 35°C on strains that were thermotolerant on lock-in assay. N2 is wild-type and TJ1052 is [*age-1(hx546)*].

RESULTS AND DISCUSSION

Large numbers of strains were screened using a "lock-in" method of assessing thermotolerance. For lock-in assays, mutant strains were compared to wild-type strains for survival of a 7-hour heat shock at 35°C (FIG. 1). Strains that appeared thermotolerant in the lock-in assays were further characterized using a standard longitudinal thermotolerance assay[6] at 35°C (FIG. 2). Using these methods, the thermotolerance of over 50 strains has been assessed. Of these, 10–15 strains have been identified as being reproducibly thermotolerant (data not shown). The genetic characterization of these mutants is ongoing.

A previous genetic screen was undertaken utilizing ethyl methanesulfonate (EMS) as a mutagen.[7] In contrast to TMP, EMS is thought to induce point mutations.[8] This screen successfully identified new ltt and Age mutants. The novel TMP-induced alleles will now be used in complementation analysis with these EMS-induced alleles. In addition, survival analysis will be undertaken on the novel TMP-induced alleles to assess whether or not the mutants exhibit the phenotypic association between ltt and Age.

We have demonstrated that a range of mutagens can be used to isolate mutations that confer increased stress resistance. Analysis of the new mutant strains will be used as part of a molecular dissection of aging and stress response in this organism.

REFERENCES

1. LITHGOW, G.J. 1996. Invertebrate gerontology: the age mutations of *Caenorhabditis elegans*. BioEssays **18:** 809–815.
2. LITHGOW, G.J. *et al.* 1995. Thermotolerance and extended lifespan conferred by single-gene mutations and induced by thermal stress. Proc. Nat. Acad. Sci. USA **92:** 7540–7544.
3. CIMINO, G.D. *et al.* 1985. Psoralens as photoactive probes of nucleic acid structure. Annu. Rev. Biochem. **54:** 1151–1193.

4. YANDELL, M.D. *et al.* 1994. Trimethylpsoralen induces small deletion mutations in *Caenorhabditis elegans*. Proc. Natl. Acad. Sci. USA **91:**1381–1385.
5. ANDO, H. *et al.* 1998. Efficient mutagenesis of zebrafish by a DNA cross-linking agent. Neurosci. Lett. **244:** 81–84.
6. LITHGOW, G.J. *et al.* 1994. Thermotolerance of a long-lived mutant of *Caenorhabditis elegans*. J. Gerontol. Biol. Sci. **49:** B270–B276.
7. WALKER, G.A. *et al.* 1998. A relationship between thermotolerance and longevity in *Caenorhabditis elegans*. J. Invest. Dermatol. Symp. Proc. **3:** 6–10.
8. ANDERSON, P. 1995. Mutagenesis. Methods Cell Biol. **48:** 31–58.

Telomere Length As a Marker of Oxidative Stress in Primary Human Fibroblast Cultures

VIOLETA SERRA,[a] TILMAN GRUNE,[b] NICOLLE SITTE,[b] GABRIELE SARETZKI,[a] AND THOMAS VON ZGLINICKI[a,c]

[a]Institute of Pathology and [b]Clinic of Physical Medicine and Rehabilitation, Charité, Humboldt University, 10098 Berlin, Germany

INTRODUCTION

Oxidative damage is regarded as a major cause of aging.[1] Telomere shortening induces aging *in vitro*, that is, replicative senescence.[2,3] Increased oxidative stress accelerates telomere shortening of human fibroblasts *in vitro*.[3,4] Telomeres in different human tissues, including peripheral leukocytes, shorten with age.[5] The variation between telomeres from different individuals is, to a large extent, genetically determined.[6] We could show that this individual variation is about the same in two cell types with a very different replicative history *in vivo*, namely, peripheral leukocytes and skin fibroblasts. Moreover, the age-corrected individual telomere length correlates very strongly with, and possibly predicts, the risk for stroke-related dementia, a major oxidative damage-dependent brain disease of the elderly.[7] To define more clearly the factors that influence the interindividual variation of telomere length, we measured the mRNA steady state levels of the antioxidant enzymes Cu/Zn superoxide dismutase (CuZnSOD), Mn superoxide dismutase (MnSOD), and cytosolic glutathione peroxidase (cGPX) in primary skin fibroblast cultures from 16 individuals and compared these expression levels with the concentration of protein carbonyls as an established indicator of oxidative stress and with the rate of telomere shortening.

MATERIALS AND METHODS

Fibroblast cultures were grown from skin biopsy specimens taken from the unexposed upper arm of 16 individuals (age range 28–91 years). They were grown in DMEM for at least 20 population doublings. Growth rates were linear over the whole range in each case and were used to extrapolate the cell number at t = 0 (explantation from the biopsy) and to calculate population doubling level (PDL). Telomere length was measured at 3–4 different time points about 5–10 PD apart as described.[4] Protein carbonyl concentrations were measured in 2–4 independent confluent cultures at a PDL of 19.4 ± 5.4 by immunoassay as described.[8,9] RNA was isolated from cells in log growth at about the same PDL using RNAeasy (Qiagen). Signal intensities on Northern blots were measured using a PhosphorImager (BioRad).

[c]Address for correspondence: Dr. Thomas von Zglinicki, Department of Gerontology, University of Newcastle, IHE, Newcastle General Hospital, Newcastle upon Tyne, NE4 6BE, UK. Phone: +44191 256 3310; fax: +44191 219 5074.

t.vonzglinicki@ncl.ac.uk

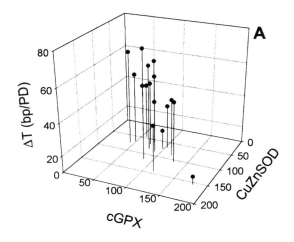

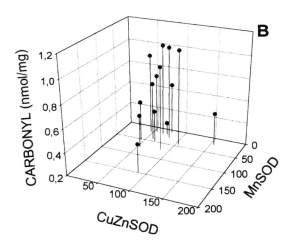

FIGURE 1. Correlation between antioxidant enzyme mRNA steady state levels and telomere shortening rate ΔT (**A**) or protein carbonyl content (**B**) in primary skin fibroblast lines derived from different donors. mRNA levels are given as ratio to GAPDH mRNA levels (in arbitrary units). Standard errors are generally below 10% for mRNA levels and carbonyls and below 25% for ΔT.

RESULTS AND DISCUSSION

The examined fibroblast lines differ not only by age of the donor, but also by their telomere length at explantation from the tissue. Although there was a slight decrease in average telomere length with donor's age (15 bp/year), the interindividual differences in telomere length are clearly larger than the age-dependent ones.[7] *In vitro*, telomeres shorten with rates between 15 and 76 bp/PD in the different cultures. These individual rates correlate with expression levels of cGPX and CuZnSOD mRNA (FIG. 1A). Higher levels of these antioxidant enzyme mRNAs are concordant with slower telomere shortening. There is no significant dependency of the telomere shortening rate on MnSOD mRNA levels (not shown).

The amount of oxidative protein damage, measured as protein carbonyl content, is also dependent on the expression of antioxidant enzyme message. However, it decreases with increasing levels of MnSOD mRNA (FIG. 1B), whereas CuZnSOD (FIG. 1B) or cGPx levels (not shown) appear to have a minor effect on carbonyl contents.

Despite the importance of posttranscriptional regulation,[10] mRNA abundance is a major factor in governing antioxidant enzyme activities of cultured fibroblasts.[11] It is interesting to note that both oxidative protein damage and telomere shortening rate decrease with expression of antioxidant enzymes. However, whereas mitochondrial MnSOD determines primarily the cellular carbonyl content, it is the combination of the two major cytoplasmic antioxidants cGPX and CuZnSOD that govern telomere shortening. As expected, expression levels of these two are positively correlated, but no significant correlation exists with MnSOD expression (not shown). Accordingly, there is only a weak positive correlation between oxidative protein damage and DNA damage, measured as telomere shortening rate (not shown).

Oxidative stress was shown previously to contribute greatly to telomere shortening due to a telomere-specific deficiency of DNA single-strand damage.[3] The correlative data presented here confirm these results.

ACKNOWLEDGMENTS

The study was supported by grants from the Deutsche Forschungsgemeinschaft and the VERUM Foundation.

REFERENCES

1. HARMAN, D. 1998. Aging: phenomena and theories. Ann. N.Y. Acad. Sci. **854:** 1–7.
2. BODNAR, A.G. *et al.* 1998. Extension of life-span by introduction of telomerase into normal human cells. Science **279:** 349–352.
3. VON ZGLINICKI, T. 1999. The role of oxidative stress in telomere length regulation and replicative senescence. Ann. N.Y. Acad. Sci. This volume.
4. VON ZGLINICKI, T. *et al.* 1995. Mild hyperoxia shortens telomeres and inhibits proliferation of human fibroblasts: a model for senescence? Exp. Cell Res. **220:** 186–193.
5. RUFER, N. *et al.* 1999. Telomere fluorescence measurements in granulocytes and T lymphocyte subsets point to a high turnover of hematopoetic stem cells and memory T cells in early childhood. J. Exp. Med. **190:** 157–167.
6. SLAGBOOM, P.E. *et al.* 1994. Genetic determination of telomere size in humans: a twin study of three age groups. Am. J. Hum. Genet. **55:** 876–882.
7. VON ZGLINICKI, T. *et al.* Blood telomere length variation might predict the risk of stroke-related dementia. J. Exp. Med. Submitted.
8. BUSS, H. *et al.* 1997. Protein carbonyl measurement by a sensitive ELISA method. Free Rad. Biol. Med. **23:** 361–366.

9. SITTE, N. *et al.* 1999. Proteasome-dependent degradation of oxidized proteins in MRC-5 fibroblasts. FEBS Lett. **440:** 399–402.
10. ALLEN, R.G. *et al.* 1999. Differences in electron transport potential, antioxidant defenses, and oxidant generation in young and senescenct fetal lung fibroblasts (WI 38). J. Cell. Physiol. **180:** 114–122.
11. SARETZKI, G. *et al.* 1998. Similar gene expression pattern in senescent and hyperoxic-treated fibroblasts. J. Gerontol. Biol. Sci. **53A:** B438–B442.

Programmed Cell Death and Senescence in Skeletal Muscle Stem Cells

KATHRYN WOODS, ANNA MARRONE, AND JANET SMITH[a]

School of Biosciences, University of Birmingham, Edgbaston, Birmingham B15 2TT, UK

INTRODUCTION

We have developed a method for the isolation and culture of clonal populations of skeletal muscle cells (SMSc).[1,2] These cells can be genetically manipulated prior to injection into host muscles, where they will either take up "satellite" cell positions, around the periphery of mature muscle fibers or will differentiate and become incorporated into the muscle fibers of their host (FIG. 1a). Genetically modified SMSc are nontumorigenic and remain quiescent as satellite cells or as part of the host muscle for at least 15 months. Quiescent SMSc can also be reisolated from their host muscle and induced to proliferate in culture. In culture, fewer than 10% of SMSc are quiescent, and they display a cell cycle length (measured by continuous BrDu labeling) that is consistent with their being stem cells (FIG. 1b). We have shown that SMSc are subject to growth factor survival[3–5] and apoptotic (FIG. 2) signals and that they will undergo programmed cell death (PCD) when subject to IGF-II withdrawal.

CELL SENESCENCE AND MUSCLE AGING

It has been known for some time that most cultured cells have a natural life span, known as the Hayflick limit,[6] beyond which they will exit from the cell cycle and begin to senesce. An exception to this rule are cells that have become transformed. It is not yet clear whether stem cells also behave in this way. Survival of cultured skeletal muscle stem cell numbers declines with age;[1,7] however, it is not known if this is due to senescence and the loss of proliferative capacity or if it is due to altered survival capacity. All these factors are likely to be governed by growth factors. Telomerase expression, telomeric length shortening, and genetic instability[8,9] are also indicators of the induction of senescence. To enable us to investigate these interactions more closely and so that we can accurately measure senescence in SMSc, we used BrDu to estimate cell cycle length and quiescence.

Using cumulative BrDu labeling (FIG. 1b), we measured the rate of S-Phase entry of two different SMSc lines. SMSc were labeled in 40 mM BrDu and sampled at regular intervals (as indicated). Total cell cycle length of SMSc, as measured by BrDu incorporation, is estimated to be 17.9 hours, of which 5.9 hours are S-phase. As might be expected of cycling stem cells, the calculated cell cycle length is relatively

[a]Author for correspondence. Phone: +(44) 121 5408; fax +(44) 121 414 5925.
j.smith.20@bham.ac.uk

(a)

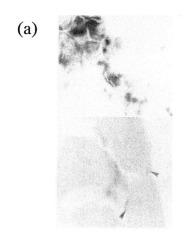

(b)

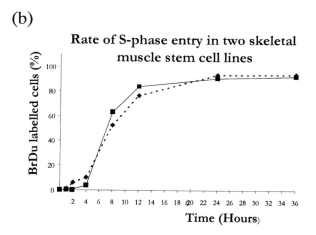

FIGURE 1. (**a**) Incorporation of the skeletal muscle stem cell line PD50A into host muscles (6 months postinjection of 4,000 cells), showing fused PD50A cells forming part of several muscle fibers as well as undifferentiated cells (*arrows*) in satellite cell positions. Positive cells express β-galactosidase, detected by histochemistry, which here shows as cytoplasmic staining. Histochemistry, genetic modification of PD50A, and its expression in host mouse muscle were achieved as described previously.[2] (**b**) Rate of S-phase entry in two skeletal muscle stem cell lines measured by continuous BrDu labeling. SMSc were cultured in 40 μM BrDu and sampled at regular intervals. Labeling was carried out and detected (using an anti-BrDu antibody from Boeringer) as described by Uzbekov et al.[11] Total cell cycle length was calculated to be 17.9 hours, of which 5.9 hours were S-Phase.

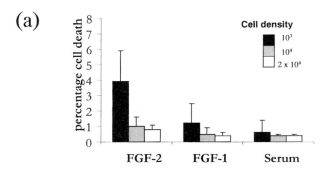

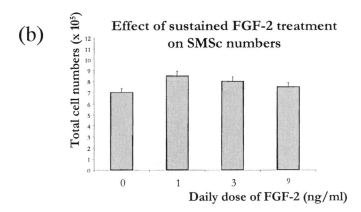

FIGURE 2. (a) FGF-2 and FGF-1 induce PCD in SMSc (PD50A). The extent of cell death is reciprocally correlated with cell plating density. Cell death was determined using trypan blue as described previously.[3] (**b**) Sustained treatment of SMSc with low doses (1–3 ng/ml) FGF-2 elevates total cell number, whereas higher doses (9 ng/ml) of FGF-2 do not. PD50A cells (7 x 10^4/cm^2) were plated out in triplicate. Growth factor (FGF-2) or carrier (PBS) was added to each dish at the indicated concentrations at plating and was replenished at 24-hour intervals. After 5 days (120 hours), cells were harvested using trypsin/EDTA and counted using a hemocytometer. Data show mean and SD at each point.

short, of which approximately a third of the time is spent in S-phase. Short cell cycle lengths with a high proportion of cells in S-phase have been commonly reported for both embryonic and stem cells.[10] In both SMSc lines assayed, S-Phase labeling was seen to plateau at around 20 hours labeling time, with 7 and 10% of cells remaining unlabeled. This small pool of cells is assumed to be quiescent (G0).

GROWTH FACTOR CONTROL OF PROGRAMMED CELL DEATH

Growth factors play an important role in the control of skeletal muscle stem cell death. We used Western blotting analysis of the cleavage products of the caspase 3 substrate poly-(ADP-ribose)-polymerase (PARP), a late stage indicator of PCD, to show that SMSc constitutively express caspases. Treatment of SMSc with the survival factor IGF-II induces inhibition of PARP cleavage within 6 hours, consistent with IGF-II acting directly upstream of this PCD pathway (A.M. and J.S., unpublished data).

We investigated the effect of FGF-1 and FGF-2 on SMSc growth and found that both can induce PCD in cultured SMSc (FIG. 2a) but that the extent of PCD observed is dependent on both the dose of FGF and the density of the cells. The effect of both FGF-1 and FGF-2 on cell death is greatest where cells are plated at lower densities (10^3/cm^2) and is less pronounced where cell densities are higher (2×10^4/cm^2). FGF induction of PCD is observable within 6 hours of treatment, suggesting a direct induction pathway. We have shown that skeletal muscle stem cell PCD is also dependent upon serum concentration. PCD is minimal at low levels of serum (1–4%) and all cell densities that we have studied (FIG. 2a; J.S., unpublished data). Very low concentrations (1 ng/ml) of FGF-2 and (to a lesser extent) FGF-1 can induce PCD in cells cultured under low serum conditions. PCD induction by FGFs at this concentration was greatest, however, when SMSc were cultured at lower (10^3/cm^2) cell densities (FIG. 2a) and was minimal when cell density was raised above 10^4 cells/ cm^2. The same effect of FGF-2 on PCD induction was also seen when cells were cultured in high serum (10%), but higher concentrations of FGFs (10 ng/ml) are required to induce an observable effect. This difference between the sensitivity of cells to FGF induction of PCD in high compared to low serum may be because in the presence of other growth factor stimuli, both FGF-1 and FGF-2 can also induce a proliferative response in SMSc. To further test this, we examined the net effect of FGF-2 on cultured SMS cell numbers (FIG. 2b). We found that low (1-3 ng/ml/day) and sustained (cells counted after 5 days) doses of FGF-2 were sufficient to raise the total cell number in skeletal muscle stem cells. However, we also saw that cell numbers were highest when SMSc were cultured in the presence of 1 ng/ml FGF-2 and that the net effect of FGF-2 at higher doses (3–9 ng/ml) was to reduce cell number. These data strongly suggest that in SMSc both PCD and the proliferative response are directly and complexly regulated by different growth factors or by different doses of the same growth factor.

DISCUSSION

We demonstrated that under conditions permissive for growth, skeletal muscle stem cells in culture are undergoing rapid turnover with a relatively short cell cycle and a very low proportion of cells in G0. We showed SMSc to be highly dependent on growth factor signals for their growth response. We studied in more detail the effects of two growth factor groups on the growth behaviors of these cells and found that IGFs (represented particularly by IGF-II) and both FGF-1 and FGF-2 have pleiotrophic effects on SMSc growth that depend largely on dose. The response of SMSc to serum is complex and suggests the presence of other (growth) factors that can act synergistically with the FGFs to inhibit or induce PCD. We also showed that the same growth factor (FGF-2) can induce both PCD and a mitogenic response in the same SMSc lines and that the net effect of such growth factors depends on dose and also on whether other (serum or growth) factors are also present. These data show that successful measurement of senescence in SMSc requires rigorous control of culture conditions and must take into account the pleiotrophic effects of growth factors on the net growth rates of these cells as well as an accurate measurement of cell cycle length.

ACKNOWLEDGMENTS

We gratefully acknowledge the support of The Royal Society (RSRG19484) and a grant from the BBRSC Sage initiative (6/SAG10075).

REFERENCES

1. SMITH, J. & P.N. SCHOFIELD. 1994. The effects of fibroblast growth factors in long term primary culture of dystrophic (mdx) mouse muscle myoblasts. Exp. Cell Res. **210:** 86–93.
2. SMITH, J. & P.N. SCHOFIELD. 1997. Stable integration of an mdx skeletal muscle cell line into dystrophic (mdx) skeletal muscle: evidence for stem cell status. Cell Growth Diff. **8:** 927–934.
3. SMITH, J., G. FOWKE & P.N. SCHOFIELD. 1995. Programmed cell death in dystrophic (mdx) muscle is inhibited by IGF-II. Cell Death Diff. **2:** 243–251.
4. SMITH, J., A. WARD, C. GOLDSMITH & R. LE DIEU. 2000. IGF-II ameliorates the dystrophic phenotype and coordinately downregulates programmed cell death. Human Gene Ther. Under revision.
5. SMITH, J. 1996. Muscle growth factors, ubiquitin and apoptosis in dystrophic muscle: apoptosis declines with age in the *mdx* mouse. B. Appl. Myol. **6:** 279–284.
6. HAYFLICK, L. 1992. Aging, longevity, and immortality *in vitro*. Exp. Gerontol. **27:** 363–368.
7. DECARY, S., V. MOULY & G.S. BUTLERBROWNE. 1996. Telomere length as a tool to monitor satellite cell amplification for cell-mediated gene therapy. Hum. Gene Ther. **7:** 1347–1350.
8. BENN, P. 1997. Aging chromosome telomeres: parallel studies with terminal repeat and telomere associated DNA probes. Mech. Age Dev. **99:** 152–166.
9. LORRIMORE, S.A., D.T. GOODHEAD & E.G. WRIGHT. 1995. The effect of p53 status on the radiosensitivity of haematopoetic stem cells. Cell Death Diff. **2:** 233–234.
10. LASKEY, R.A., A.D. MILLS & N.R. MORRIS. 1977. Assembly of SV40 chromatin in a cell-free system from *Xenopus* eggs. Cell **10:** 237–243.
11. UZBEKOV, R., I. CHARTRAIN, M. PHILIPPE & Y. ARLOT-BONNEMAINS. 1998. Cell cycle analysis and synchonisation of the Xenopus cell line XL2. Exp. Cell Res. **242:** 60–68.

Study of the H1 Linker Histone Variant, H1o, during the *in Vitro* Aging of Human Diploid Fibroblasts

D.S. TSAPALI, K.E. SEKERI-PATARYAS, AND T.G. SOURLINGAS[a]

National Center for Scientific Research, "DEMOKRITOS" Aghia Paraskevi, 153 10, Athens, Greece

INTRODUCTION

Many senescence-associated changes that are important components of *in vitro* aging, such as oxidative stress, telomere shortening, and expression of certain genes, have been intensively studied during the past years. Less well studied is another characteristic of *in vitro* aging, namely, its apparent similarity to differentiation. One important area of research into the similarities between aging and differentiation is the study of changes in the constitution of chromatin, which take place during both of these biological processes. Changes in the histone variant constitution of chromatin, as well as changes in their posttranslational modifications, must take place for chromatin remodeling to occur during differentiation.[1] The H1 linker histone variant H1o is closely associated with terminal differentiation in many mammalian cell lines, and its accumulation in chromatin has been linked to transcriptional changes occurring during differentiation.[2] H1o was first detected in tissues exhibiting a low rate of cell division. Its accumulation can be induced in cultured cells upon growth inhibition. Conversely, the amount of H1o in the chromatin of quiescent cells is reduced upon resumption of proliferation.[3] H1o was also found to accumulate in various cell systems that can be induced to differentiate using agents such as sodium butyrate, a histone deacetylase inhibitor.[4,5] Expression of the H1o gene, which occurs when cells are committed to a differentiation program, has also been associated with core histone acetylation.[5,6]

In light of the foregoing, the aim of this study was to analyze and compare the synthesis rate of H1o as a function of increasing cumulative population doublings (CPDs) and during the postmitotic senescent state in the *in vitro* model aging cell system of human diploid fibroblasts (HDFs; lung embryonic, Flow 2002).

RESULTS AND DISCUSSION

Analysis of H1o was accomplished using HDF cell cultures of different CPDs synchronized by serum deprivation in the Go, G1, and S phases of the cell cycle. Initially, cell cycle kinetics of [3H]thymidine incorporation were ascertained in the synchronized HDF populations of different *in vitro* ages (i.e., CPDs 20, 30, 40, and 50),

[a]Corresponding author. Phone: +301 650 3571/2; fax: +301 651-1767.
ksek@mail.demokritos.gr

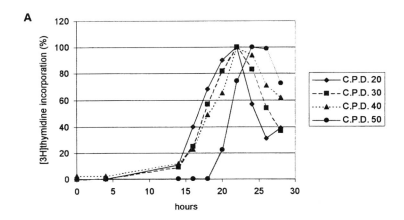

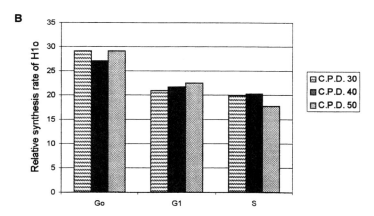

FIGURE 1. Rate of DNA synthesis during the cell cycle of HDF cell populations synchronized by serum deprivation. Cells inoculated at a cell density of 10^4 cells/cm^2 were incubated at 37°C with [^{3}H]thymidine for 120 minutes. Duplicate aliquots were used for each time point. Results are expressed as the percent incorporation, where the peak of DNA synthesis, occurring 22 hours after readdition of serum, is expressed as 100% (**A**). H1o relative synthesis rate of synchronized HDF cell populations. See text for details (**B**).

most of which would, in continuation, be used in the analysis of H1o. Results from the experiments just described (FIG. 1A) showed that these HDF cell populations of different CPDs (i.e., 20, 30, 40, and 50) have similar cell cycle dynamics as far as the ease with which they can be synchronized and the levels in which they can incorporate radiolabeled thymidine. This shows that the majority of cells in these four cell populations, including CPD 50, are mitotically active.

In continuation, the relative synthesis rate of H1o with respect to the entire H1 fraction was ascertained during the Go, late G1 (12 hours after readdition of serum),

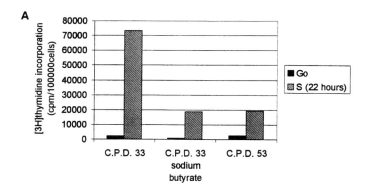

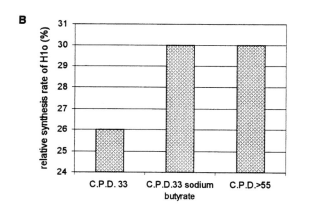

FIGURE 2. [³H]Thymidine incorporation of HDF cell populations. Samples were obtained from CPD 33 cell monolayers in the absence and/or continued presence of 0.5 mM sodium butyrate and CPD 53 monolayers (10^4 cells/cm²) synchronized in the Go and S phases of the cell cycle (**A**). Relative synthesis rate of H1o expressed as the percent synthesis with respect to the entire H1 fraction of HDF cell monolayers of CPD 33 in exponential growth, CPD 33 in the continued presence of 0.5 mM sodium butyrate, and CPD > 55. See text for details (**B**).

and S phases (22 hours after readdition of serum) of the cell cycle of cell populations with CPDs 30, 40, and 50 (FIG. 1B). Cell monolayers were incubated for 4 hours (i.e., in the serum-deprived Go phase, 8–12 hours after restimulation and 18–22 hours after restimulation) at 37°C with radioactive protein precursers (¹⁴C-arginine and ¹⁴C-lysine). After HCl extraction of the total histone fraction, H1o was separated from the rest of the H1 fraction by SDS-PAGE. The exact electrophoretic position of H1o was ascertained by Western analysis using an H1o monoclonal antibody[7] (generously given to us by Dr. S. Khochbin). In continuation, the H1o band was excised, solubilized, and incorporated radioactivity quantitated by liquid scintillation spectrometry. FIGURE 1B shows the results of this study. The relative synthesis rate of H1o is expressed as the percent incorporation relative to the incorporated radio-activity of the entire H1 fraction. From FIGURE 1B it can be seen that the synthesis rate of H1o during the Go phase is higher than that found during the late G1 and/or S phase of the cell cycle. This was observed at all ages studied (CPDs 30, 40, and 50). In other words, cell cycle phase dependent- changes in the H1o synthesis rate were observed which, though, were not found to change with increasing age in mitotically active HDF cell populations.

Because cells in terminally differentiated tissues are in a nonmitotic state, the researchers of this investigation wished to analyze the H1o relative synthesis rate in postmitotic HDF cell populations. Rapid *in vitro* aging as well as obtaining postmitotic cells in sufficient amounts for H1o analysis was accomplished either by increasing the split ratios (cell populations of CPD 54 were successively passaged two to three times using a 1:4 split ratio instead of the 1:2 split ratio normally used during routine subculturing) during each passage or by using the agent sodium butyrate (cell populations of CPD 28 were routinely subcultured once a week for 6 weeks using a 1:2 split ratio in the continued presence of 0.5 mM sodium butyrate) known to induce differentiation and lately found to bring about an irreversible senescent-like state in HDFs.[8,9] A population was characterized as being postmitotic when its monolayers were no longer able to reach confluency, even after an extended in-culture period of time, and by its very low incorporation of radiolabeled thymidine (FIG. 2A). FIGURE 2B shows the H1o relative synthesis rate of the aged HDF cell populations just described as compared to young HDF populations in exponential growth (CPD 33). It can be seen that both of these postmitotic cell populations, which have reached senescence either by increasing their split ratios (CPD > 55) or after sodium butyrate treatment, show an increased synthesis of H1o as compared to that of actively dividing cells. This increased synthesis of H1o does not follow the normal cell cycle synthesis pattern observed for this linker histone variant during the late G1 phase of cycling cells (FIG. 1B). Despite the fact that senescent populations are known to be in an arrested late-G1–like state, many cell cycle related events, crucial to the cell's forthcoming entrance into the S phase, fail to occur, leading to the establishment of the senescent state.[10] The synthetic behavior of this unique linker histone variant reported in this communication may be another manifestation of senescence. Experiments, already underway, assessing the transcriptional levels of the H1o gene will shed further light regarding these observations.

ACKNOWLEDGMENT

We wish to thank our laboratory technician, Kalliope Kalokyri, for invaluable technical assistance throughout the course of this work.

REFERENCES

1. VAN HOLDE, K. & J. ZLATANOVA. 1996. Chromatin architectural proteins and transcription factors: a structural connection. BioEssays **18:** 697–700.
2. DONG, Y., L. DAKAI & A.I. SKOULTCHI. 1995. An upstream control region required for inducible transcription of the mouse H1o histone gene during terminal differentiation. Mol. Cell Biol. **15:** 1889–1900.

3. LEA, M.A. 1987. Relationship of H1o histone to differentiation and cancer. Cancer Biochem. Biophys. **9:** 199–209.
4. Chabanas, A., E. Khoury, P. Goeltz *et al.* 1985. Effects of butyric acid on cell cycle regulation and induction of histone H1o in mouse cells and tissue culture. Inducibility of H1o in late S-G2 phase of the cell cycle. J. Mol. Biol. **183:** 141–151.
5. SEIGNEURIN, D., D. GRUNWALD, J.J. LAWRENCE & S. KHOCHBIN. 1995. Developmentally regulated chromatin acetylation and histone H1o accumulation. Int. J. Dev. Biol. **39:** 597–603.
6. GIRARDOT, V., T. RABILLOUD, M. YOSHIDA *et al.* 1994. Relationship between core histone acetylation and histone H1o gene activity. Eur. J. Biochem. **224:** 885–892.
7. DOUSSON, S., C. GORKE, C. GILLY & J. J. LAWRENCE. 1989. Histone H1o mapping using monoclonal antibodies. Eur. J. Immunol. **19:** 1123–1129.
8. OGRYZKO, V.V., T.H. HIRAI, V.R. RUSSANOVA *et al.* 1996. Human fibroblast commitment to a senescence-like state in response to histone deacetylase inhibitors is cell cycle dependent. Mol. Cell Biol. **16:** 5210–5218.
9. XIAO, H., T. HASEGAWA, O. MIYAISHI *et al.* 1997. Sodium butyrate induces NIH3T3 cells to senescence-like state and enhances promoter activity of p21[WAF/CIP1] in p53-independent manner. Biochem. Biophys. Res. Commun. **237:** 457–460.
10. PIGNOLO, R.J., B.G. MARTIN, J.H. HORTON *et al.* 1998. The pathway of cell senescence: WI-38 cells arrest in late G1 and are unable to traverse the cell cycle from a true Go state. Exp. Gerontol. **33:** 67–80.

Aging-Related Muscle Dysfunction

Failure of Adaptation to Oxidative Stress?

S. SPIERS,[a] F. McARDLE, AND M. J. JACKSON

Cell Pathophysiology Group, University of Liverpool, Department of Medicine, UCD5, Duncan Building, Liverpool, L69 3GA, UK

INTRODUCTION

Oxygen-free radicals and oxidant-induced damage to cellular molecules have been implicated as important contributors to the aging process. In aged muscle, there is an increased content of products of oxidative damage to proteins and DNA[1,2] together with a marked increase in the number and variety of mitochondrial DNA rearrangements.[3]

Mammalian cells respond to oxidative stress by variations in the rate of cell growth, changes in cell cycle length, and marked responses in resistance to oxidative stress.[4] In muscle there is a transient response to contractile activity by increased expression of several stress or heat shock proteins (HSPs) and by increasing activity of superoxide dismutase and catalase enzymes.[5] Evidence also exists that aging results in a reduced ability to express HSPs in response to stress.[6]

The decline in skeletal muscle function that occurs with age appears to be due primarily to a loss of muscle fibers. Reduction in fiber number may be related to an increased susceptibility of aged muscle to contraction-induced damage and a reduced rate of regeneration following damage.[7] Muscles from aged animals take significantly longer to recover from exercise involving lengthening contractions than do muscles from adult animals and do not show the same adaptive response to preventing muscle damage induced by further exercise protocols.[8]

Oxidative stress therefore plays an important role in the aging process, and studies in nonmammalian systems suggest that a reduction in this stress may reduce age-related tissue dysfunction. It is also clear that cells respond to episodes of oxidative stress to prevent further tissue damage and that failure of adaptation is a feature of the aging process.

METHODS AND RESULTS

Our aim is to define the adaptive response in muscle to exercise-induced oxidative stress and to determine whether this is modified with aging. The adaptive response of skeletal muscle cells will be characterized *in vitro* using mRNA differential display techniques. cDNAs obtained from this study will then be utilized to investigate the adaptive response of young and aged murine skeletal muscle *in vivo*.

[a]Phone: 0151-706-4089; fax: 0151-706-5952.
sspiers@liv.ac.uk

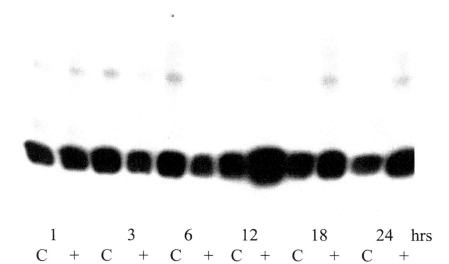

FIGURE 1. HSP25 expression in C_2C_{12} myotubes. Five-day-old C_2C_{12} myotubes were exposed to a single bolus of 0 or 100 mM H_2O_2. Cells were harvested at the indicated times, sonicated, and analyzed by Western blotting for HSP25 expression. C, control (0 μM H_2O_2); +, stressed (100 μM H_2O_2).

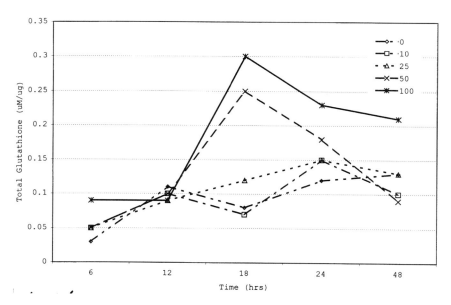

FIGURE 2. Total glutathione in H_2O_2 stressed C_2C_{12} myotubes. Five-day-old C_2C_{12} myotubes were exposed to a single bolus of H_2O_2 (0–100 μM) and harvested at 6, 12, 18, 24, and 48 hours. Cells were sonicated prior to analysis for total glutathione.

Preliminary studies have been undertaken to characterize the adaptive response of C_2C_{12} myotubes to hydrogen peroxide-induced oxidative stress. Analysis of HSP25 expression demonstrates a significant increase in protein levels 12 hours poststress (FIG. 1). Cellular glutathione levels also increase in a dose-dependent manner, peaking 18 hours poststress (FIG. 2).

Differential display results support these data with the identification of polymerase chain reaction (PCR) products with homology to the enzyme γ-glutamylcysteine synthetase and NF-κB essential modulator.

DISCUSSION

Our results indicate that the application of mild nondamaging oxidative stress to myotubes results in an adaptive response. Other workers have demonstrated an HSP25-mediated rise in cellular glutathione in response to hydrogen peroxide-induced stress.[9] This response appears to be transient, and glutathione levels return to control values by 48 hours.

Preliminary differential display studies demonstrate early changes in gene expression following stress. Gamma-glutamylcysteine synthetase is involved in glutathione synthesis, whilst potential enhanced activation of the transcription factor NF-κB is of interest in that NF-κB is known to enhance the cellular response to oxidative stress.[10]

ACKNOWLEDGMENT

We thank the Biotechnology and Biological Sciences Research Council (BBSRC) for their financial support.

REFERENCES

1. SOHALL, R.S. *et al.* 1994. Mech. Ageing Dev. **74:** 121–133.
2. SOHALL, R.S. *et al.* 1994. Mech. Ageing Dev. **76:** 215–224.
3. MELOV, S. *et al.* 1995. Nucleic Acid Res. **23:** 4122–4126.
4. WIESE, A.G. *et al.* 1995. Arch. Biochem. Biophys. **318:** 231–240.
5. McARDLE, A. *et al.* 1997. J. Physiol. **499:** 9P (abstr.).
6. LIU, A.Y. *et al.* 1996. *In* Stress Inducible Cellular Responses. U.Feige. R.I.
7. FAULKNER, J.A. *et al.* 1990. Ann. Rev. Gerentol. Geriatr. V.J. Cristofalo & M.P. Lawton, Eds.:147–166. Springer. New York.
8. McBRIDE, T.A. *et al.* 1995. Mech. Ageing Dev. **83:** 185–200.
9. PREVILLE, X. *et al.* 1999. Exp. Cell Res. **247:** 61–78.
10. HELENIUS, M. *et al.* 1996. J. Mol. Cell Cardiol. **28:** 487–498.

A Theoretical Model for *"in machina"* Experiments on Immunosenescence

SILVANA VALENSIN[a] AND GIANNI DI CARO[b]

[a]*Department of Experimental Pathology, University of Bologna, Via S. Giacomo 12, 40127 Bologna, Italy*

[b]*IRIDIA, Universite Libre de Bruxelles, 50 Av. F. Roosevelt, 1050 Bruxelles, Belgium*

INTRODUCTION

Immunosenescence refers to the changes that occur in the immune system (IS) with age. Old literature suggests that almost all immune responses undergo profound deterioration with age, but recent data indicate that this assumption is inappropriate. Indeed, with age some parameters decrease, whereas others increase or remain unchanged; therefore, immunosenescence is a phenomenon much more complex than previously thought.[1] Based on these data, a remodeling theory of immunosenescence was proposed, which suggests that the body undergoes a continuous process of adaptation to cope with external and internal damaging agents. In particular, innate immunity appears to be well preserved and even upregulated in elderly humans and animals,[2] whereas adaptive immunity undergoes profound deterioration (shrinkage of T-cell repertoire, depletion of naïve T cells, and progressive accumulation of memory T cells, among others). We present here a theoretical computational model that has been designed basically to simulate the IS (Artificial Immune System) and that has been implemented to investigate the physiologic behavior of the IS and its changes over time by "in machina" experiments.[3]

MAJOR CHARACTERISTICS OF ABISS (AGENT BASED IMMUNE SYSTEM SIMULATOR)

The most relevant characteristics of ABISS are the following:

1. It is implemented using an object-oriented programming language, the C++, particularly suitable to represent complex systems, such as the IS, in a relatively easy way. Moreover, the choice of an OO programming language allows easy updating of the program.

2. It implements in deep immunologic detail the following immune entities: B cells, T helper (Th1 and Th2) and cytotoxic T lymphocytes, macrophages, dendritic cells, NK cells, antibodies and antigens, and their interactions. Such interactions involve also the role of cytokines, MHC-I and MHC-II complexes, costimulator factors, ligands, surface markers, and surface receptors (a detailed example of such interactions is given in FIG. 1).

[a]Phone: +39 051 2094741; fax: +39 051 2094747.
valensin@mangrovia.unibo.it

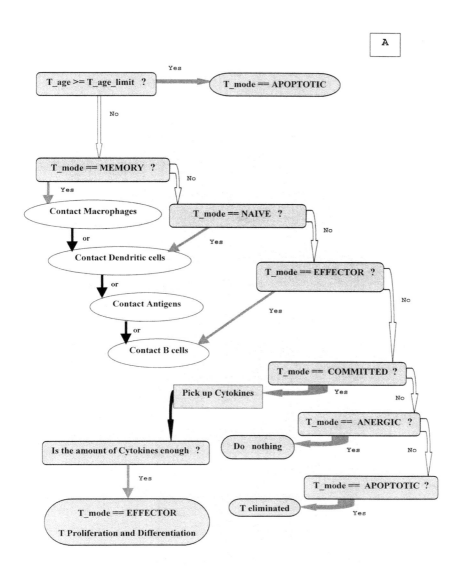

FIGURE 1. Some implementation details: interactions between Th lymphocytes and other immune cells, and rules for state changes (same box color refers to the same type of cell; *yes* and *no arrows* indicate the flow direction when the condition in the box *is* or *is not* satisfied). (A) Flow chart of Th-to-cell possible interactions based on the Th status; (B) flow chart of rules mastering the Th memory and naive state changes when they interact with a dendritic cell.

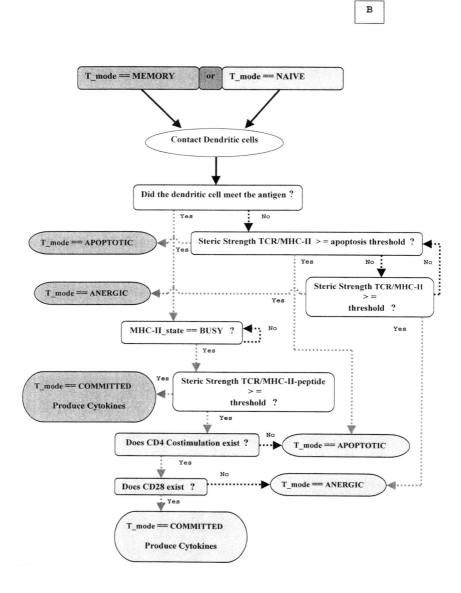

FIGURE 1B. *See legend on previous page.*

3. It is designed assigning a class to each immunologic element, that is, a general framework describing variables (specific attributes of the elements) and methods (possible actions the element can carry out within the environment). All elements belong to macrocells, that is, big containers of immunologic interacting items arranged as a 2D grid (tessellation of a 2D artificial immunologic space), among which elements can move if some conditions are satisfied. Accordingly, elements perform: direct element-to-element local interactions, higher level interactions (occurring among groups of them), and indirect reciprocal interactions through the local environment of the macrocell they share (for instance, the type and amount of available cytokines).

4. It has a high degree of flexibility and adaptability because most of the parameters to set up in order to perform a simulation must be fixed by the user, and because the aforementioned entities must be switched on and off, depending on the experimental requirements. Both features enable different degrees of complexity and variability in the system that are exploited for the study of the IS dynamics and metadynamics at different levels of complexity.

5. It allows complex and extensive "in machina" experiments by which changes over time of the real IS are envisaged by means of artificial longitudinal studies.

REFERENCES

1. FRANCESCHI, C. *et al*. 1995. The immunology of exceptional individuals: the lesson of centenarians. Immunol. Today **16:** 12–16.
2. OTTAVIANI, E., S. VALENSIN & C. FRANCESCHI. 1998. The neuro-immunological interface in an evolutionary perspective: the dynamic relationship between effector and recognition systems. Front. Biosci. **3:** 431–435.
3. VALENSIN, S. 1999. A Computational Model for the Immune System and Its Ageing. M. Sci. Thesis. The University of Manchester, Manchester, UK

Index of Contributors

Agiostratidou, G., 133–142

Bandyopadhyay, D., 71–84
Basun, H., 295–298
Beekman, M., 50–63
Benne, R., 267–281
Blakely, E.L., 199–207
Bohr, V.A., 167–179
Bonafè, M., 208–218, 244–254
Borg, J., 226–243
Borghouts, C., 31–39, 291–294
Bornemann, K.D., 260–266
Borthwick, G.M., 199–207
Brierley, E.J., 199–207
Bulteau, A.-L., 143–154
Bürkle, A., 126–132
Burton, M., 306–309

Cantatore, P., 299–301
Caragounis, A., 226–243
Carrieri, G., 208–218
Chainiaux, F., 85–98, 302–305, 306–309
Chen, Q.M., 111–125, 306–309
Chondrogianni, N., 133–142, 143–154
Conconi, M., 143–154
Corder, E.H., 295–298
Corder, L.S., 295–298
Cormio, A., 299–301
Cottrell, D.A., 199–207
Cypser, J.R., 40–49

De Benedictis, G., 208–218, 244–254
De Luca, M., 244–254
Demoliou-Mason, C., 321–323
Di Caro, G., 344–347
Dierick, J.-F., 85–98, 302–305, 306–309
Drosopoulos, N., 133–142
Dumont, P., 85–98, 306–309

Eliaers, F., 85–98, 302–305, 306–309

Felzani, G., 299–301
Fischer, D.F., 267–281
Francavilla, A., 299–301
Franceschi, C., 208–218, 244–254
Fratiglioni, L., 295–298
Friguet, B., 143–154
Frippiat, C., 85–98, 306–309

Gadaleta, M.N., 299–301
Gonos, E.S., 133–142
Grune, T., 327–330
Guglielmi, F.W., 299–301
Guo, Z., 295–298

Handris, P., 155–166
Heffernan, D.R., 226–243
Heijmans, B.T., 50–63
Hol, E.M., 267–281
Hughes, A., 321–323

Ince, P., 199–207
Islam, M., 226–243

Jackson, M.J., 341–343
Jaruga, E., 310–314
Jazwinski, S.M., 21–30
Jenkins, N.L., 319–320, 324–326
Johnson, M.A., 199–207
Johnson, T.E., 40–49

Kawas, C.H., 255–259
Kirkwood, T.B.L., 14–20, 315–318
Kletsas, D., 155–166, 321–323
Kopsidas, G., 226–243
Kotsota, V., 133–142
Kovalenko, S.A., 226–243
Kramarova, L., 226–243

Lannfelt, L., 295–298
Larsen, P.M., 302–305
Lezza, A.M.S., 299–301
Linnane, A.W., 226–243
Lithgow, G.J., 319–320, 324–326

Manton, K.G., 295–298
Marangi, L., 299–301
Marrone, A., 331–335
Martin, G.M., 1–13
Martin, K., 315–318
Martin, L.J., 255–259
McArdle, F., 341–343
McColl, G., 319–320
Medrano, E.E., 71–84
Melov, S., 219–225
Merry, B.J., 180–198
Meulenbelt, I., 50–63
Murakami, S., 40–49

Nehlin, J.O., 167–179

Olivieri, F., 244–254
Osiewacz, H.D., 31–39, 291–294
Ottaviani, E., 155–166, 244–254

Pascal, T., 85–98, 302–305, 306–309
Petropoulos, I., 143–154
Petropoulou, C., 133–142
Potten, C.S., 315–318
Pratsinis, H., 155–166, 321–323

Radzieszewska, E., 310–314
Rattan, S.I.S., 282–290
Remacle, J., 85–98, 302–305, 306–309
Roepstorff, P., 302–305

Sampayo, J.N., 324–326
Saretzki, G., 327–330
Schächter, F., 64–70

Sekeri-Pataryas, K.E., 336–340
Serra, V., 327–330
Sevaslidou, E., 155–166
Sikora, E., 310–314
Simoes, D., 133–142
Sitte, N., 327–330
Skierski, J., 310–314
Skovgaard, G.L., 167–179
Slagboom, P.E., 50–63
Sluse, F., 85–98
Smith, J., 331–335
Sourlingas, T.G., 336–340
Spiers, S., 341–343
Stathakos, D., 155–166
Staufenbiel, M., 260–266
Stojanovski, D., 226–243

Taylor, G.A., 199–207
Tedesco, P.M., 40–49
Toussaint, O., 85–98, 302–305, 306–309
Troncoso, J.C., 255–259
Tsapali, D.S., 336–340
Turnbull, D.M., 199–207

Valensin, S., 244–254, 344–347
van Leeuwen, F.W., 267–281
Varcasia, O., 208–218
Vecchiet, J., 299–301
Viitanen, M., 295–298
von Zglinicki, T., 99–110, 327–330

Walker, D.W., 319–320
West, M.J., 255–259
Westendorp, R.G.J., 50–63
Winblad, B., 295–298
Woods, K., 331–335

Yarovaya, N., 226–243

Zervolea, I., 155–166